Introduction to Graph Signal Processing

An intuitive and accessible text explaining the fundamentals and applications of graph signal processing. Requiring only an elementary understanding of linear algebra, it covers both basic and advanced topics, including node domain processing, graph signal frequency, sampling, and graph signal representations, as well as how to choose a graph. Understand the basic insights behind key concepts and learn how graphs can be associated with a range of specific applications across physical, biological, and social networks, distributed sensor networks, image and video processing, and machine learning. With numerous exercises and Matlab examples to help the reader put knowledge into practice, and a solutions manual available online for instructors, this unique text is essential reading for graduate and senior undergraduate students taking courses on graph signal processing, signal processing, information processing, and data analysis, as well as researchers and industry professionals.

Antonio Ortega is a professor of electrical and computer engineering at the University of Southern California, and a fellow of the IEEE.

Introduction to Graph Signal Processing

ANTONIO ORTEGA

University of Southern California

CAMBRIDGE
UNIVERSITY PRESS

CAMBRIDGE
UNIVERSITY PRESS

University Printing House, Cambridge CB2 8BS, United Kingdom

One Liberty Plaza, 20th Floor, New York, NY 10006, USA

477 Williamstown Road, Port Melbourne, VIC 3207, Australia

314–321, 3rd Floor, Plot 3, Splendor Forum, Jasola District Centre,
New Delhi – 110025, India

103 Penang Road, #05–06/07, Visioncrest Commercial, Singapore 238467

Cambridge University Press is part of the University of Cambridge.

It furthers the University's mission by disseminating knowledge in the pursuit of
education, learning, and research at the highest international levels of excellence.

www.cambridge.org
Information on this title: www.cambridge.org/9781108428132
DOI: 10.1017/9781108552349

First published 2022

A catalogue record for this publication is available from the British Library.

Library of Congress Cataloging-in-Publication Data
Names: Ortega, Antonio, 1965– author.
Title: Introduction to graph signal processing / Antonio Ortega, University of Southern California.
Description: New York, NY : Cambridge University Press, 2021. |
    Includes bibliographical references and index.
Identifiers: LCCN 2021038900 | ISBN 9781108428132 (hardback)
Subjects: LCSH: Signal processing. | BISAC: TECHNOLOGY & ENGINEERING /
    Signals & Signal Processing | TECHNOLOGY & ENGINEERING / Signals & Signal Processing
Classification: LCC TK5102.9 .O77 2021 | DDC 621.382/2–dc23
LC record available at https://lccn.loc.gov/2021038900

ISBN 978-1-108-42813-2 Hardback

To Mineyo and Naoto

Contents

Preface

This book grew out of my own research with students and collaborators, and evolved into a set of notes developed for a special topics course on graph signal processing at the University of Southern California. I have taught several versions of this material as a standalone class as well as in combination with advanced topics in signal processing (wavelets and extensions, dictionary representations and compressed sensing). In the meantime, much has changed since the first time I offered this class in Fall 2013. Graph signal processing (GSP) has grown as a field, understanding of specific problems has improved, and more applications have been considered. Both my experience in teaching this material and the evolution of the field informed the choices I made in writing this book.

First, while this is a book about graph signal processing, my goal has been to make the book accessible to readers who do not have any signal processing background. The main assumption I make is that readers will have taken an elementary linear algebra course. Appendix A provides a review of elementary concepts, presented from a signal representation perspective, and can be used to review this material.

Second, from my experience teaching this material to masters and PhD students, and from feedback I received from undergraduates who read early drafts of the book, I think the main challenge for those studying this topic for the first time is to understand the basic insights that make it possible to *use* the concepts. Without those insights, even if the mathematical ideas are understood, it is difficult to apply them. For this reason, I have spent time trying to develop intuition about the key concepts, lingering on ideas that are likely to be obvious to active researchers but hoping that they will prove to be useful to those new to the area. In short, whether or not I have succeeded in these goals, the word "Introduction" certainly deserves its place on the title.

Finally, by design, this book does not aim to provide a comprehensive and detailed survey of all recent research. In a rapidly evolving field this is difficult to do: much has been published but competing methods have not been compared, connections have not been made, the dust hasn't quite settled yet. I have chosen to summarize, classify and make connections where possible, but many details and approaches are left out. For advanced topics covered in this book (sampling, graph learning, signal representations) there are recent overview papers that provide a more detailed literature survey. Similarly, GSP is being used in a growing number of applications, and only a small subset of those are introduced in this book.

How to use this book   There are of course many different ways to use the material in this book for teaching. I describe some possible scenarios.

- For a semester-long advanced undergraduate class, Chapter 1 can be followed by the first three sections of Appendix A, and then the rest of the chapters. The remaining sections in Appendix A can be introduced before tackling Chapter 5. Additional programming assignments can be included to help students with limited Matlab experience.
- I have used this material as part of a semester-long graduate level class covering other topics (e.g., filterbanks, various types of wavelets and compressed sensing) and colleagues at other universities have used it in a similar way. For this type of course, the material in Appendix A can be used as an introduction to both GSP and to other advanced signal processing topics.
- Finally, a semester-long GSP class for advanced graduate students can complement the material in this book with reading from recent literature and advanced research oriented projects.

Exercises   Since Chapters 1–3 and the two appendices cover basic concepts they contain exercises, while for Chapters 4–7, which deal with more advanced topics, students could be asked to read some of the published literature and work on a class project.

Matlab examples   There are sections in Appendix B corresponding to Chapters 1–6, so this appendix can be used to complement each of those chapters, allowing students to get a more hands-on experience through Matlab code examples. Alternatively, Appendix B can be used as a standalone introduction to GraSP.

The GraSP toolbox is freely available at https://www.grasp-toolbox.org/. The source code for all examples and supplementary materials is available on the book's web page (http://www.graph-signal-processing-book.org/).

Acknowledgments

Interest in graph signal processing (GSP) has grown out of research on multiple applications where observed signals can be associated with an underlying graph. The journey that led me to completing this book started, more than 10 years ago, with the study of the representation and compression of signals captured by sensor networks. I often get asked about good applications for GSP, as if the results in this field had emerged out of purely theoretical research. In fact, much of what is described here, and in particular most of my own work on this topic, has been motivated by practical applications of GSP. While applications are important as a motivation, the reader should not be looking here for detailed solutions to problems that arise in specific applications. Instead, the main goal of this book is to develop the mathematical tools and insights that can allow us to think about some of these problems in terms of the processing of signals on graphs.

This book would not have been possible without the help and support of many people. First and foremost I would like to thank all of my current and former PhD students and postdoctoral fellows at the University of Southern California (USC). A brief note in these acknowledgments can hardly do justice to the importance of their contributions. Among my former PhD students I would like to thank in particular Alexandre Ciancio and Godwin Shen, who focused on transforms for sensor networks and developed methods for transforms over trees; Sunil Narang, who extended these ideas to graphs and introduced critically sampled graph filterbanks; Wooshik Kim, Yung-Hsuan (Jessie) Chao, Hilmi Egilmez and Eduardo Pavez, who developed new graph constructions and studied a number of image and video applications; Jiun-Yu (Joanne) Kao and Amin Rezapour, who took GSP methods into new and interesting application domains; and Akshay Gadde and Aamir Anis, who developed new methods for graph signal sampling and its application to machine learning. Ongoing and recent work with some of my current students, including Pratyusha Das, Keng-Shih Lu, Ajinkya Jayawant and Sarath Shekkizhar, as well as other work with Shay Deutsch, Alexander Serrano, Yoon Hak Kim, Lingyan Sheng, Sungwon Lee and Yongzhe Wang, also contributed to my research in this area. Finally, I would like to acknowledge the contributions of undergraduate and graduate students from various universities who visited USC and collaborated with my group, and in particular Eduardo Martínez-Enríquez from Universidad Carlos III, Madrid, and a series of students from Universitat Politècnica de Catalunya, Barcelona, and in particular Xavier Perez-Trufero, Apostol Gjika, Eduard Sanou, Javier Maroto, Victor González and Jùlia Barrufet, as well as David Bonet, who also provided comments on the manuscript.

Special thanks go to Benjamin Girault. Benjamin started the development of the GraSP Matlab toolbox during his PhD studies and has continued to grow, develop, and maintain it since joining USC as a postdoc. Without Benjamin's contribution, including Appendix B, a comprehensive introduction to GraSP, and numerous code examples, this book would be a lot less useful to students and practitioners. Benjamin's contributions go far beyond this one chapter. He has been a close collaborator, his ideas have fundamentally shaped several parts of this book and he has also read multiple chapter drafts, some of them more than once, and sometimes early versions that were far from being ready. His comments on various versions of the manuscript were thoughtful and detailed, and always prompted me to go deeper into the material.

Throughout my time at USC I have been fortunate to work with many outstanding colleagues in the school of engineering and beyond. Among those who have collaborated more directly in research related to this book, I would like to mention Salman Avistimehr, Bhaskar Krishnamachari, Urbashi (Ubli) Mitra, Shrikanth (Shri) Narayanan and Cyrus Shahabi, as well as some of their students and postdocs, among others Sundeep Pattem, Mahesh Sathiamoorthy, Eyal En Gad, Aly El Gamal, Basak Guler, Nicolò Michelusi and Marco Levorato.

One of the great pleasures of academic research is having the opportunity to collaborate with researchers around the world, both remotely and through some very enjoyable visits. Some of these collaborations have had a major role in shaping my research in this area. During several visits to École polytechnique fédérale de Lausanne (EPFL), discussions with Pascal Frossard, Pierre Vandergheynst and David Shuman led to an overview paper and to additional collaboration with Thomas Maugey, Xiaowen Dong and Dorina Thanou, among others. It was during an extended sabbatical visit to National Institute of Informatics (NII) in Tokyo, Japan, that I started drafting this book, and I would like to thank my long-time collaborator Gene Cheung (then at NII, now at York University), and his students Wei Xu, Jiahao Pang and Jin Zeng for many productive discussions and a very fruitful collaboration. A short visit to the University of New South Wales (UNSW) in Sydney, Australia, and many discussions with David Taubman are also acknowledged. Since 2015 I have also visited regularly the Tokyo University of Agriculture and Technology (TUAT), and various parts of the manuscript were developed there. I am very grateful to TUAT and my hosts Yuichi Tanaka and Toshihisa Tanaka and their students, in particular Akie Sakiyama and Koki Yamada, for hosting me and for their collaboration. While most of this collaboration has been remote, I would also like to thank Vincent Gripon, and his students Carlos Lassance and Myriam Bontonou, at Institut Mines-Télécom (IMT) Atlantique and David Tay at Deakin University for their collaboration.

My work in this area has benefited greatly from collaboration and funding from several companies, and in particular I would like to thank Anthony Vetro, Dong Tian and Hassan Mansour at Mitsubishi Electric Research Labs (MERL), Phil Chou, Debargha Mukherjee, Yue Chen at Google and Amir Said at Qualcomm.

My research in this area has been supported by multiple sources over the years and this funding is most gratefully acknowledged. In particular, I would like to acknowledge multiple grants from the National Science Foundation and funding from the National

Aeronautics and Space Administration (NASA), from the Defense Advanced Research Project Agency (DARPA) and several companies, including Samsung, LG Electronics, Google, Mitsubishi Electric, Qualcomm, Tencent and KDDI. My summer and sabbatical visits were funded by the Japan Society for the Promotion of Science, NII, UNSW, TUAT and EPFL.

A first version of this manuscript was based on class notes for a special topics course on GSP at USC, and portions of successive drafts were used in teaching a class that included GSP along with wavelets. I am thankful to students in those classes for their questions and comments. Several students who took my undergraduate linear algebra class gamely volunteered to test my hypothesis that this material could be made accessible to undergraduate students with only basic linear algebra. Time will tell whether this is possible, or a good idea, but I am grateful to them, in particular Alex Vilesov, Reshma Kopparapu, Pengfei Chang, Lorand Cheng and Keshav Sriram, for their comments and questions and Catherine (Cami) Amein for her comments and for her wonderful cover illustration.

Several people helped me with detailed comments on various parts of the manuscript. In addition to Benjamin Girault, I would like to thank Yuichi Tanaka, David Tay, Hilmi Egilmez and Eduardo Pavez. Baltasar Beferull-Lozano and Gene Cheung both used the material in their teaching and provided comments. I am also thankful to the team at Cambridge University Press, including Julie Lancashire, Julia Ford, and Sarah Strange, for their inexhaustible patience and their support throughout the years it took for this book to be completed, Susan Parkinson for her detailed reading and many helpful suggestions and Sam Fearnley for managing the final production.

As of the time of this writing, 2020 and 2021 have been often described as "interesting" and "unusual" years. While not much else has gone according to plan in the last year, I can at least look at this book as a small (if late) positive outcome. I thank my wife Mineyo and our son Naoto for supporting me as I completed this project over the last few years, for pretending to believe my repeated (and highly unreliable) claims that the book was almost done and for making this year of work from home interesting and unusual, but in a very good way.

Notation

Vectors are written in boldface lowercase, $\mathbf{x}$, while matrices are in capital boldface, $\mathbf{A}, \mathbf{B}$. For graph matrices we use calligraphic letters if the matrix is normalized, so for example we would write $\mathcal{L}$ instead of $\mathbf{L}$. A summary of other specific notation is given below.

$S_1 \cap S_2$	Intersection of sets S_1 and S_2
$S_1 \cup S_2$	Union of sets S_1 and S_2
$\lvert S \rvert$	Number of elements in set S
$\mathbf{I}$	Identity matrix
$\mathbf{J}$	Exchange matrix – all ones on the anti-diagonal
$\mathbf{1}$	Vector with all entries equal to 1
$\mathbf{A}$	Adjacency matrix
$\mathbf{B}$	Incidence matrix
$\mathbf{E}$	Matrix of self-loop weights
$\mathbf{L}$	Combinatorial graph Laplacian
$\mathcal{A}$	Symmetric normalized adjacency matrix
$\mathcal{L}$	Symmetric normalized Laplacian
$\mathcal{Q}$	Row normalized adjacency matrix
$\mathcal{P}$	Column normalized adjacency matrix
$\mathcal{T}$	Random walk Laplacian
$\mathbf{S}$	Sample covariance matrix
$\mathbf{Q}$	Precision matrix
$i \sim j$	Nodes i and j are connected
$\mathcal{I}(A)$	Indicator function (1 if A is true, 0 otherwise)
$\mathbf{Z}$	Generic one-hop fundamental graph operator
$\Delta_{\mathbf{S}}(\mathbf{x}) = \mathbf{x}^{\mathsf{T}} \mathbf{S} \mathbf{x}$	Variation operator
$\mathcal{N}(i)$	Set of nodes connected to node i
$\mathcal{N}_k(i)$	Set of nodes in the k-hop neighborhood of node i
$p_{\min}(\mathbf{Z})$	Minimal polynomial of $\mathbf{Z}$
$p_{\mathrm{c}}(\mathbf{Z})$	Characteristic polynomial of $\mathbf{Z}$
$p_x(\mathbf{Z})$	Minimal polynomial of $\mathbf{Z}$ for vector $\mathbf{x}$

1 Introduction

Graph signal processing (GSP), or **signal processing on graphs,** is the study of how to analyze and process data associated with graphs. Graphs can represent networks, including both physical networks (the Internet, sensor networks, electrical grids or the brain) and information networks (the world wide web, Wikipedia or an online social network). Graphs can also be used to represent data (pixels in an image, or a training dataset used for machine learning). In this chapter we start by defining graphs and graph signals (Section 1.1). We develop some intuition about how to quantify signal variation on a graph, which will later be used to introduce the concept of graph signal frequency. We give examples of simple operations that can be performed on graphs, such as filtering or sampling. An important feature of GSP is that processing tools can be adapted to graph characteristics (Section 1.2). We present examples of graphs encountered in conventional signal processing (Section 1.3) and this is followed by a discussion of graphs arising in some representative applications (Section 1.4) and a brief overview of mathematical models for graphs (Section 1.5). We conclude the chapter with a roadmap to the rest of the book (Section 1.6).

1.1 From Signals to Graph Signals

1.1.1 Graphs

A **graph** consists of a series of **nodes**, or vertices, whose relations are captured by **edges** (see Figure 1.1). Two nodes are said to be **connected** when there exists an edge between them. Figure 1.1 shows a fully connected three-node graph, where all nodes connect with each other. Depending on the application, nodes and edges may correspond to components in a physical network (e.g., two computers connected by a communication link), entries in an information network (e.g., two linked Wikipedia pages) or different items in a dataset (e.g., two similar items used for training a machine learning system).

More formally, we define a graph as a set of nodes or vertices, $\mathcal{V} = \{v_1, v_2, \ldots, v_N\}$ or $\mathcal{V} = \{1, 2, \ldots, N\}^1$, and a set of edges, $\mathcal{E} = \{e_1, e_2, \ldots, e_M\}$ or $\mathcal{E} = \{1, 2, \ldots, M\}$. We may also denote an edge as $e = v_i v_j$ if it connects nodes i and j. The edges can be **weighted**, so that a scalar value is associated with each edge, or **unweighted**, i.e., all edges have weight equal to 1. With each node we associate a **degree**, the total weight

---

[1] We will use the terms vertex and node interchangeably throughout this book.

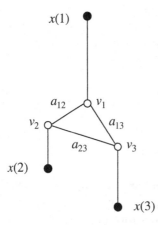

Figure 1.1 Simple example graph with three nodes, v_1, v_2, v_3, and edge weights a_{12}, a_{13}, a_{23}. There is a graph signal associated with this graph, with values $x(1), x(2), x(3)$ respectively at nodes v_1, v_2 and v_3. We will consider weights that are positive and capture the proximity or similarity between the nodes. For example, if $a_{12} > a_{13}$ then the first node, v_1, is closer (or more similar) to the second node, v_2, than to the third node, v_3.

of the edges connecting to that node. Some graphs are **directed**: only one of $v_i v_j$ and $v_j v_i$ may exist, or they may both exist but have different edge weights. Other graphs are **undirected**: $v_i v_j$ and $v_j v_i$ both exist and both edges have the same weight if the graph is weighted. The choice between a directed or an undirected graph will depend on the application, as shown in the following examples.

Example 1.1   Directed graph   If A and B represent two locations on a map, and we choose an edge weight based on distance, it makes sense for the edge to be undirected, since the distance between A and B is the same as the distance between B and A. On the other hand, if we are considering the distance between those two points following an existing road network, the two distances may be different, as illustrated in Figure 1.2.

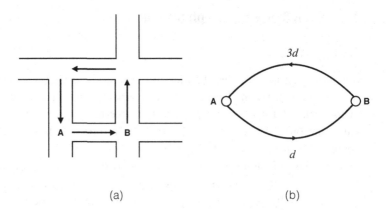

(a)   (b)

Figure 1.2 Directed road network graph. (a) City area with one-way streets, where driving distances on the road network are not symmetric. (b) Corresponding directed graph with different weights in each direction.

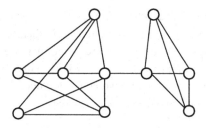

Figure 1.3 Undirected social network graph: Users have to agree to connect to each other and thus connections are undirected. It is common for connected users to share multiple common connections.

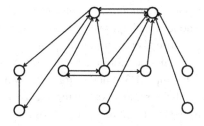

Figure 1.4 Example of a directed social graph. Some users attract many followers while others do not have any. Sometimes there are connections in both directions.

Example 1.2 Undirected graph In some social networks two users have to agree to be connected, while in others one user might follow another, but not the other way around. The first type of social network leads to an undirected graph (Figure 1.3), while the second should be represented by a directed graph (Figure 1.4). The questions of interest and properties are in general quite different for these two types of social graphs. In an undirected social graph connected users often share common connections which leads to a "clustering" behavior: we can observe groups of users with strong connections. In the directed social network case, a typical characteristic is for some users to have many connections, while most users have very few.

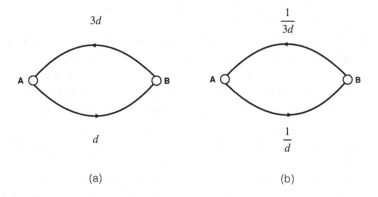

Figure 1.5 (a) Directed distance version of Figure 1.2, where larger weights indicate longer distance. (b) In a similarity graph, where, as in this example, weights are inversely proportional to distance, nodes that are closer are more likely to be similar.

Example 1.3   Distance graphs and similarity graphs   In Figure 1.2, edge weights represent distance, i.e., a larger weight corresponds to nodes that are farther apart. In contrast, in a similarity graph, larger weights correspond to greater similarity and smaller distance, as shown in Figure 1.5. Throughout most of this book we will use similarity graphs, with the assumption that signals observed at two connected nodes will tend to have similar values if the edge weight between those nodes is large.

Most weighted graphs studied in this book are *similarity* graphs, where the (positive) edge weights quantify how similar two nodes are (larger weights mean greater similarity). For example, a larger weight may be chosen between two nodes in a sensor network if those two nodes are close to each other, and therefore their measurements are more likely to be similar. This is illustrated in Example 1.3, where a graph has weights representing distance, while a second graph with the same topology represents similarity.

More generally, the meaning given to the existence of an edge between two nodes, i and j, depends on the application. A edge could be used to indicate that "i is at certain distance from j" or that "i may happen if j happened" or simply that "A and B are similar". Thus, graph representations are general. They provide a language and associated mathematical tools for very different problems and applications.

1.1.2   Frequency Analysis of Graph Signals

Graph signals   This book is about processing signals defined on graphs. Assume that each node i in a graph has a scalar value $x(i)$ associated with it.[2] The aggregation of all these scalar values into a vector $\mathbf{x}$ of dimension $|\mathcal{V}| = N$ will be called a **graph signal**. As an example, the graph signal in Figure 1.1 takes values $x(1), x(2), x(3)$ corresponding to nodes v_1, v_2, v_3, respectively. For a given graph, there can be many different graph signals. As an example, a set of sensors can make different daily measurements, each leading to a different graph signal.

What we define to be a graph signal depends on the application: in a graph representing physical sensors, graph signals could be measurements (e.g., temperature); in a graph representing a machine learning dataset, the signal might represent the label information; a graph associated with an image may have pixel intensities as graph signals; and so on. Concrete examples of graph signals are discussed in Section 1.4.

Frequency and variation   Frequency is a key concept in conventional signal processing. Informally, terms such as "low frequency" or "high frequency" are commonly used in everyday conversation when referring to music and audio signals. A more mathematical definition of frequency, with complex sinusoids representing elementary frequencies, would be familiar to many science and engineering students. The definition of frequency, the number of cycles per second, indicates that, loosely speaking, higher frequency signals change faster, i.e., have greater variation, than signals with lower

---

[2] Generalization to the case where vectors are associated with each node is possible, but is not explicitly considered in this book.

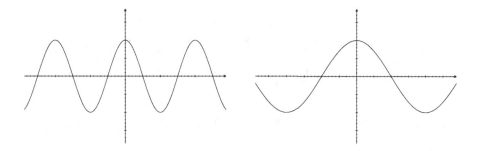

Figure 1.6 The sinusoid on the left has higher frequency than than on the right (the respective periods are 2 and 4). This can also be viewed in terms of variation. For example, selecting a small interval length Δt, say $\Delta t = 0.1$, we can observe that with any interval $[t, t + \Delta t]$ there is in general more change, i.e., more variation, for the sinusoid on the left. For graphs with arbitrary connections between edges and nodes, the focus of our interest, we will not be able to easily plot a graph signal or define frequencies in terms of sinusoids, but we will use the notion of variation to define elementary frequencies.

frequencies. This is illustrated in Figure 1.6. Since it is not possible, except for very particular cases, to have a meaningful definition of sinusoids for a graph with arbitrary nodes and edges, we define the concept of frequency for graph signals by quantifying the **graph signal variation**: greater variation on the graph leads to higher frequency.

Defining variation for time signals   Before considering graph signals, let us take a closer look at a conventional discrete-time signal, where each sample is associated with a point in time (e.g., speech, audio). We can reason in a similar way with images, where each pixel represents color or intensity in space. Taking the time-based signal of Figure 1.7(a) as an example, we can quantify how fast the signal changes, i.e., how large its **variation** is. Let us do this by computing the difference between two consecutive samples as $|x(k+1)-x(k)|$, where the absolute value is used since increases and decreases are equivalent in terms of variation. Notice that if we multiply all the samples $x(k)$ by some constant C, the overall variation would be multiplied by C. Thus, if we wish to compare two different signals in terms of variation we need to normalize the signals or, equivalently, normalize the variation. Then define a normalized **total variation**:

$$TV_t = \frac{1}{\sum_k |x(k)|} \sum_k |x(k + 1) - x(k)|. \tag{1.1}$$

Notice that in Figure 1.7(a) the samples are equispaced, as is the case for conventional signals. Thus, in an audio signal with N samples per second, the interval between two consecutive samples is $1/N$ seconds. Similarly, the distance between neighboring pixels in an image is always the same.

Variation for irregularly spaced time signals   In order to develop a metric of variation for graph signals, let us consider now the case where we have samples at arbitrary

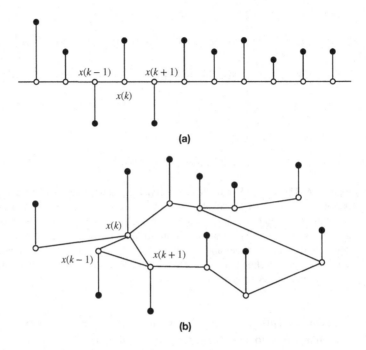

Figure 1.7 Variation and frequency in the time and node domains. The same signal seen as (a) observations in time and (b) observations placed on a graph. The signal in (a) can be viewed as being positioned on a line graph, while in (b) we have an arbitrary graph.

times $k_1, k_2, \ldots$ (e.g., due to missing data in the observations). Then, our definition can be adapted to measure "variation per unit time," that is:

$$\frac{1}{(k_{i+1} - k_i)}|x(k_{i+1}) - x(k_i)|$$

where the variation between samples further apart in time counts less. Then we can define the **normalized total variation** as

$$TV_t = \frac{1}{\sum_i |x(k_i)|} \sum_i \frac{1}{(k_{i+1} - k_i)}|x(k_{i+1}) - x(k_i)|, \qquad (1.2)$$

which gives us (1.1) when $k_{i+1} - k_i = 1$ for all i.

Defining variation for graph signals   To define variation for graph signals, let us follow the same principles we used to define (1.2): (i) compute the absolute difference between the signal value at *neighboring nodes* and (ii) scale this measure of variation by the *distance between nodes*.

Referring to the graph example in Figure 1.7(b), note that the node index has no meaning, i.e., the difference between the indices of two nodes v_1 and v_2 (i.e., $1 - 2$) cannot be interpreted as a time, as in the example in Figure 1.7(a). The example in Figure 1.5(b) showed that one can define a notion of the **similarity** between nodes as the inverse of their distance. Thus, rather than divide variations by distance, as in (1.2), let us *multiply them by the similarity*. Then the definitions of variation for graphs will be based on elementary operators such as

$$a_{12}|x(v_1) - x(v_2)| = \frac{1}{d_{12}}|x(v_1) - x(v_2)|$$

where the difference between the signals at connected nodes v_1 and v_2 is weighted by their similarity, a_{12}. Note that when distance becomes arbitrarily small its reciprocal becomes arbitrarily large. In practice, however, similarity is often derived from distance using a kernel function, such as an exponential, so that the maximum similarity is 1 as the distance between nodes becomes small.

With this definition the variation can be quantified for any graph by considering only nodes that are neighbors (i.e., such that their similarity is non-zero). Thus, the metric of (1.2) can be extended in a natural way as follows:

$$TV_g = \frac{1}{\sum_i |x(v_i)|} \sum_{i \sim j} a_{ij}|x(v_i) - x(v_j)|, \tag{1.3}$$

where the sum is over all pairs of nodes i and j that are connected, denoted by $i \sim j$, corresponding to $a_{ij} \neq 0$. The intuition behind this definition is illustrated in the following example. The problem of defining frequencies for graph signals will be studied in detail in Chapter 3.

Example 1.4   Time variation and graph variation   Compare Figure 1.7(b) with Figure 1.7(a). The nodes v_k, v_{k-1}, v_{k+1} have the same values in Figure 1.7(a) and in Figure 1.7(b). But in Figure 1.7(b) the connections between these nodes (and between those and other nodes) are somewhat more complicated. For example, unlike in Figure 1.7(a), nodes v_{k-1} and v_{k+1} are connected. Also, node v_k is connected to two other nodes apart from v_{k-1} and v_{k+1}, and since these other nodes have values closer to $x(k)$ the frequency of the signal on the graph of Figure 1.7(b) may be lower since the variation is smaller.

Even a simple definition of graph variation such as that of (1.3) can be useful in practice. For example, if we assume that data variation should be small under normal circumstances, the definition (1.3) can be used to detect anomalies corresponding to excessive variation. This is illustrated by the following example.

Example 1.5   Sensor networks – anomaly detection   Consider a set of temperature sensors inside a large factory. The goal is to use temperature measurements to determine whether there may be problems with some of the equipment. As an example, equipment overheating could be observed through locally higher temperature measurements.

Construct a graph where each node represents a sensor (and measurement). In order to detect an "anomaly" we can compare temperatures in different parts of the factory floor and try to identify unexpected behavior. Each graph node is associated with a sensor, so we need to decide how the nodes should be connected, i.e., how to choose edge weights between pairs of nodes.

We may consider the situation to be anomalous if a sensor measurement is very different from that of its close neighbors. Intuitively, it may be normal to measure very different temperature in areas far away from each other, but if two sensors are nearby and the temperature is much higher in one sensor, this could be an indication of equipment (or sensor) malfunction. Thus, we can consider a graph with edges having weights that are decreasing functions of the distance, and where only nodes close enough to each other will be considered to be connected. A simple detection strategy may then involve comparing the signal values at a node with those in the immediate neighborhood.

1.1.3   Filtering and Sampling Graph Signals

Conventional signals can be represented as a linear combination of elementary signals representing frequencies (e.g., complex sinusoids in the case of the Fourier transform). These frequencies can be ordered, so that we can talk about high and low frequencies. Similarly, we can develop graph signal processing for a given graph by defining frequency in a way that takes into account the characteristics of that graph.

This idea of quantifying signal variation across nodes will lead us to introduce frequency representations for graph signals, with high frequencies corresponding to fast signal variation across connected nodes. This notion of frequency will allow us to define tools to analyze, filter, transform and sample graph signals in a way that takes into account the signal variation on the graph. In terms that would be familiar to a signal processing practitioner, these tools will allow us to talk about "low pass" graph signals, with corresponding "low pass" filters. To address the sampling problem, we would like to be able to define the required level of "smoothness" that a signal is required to have in order for it to be recovered from a set of signal samples (i.e., observations made at a subset of nodes).

Example 1.6   Sensor networks – denoising   Considering again Example 1.5, we can formulate the problem of denoising the observed data using the concept of the graph signal frequency. Removing noise that affects measurements requires making assumptions about the signal measured and the measurement noise. For example, if we expect the temperature measurements to be similar across neighboring nodes we may use this to design a low pass filter that can reduce noise in the temperature measurements before processing them. A simple example of this could be a filter that updates the value at each node using a weighted average of the values at neighboring nodes.

Example 1.7   Sensor network – power management   Assume that it is necessary to turn off some sensors to reduce energy consumption. We can formulate the problem of selecting which sensors to turn off on the basis of the graph signal frequency, by viewing this as a sampling problem, Then, the goal is to select a subset of sensors under the assumption that the values at sensors that were not observed can be estimated using interpolating filters. Graph signal sampling will take into account the relative position of the nodes (the graph structure) to select the most informative nodes (sensors).

1.1.4 Graph Signal Processing versus Vector Processing

For a graph with N nodes we can combine all the $x(v)$ into a single vector $\mathbf{x} \in \mathbb{R}^N$. Thus, it is worth asking why we could not simply work with $\mathbf{x}$ as a vector in $\mathbb{R}^N$ and apply existing methods from linear algebra to transform and analyze this vector.

To answer this question, first note that for regular domain signals the sample indices are meaningful, e.g., in the time domain we know that sample $x(k + 1)$ comes after sample $x(k)$. In contrast, for graph signals the indices associated with each node are arbitrary. Thus, for two nodes i and j, the indices i and j themselves do not matter; what matters is whether there is an edge between those nodes. We can change the labels but, as long as the connections in the graph remain the same, the tools we develop for graph signal analysis will be the same (a simple permutation of the input signal). In general, for a graph with arbitrary connectivity, there is no obvious way to use a standard transform such as the DFT, given that there are many different ways of mapping a graph signal into a vector along a line. For example, given the signal of Figure 1.7(b) there are many ways of mapping it into a vector along a line as in Figure 1.7(a). Second, the same signal may have very different interpretations depending on how the nodes are connected. This is really the main motivation for graph signal processing and can be illustrated with a simple example.

Example 1.8   Consider the two graphs G and G' in Figure 1.7(a) and (b). The same graph signal is associated with both graphs. For which of those two graphs does the signal have higher variation?

Solution
Comparing Figure 1.7(a) and Figure 1.7(b), we observe that there are only two nodes with negative values. Note that those two nodes have only one connection to others with positive values in Figure 1.7(a), while several such connections exist in Figure 1.7(b). From this, we may infer that the signal has a higher frequency for the graph of Figure 1.7(b).

From Example 1.8 we see that the same graph signal can be associated with two different graphs (with the same nodes but different edges). Processing this signal as a graph signal is different from treating it as a vector because the tools we use for processing (e.g., our definition of frequency) depend on the graph. This idea is further developed next.

1.2   GSP Tools Adapt to Graph Characteristics

An important feature of graph signal processing is that it provides a *common* framework to study systems that are fundamentally different in their behavior and properties.

1.2.1   Selecting the Right Graph for a Task

In data science applications the goal is to extract useful information from data. To that end, GSP methods process and analyze data taking into account the topology of an underlying graph. However, a given signal **x** can be interpreted in different ways depending on the graph with which it is associated, as seen in Example 1.8.

In some cases the choice of graph is obvious given the application, while in others choosing the "right" graph is key to achieving meaningful results. We will discuss both types of scenarios, including techniques to optimize the graph selection, in Chapter 6. The problem of selecting a graph for a specific problem is illustrated in Example 1.9.

Example 1.9   In Example 1.5 assume that the temperature sensors are deployed inside a building, for which we have a floor plan. A straightforward definition of a graph would be simply based on the physical distances between the sensors, as discussed in Example 1.5. Given that variation (and thus frequency) depends on how the graph is connected, suggest alternative ways in which a graph could be built in order to provide a better analysis of the temperature within the building.

Solution
In Example 1.5 we were trying to detect anomalous behavior by identifying sensors whose temperature measurements are very different from those of neighboring sensors, since we expect that temperatures will generally be uniform within a room but may change from room to room. A possible solution is to select edge weights that decrease with distance but also take into account the presence of walls in between the corresponding sensors.

1.2.2   Graph Diversity: How General Are GSP Tools?

Whether a graph is given or has to be selected, we can encounter significant diversity in the graph properties. Graph sizes can vary significantly, from hundreds of nodes (e.g., in a sensor network) to millions of nodes (e.g., in an online social network). Similarly, the graph topology can be very regular (all nodes have the same number of connections) or highly irregular (some nodes have orders of magnitude more connections than others). Graphs representing online social networks can be highly irregular (e.g., some users have millions of followers, others few), while other graphs can be designed to be very regular (e.g., a K-nearest-neighbor graph connects each point in space to exactly the K closest points in space). Graphs can be directed or undirected (see Example 1.3), weighted or unweighted. Thus, even though we will be defining a single set of mathematical tools to work with many different types of graphs we should always keep in mind the following remark.

> *Remark* 1.1   In this book we develop tools that are meant to be generic and applicable to graphs with different characteristics. However, interpretation and insights derived from processing may be quite different from case to case. The tools may be generic, but their interpretation is highly dependent on graph properties.

Throughout this book, our goal will be to strike a balance between generality and specific behavior, by describing the general ideas first and then explaining how they apply to specific graphs. In what follows we present several examples of graphs and graph signals to illustrate the diversity of cases we may encounter. These examples will be used again in later chapters to provide more insights about the concepts we introduce.

For each of these examples we explain what the nodes and edges are and provide examples of graph signals of interest. We also include in our discussion the **degree distribution**, which measures how different the nodes in the graph are from each other. This is based on computing the degree of each node (its total number of edges if the graph is unweighted, or the sum of all its edge weights if the graph is weighted) and then characterizing how the degree changes across the graph. For example, a regular graph would be such that all its nodes have (approximately) the same degree. The examples in the following sections were generated with Matlab using the GraSP toolbox, which will be used throughout the book and is introduced in more detail in Appendix B. The code used to generate many figures in the book is available from the book's web page.[3].

1.3   Graphs in Classical Signal Processing

We start by introducing graphs that are closely linked to conventional signal processing, namely, path,[4] cycle and grid graphs, which correspond to finite 1D and 2D discrete signals. Frequency representations for these graphs match exactly those available for the corresponding signals (1D and 2D) studied in classical signal processing. This allows us to view GSP tools as a "natural" generalization of existing methods.

1.3.1   Path Graphs, Cycle Graphs and Discrete-Time Signals

Discrete-time (and finite-length) signals can be interpreted as graph signals. We do this by associating one graph node to each time instant, and letting a sample in time be the signal associated with the corresponding node. For a finite-length signal we can choose the path graph of Figure 1.8, generated using Code Example 1.1. If the similarity between neighboring nodes in Figure 1.8 is 1, then the total variation of (1.2) and the

---

[3] http://www.graph-signal-processing-book.org/

[4] We use the term path graph, rather than line graph. In graph theory, the term line graph is reserved for a graph derived from an existing graph, where edges in the original graph become nodes in the line graph.

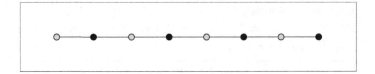

Figure 1.8 Define a path graph with N nodes and with all edge weights equal to 1. Then, a length-N time signal, where consecutive samples in time are associated with consecutive nodes on the graph, can be viewed as a graph signal. In this case the *graph frequency* definitions will correspond to the conventional definitions for time signals.

Code Example 1.1 `Matlab` code used to generate Figure 1.8

```
1  path10 = grasp_non_directed_path(8);
2  signal10 = [200 10 200 10 200 10 200 10]';
3  grasp_show_graph(gca, path8,...
4                  'node_values', signal8,...
5                  'color_map', 'gray',...
6                  'value_scale', [0 255]);
7  ylim([4 6]);
8
9  save_figure('path8');
```

graph total variation of (1.3) are equal. The graph frequency definitions that we will develop in Chapter 3 can be shown to correspond exactly to the discrete cosine transform (DCT)[1].

The same finite-length signal can also be mapped to a cycle graph (Figure 1.9). Since two nodes connected in the graph are consecutive in time, a cycle can be viewed as a representation of an infinite-length periodic signal with period N, the number of nodes. The graph of Figure 1.8 is almost regular, i.e., it has the same connectivity for each node except the two end nodes, while the cycle graph of Figure 1.9 is exactly regular. In this case, as will be seen in Chapter 3, the frequency representation associated with the graph is the discrete Fourier transform (DFT).

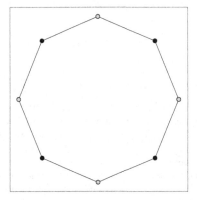

Figure 1.9 Cycle graph with same graph signal as in Figure 1.8. Note that here the two end-points of Figure 1.8 are connected. This is equivalent to viewing this signal as containing eight samples of a periodic signal with period eight.

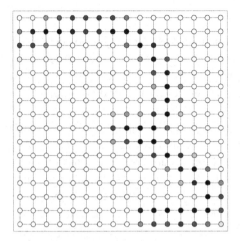

Figure 1.10 Grid graph representing an image from the USPS dataset. Each node corresponds to a pixel, and the signal at each node is the intensity of the corresponding pixel. In a signal-independent graph all edge weights can be equal.

The same signal is associated with two different graphs in Figure 1.8 and Figure 1.9, leading to different interpretations. Specifically, on the path graph of Figure 1.8, nodes 1 and N are not connected, so that the difference between $x(v_1)$ and $x(v_N)$ does not affect the total variation (1.3). In contrast, on the cycle graph of Figure 1.9, nodes 1 and N are connected (corresponding to the connection of consecutive cycles in a periodic signal) and thus $|x(v_1) - x(v_N)|$ will contribute to the total variation in (1.3). Comparing these two graph choices also reminds us that conventional signal processing relies on having multiple different representations for the same signal (e.g., DCT and DFT, among many others), with the choice of representation dependent on the specific application.

In Box 1.1 we summarize the basic ideas of 1D signals and their associated graphs. Throughout this chapter we will use similar boxes to summarize the properties of other graphs of interest. Each box includes a description of what the nodes are, and how their edges and edge weights may be chosen. We also mention whether the graph is regular (each node has roughly the same number of neighbors), describe the signals of interest and the typical scale of those graphs and give example applications.

Box 1.1   1D signals

- **Nodes**: one per signal sample
- **Edges**: only between samples that are neighbors in time
- **Degree distribution**: regular (each node has two neighbors) or nearly regular (in the line graph case, the two end nodes only have one neighbor each)
- **Signal**: the values of each sample in time
- **Scale**: each finite set of samples can be viewed as a "window" for time signal analysis; typical windows could be a few hundred to a few thousand samples
- **Applications**: audio, speech processing, for example

1.3.2 Images and Grid Graphs

Two-dimensional images can be interpreted as graph signals by defining a regular grid graph, where each node, corresponding to a pixel, can be connected to its four immediate neighbors. Pixels at the image boundaries can have two or three neighbors, as shown in Figure 1.10. The frequency modes for this grid graph are obtained from the 2D separable DCT.

As an alternative, we can also consider a grid graph that is exactly regular (all nodes have the same number of neighbors) by connecting row and column path graphs as cycles. Then each row (or column) of the grid graph corresponds to a path graph (similar to Figure 1.8) and becomes connected as a cycle (similar to Figure 1.9). Thus we connect the left and right nodes of each horizontal path graph, and similarly the top and bottom nodes of each vertical path graph. Mathematically, assume that pixels are indexed by $x(i, j)$ where $i, j = 0, \ldots, N - 1$. Then in a 4-connected graph, a given pixel would be connected to four neighbors: $x((i+1) \bmod N, j)$, $x(i, (j+1) \bmod N)$, $x((i-1) \bmod N, j)$, and $x(i, (j-1) \bmod N)$, where the modulo operation allows nodes along the top row and right column to be connected to the corresponding nodes along the bottom row and left column, respectively. For example, the pixel at the top right corner will have as a neighbor to its "right" the pixel at the top left corner, and its neighbor "above" will be the bottom right pixel. The frequency representation for this graph is the 2D separable DFT.

Box 1.2   2D images

- **Nodes**: one per pixel
- **Edges**: only between pixels that are neighbors in space
- **Degree distribution**: regular
- **Signal**: intensity (or color information) at each pixel
- **Scale**: total number of nodes is equal to the number of pixels (typically in the order of millions)
- **Applications**: image processing

1.3.3 Path and Grid Graphs with Unequal Weights

While signals associated with the graphs of Box 1.1 and Box 1.2 can already be analyzed using existing signal processing methods, these examples are still important. These representations lead to graph-based tools corresponding to well-known transforms, such as the 8×8 DCT for the path graph of Figure 1.8, which has been widely used, for example in the JPEG image compression standard [2]. But slight changes to those weights lead to different transformations (see Section 7.4). As graphs deviate from the properties used in conventional signal processing (e.g., when the edge weights are no longer equal), we can observe how the corresponding signal representations change; this provides insights about how the tools we develop adapt to changes in graph topology. For example, if all edge weights on the path graph are equal to 1 except for $a_{i,i+1}$ then this serves to indicate that samples up to i are somewhat decoupled from those after i. In the context of

Table 1.1 Classes of problems of interest

Category	Nodes	Links	Example
Physical networks	Devices	Communication	Sensor networks
Information networks	Items	Links	Web
Machine learning	Data examples	Similarity	Semi-supervised learning
Complex systems	System state	System transitions	Reinforcement learning
Social networks	People	Relationships	Online social networks

image processing this can be used to signal the presence of an image contour between i and $i + 1$. This simple observation can be seen to be a basic principle behind bilateral filtering, a highly popular image-dependent image filter [3].

1.4 Graphs Everywhere?

As technology for sensing, computing and communicating continues to improve, we are becoming increasingly reliant on a series of very large scale networks: the Internet, which connects computers and phones as well as a rapidly growing number of devices and systems (the Internet of Things); large information networks such as the web or online social networks. Even networks that have existed for decades (e.g., transportation or electrical networks) are now more complex and increasingly a focus of data-driven optimization. In what follows we describe examples of systems that can be understood and analyzed by using a graph representation, and to which GSP tools can potentially be applied. Our aim is not to be exhaustive, but rather to illustrate the wide variety of applications that can be considered. We divide these examples into several categories, which we discuss next (see Table 1.1).

1.4.1 Physical Networks

A physical network is a system where there are devices or components that can be represented as nodes in the graph model. The edges between two nodes are a function of distance or represent a physical communication link. For example, in a transportation network, the nodes may represent hubs or intersections while the links (edges) may correspond to roads or train tracks. In an electric network the nodes may include power generators and homes, while the links would represent transmission lines. In a communication network such as the Internet, each node may be a computer or other connected device, while each physical link between devices would represent a communication link.

The transportation road network in Figure 1.11 demonstrates one key feature of many of these physical networks: node position corresponds to an actual location in space. In some cases, e.g., sensor networks such as that of Figure 1.12, there are no obvious links between the nodes (the sensors), and thus one may construct a graph as a function of distance. In others, e.g., road networks, a better choice might be to select edge weights

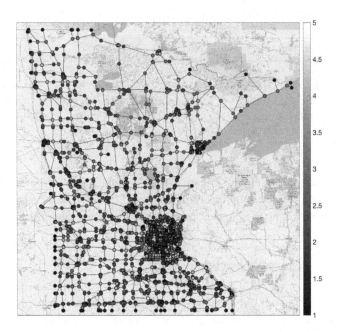

Figure 1.11 Minnesota road network graph. The signal value associated with each node on the graph is the corresponding node degree. Note that the node degrees in the graph are low, reflecting the physical limitations of the road network.

that are a function of the distance between nodes when following the network (i.e., the distance along the road, rather than the distance between nodes). This was illustrated in Figure 1.2. Finally, in communications networks, e.g., the Internet, the exact position of the nodes may be known but may not be relevant to define a graph representation. Instead, the capacity of the communication links may be more important. When analyzing a physical network, the properties of the graph topology may be constrained by those of the system itself. For example, in a realistic road network the number of roads leading to an intersection cannot be arbitrarily large, and thus the corresponding graphs will have low degree (see Figure 1.11). Also, as shown in Figure 1.2, one-way streets in a transportation network lead to directed graphs.

Box 1.3 Sensor network

- **Nodes**: one per sensor
- **Edges**: only between sensors that are within a certain range of each other (e.g., radio range)
- **Degree distribution**: can be regular, e.g., k-nearest-neighbor graph
- **Signal**: a set of sensor measurements, for example of temperature
- **Scale**: dozens to thousands of nodes
- **Applications**: environmental monitoring, anomaly detection

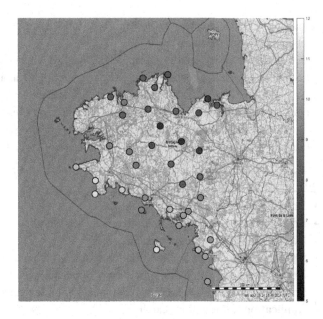

Figure 1.12 Sensor network example: weather stations. Note that here we know only the positions of the sensors. We could create a graph by connecting each sensor to its nearest neighbors. A graph signal is a set of temperatures (Celsius) measured at each weather station at a given time.

Sensor networks A sensor network is a concrete example of a physical system (see Box 1.3). A variety of systems can be described as sensor networks, including sets of distributed temperature sensors (indoors or outdoors), surveillance cameras or even devices carried by users, such as cellphones. Sensor networks are often deployed to sample information from the real world, e.g., temperature. Thus, the nodes are likely to have a location in space, and edge weights between nodes can be computed as a function of distance. The example of Figure 1.12 shows a regional network of weather sensors. Sensors can be deployed at different scales, within a building or on a bridge, across a city, or a region, and can gather many different types of measurements (temperature, air quality, vibrations, etc.) The processing of data obtained from sensor networks is an important motivating application for GSP. Data acquired by a sensor network is very often irregularly spaced. The methods we consider here will allow us to take into account sensor location as part of the processing, as illustrated in Examples 1.5, 1.6 and 1.7.

1.4.2 Information Networks

Organization of information as a network or graph is not new (e.g., encyclopedias are organized with entries and cross-references). The Internet has made this way of organizing information more explicit (and easier to navigate). Indeed, one of the most popular sources of information is Wikipedia, which is in itself an online version of an encyclopedia. In typical information networks each node represents an item of information (a web page, a Wikipedia entry) while (directed) links correspond to linking and cross-referencing between the items.

The web can be easily seen as a graph (see Box 1.4) where each page corresponds to a node, and the graph is directed, since one page can link to another without the linking being reciprocal. This graph structure has often been used to analyze page content. As an example, we can consider blogs dealing with specific topics (e.g., politics) and establish how they link to each other. Then a graph signal may be created at each node (blog) by creating a histogram of the frequency of appearances of certain keywords in that blog.

Box 1.4   The web

- **Nodes**: one per web page
- **Edges**: between web pages that reference each other
- **Degree distribution**: highly irregular
- **Signal**: some quantitative information extracted from a particular web page or blog (e.g., a distribution of the frequency of occurrence of certain keywords)
- **Scale**: the number of web pages is estimated to be several billion
- **Application**: information search

Note that graph-based representations of the web were key to the original PageRank search algorithm, where the essential idea was that the most relevant pages in a search would be those more likely to be traversed through a random walk of all pages that contain the search term [4].

1.4.3   Machine Learning

Supervised classification is a machine learning problem where the goal is to assign labels to data. For example, we may have a collection of images belonging to some categories (dogs or cats, say) and we would like to automate the process of determining to which category a new image belongs. An initial step in this design is to collect a representative set of labeled images, i.e., a training set containing examples of all categories under consideration. Then it is useful to consider the **similarity graph** associated with this training set. To do this, each image is mapped to a feature vector, which could be formed by the pixel values, or some information derived from the pixel values. In this graph each node corresponds to a data point, i.e., one image in the training set, and the edge weights are a function of the similarity between two data points, i.e., how similar the two images are in the chosen feature space. A typical approach to define a similarity graph is to associate a function of the form

$$a_{ij} = \exp(-d(\mathbf{x}_i, \mathbf{x}_j)^2 / \sigma^2)$$

with the distance between two data points $\mathbf{x}_i$ and $\mathbf{x}_j$. Thus, maximum similarity ($a_{ij} \approx 1$) is achieved when the two objects (for example, images) are close to each other. If the choice of distance metric is good, we expect the similarity between objects in different classes to be low (and $a_{ij} \approx 0$ if $d(\mathbf{x}_i, \mathbf{x}_j) \gg \sigma^2$).

> **Box 1.5   Learning: similarity graph**
>
> - **Nodes**: one per data point used in the learning process
> - **Edges**: between data points as a function of their distance in feature space
> - **Degree distribution**: can be regular, e.g., k-nearest-neighbor graphs
> - **Signal**: label for each data point
> - **Scale**: the number of nodes corresponds to the size of the training set, which could be in the order of millions of points for some modern datasets
> - **Application**: classifier design, semi-supervised learning

Graph-based methods have been used for unsupervised clustering, where the goal is to find a natural way to group data points having the same label. Notice that if a good feature space has been chosen (and the classification problem is relatively simple) one would expect neighboring nodes on the graph to have the same label. We will view this as a "smoothness" associated with the label signal. We will introduce this notion more formally and apply it to learning in Chapter 6.

1.4.4   Analyzing Complex Systems

In a complex system, many actions and events can affect overall behavior and performance. Examples of these systems include communication or transportation networks, a manufacturing plant, and a set of autonomous agents. A common problem in such systems is to take actions, often in a distributed way, in order to achieve some overarching performance goals. Note that, even in cases where there is an underlying physical system, e.g., a transportation network, we are simply interested in the logical graph describing the state of the system. As an example, in a physical communication network each node would correspond to router or queue, and a signal may represent the number of packets queuing. However, the state networks we consider here will have one state (node) corresponding to each possible state of the queue.

Complex systems can often be described by a series of state variables such that environmental changes or actions performed by a controller are meant to achieve desired changes to a state variable. For example, these state variables could be discrete, e.g., a counter that registers the number of occurrences of an event and resets to zero. These complex systems are in general non-deterministic, so that for a given action we cannot guarantee that the state of the system will evolve in a predetermined way.

We can create a directed graph where each node represents one possible system state, and directed edges correspond to possible transitions between states. For state variables represented by a counter, each node is associated with a possible counter value and each edge represents possible changes to the counter. If the counter value can only increase or decrease by 1, the corresponding graph will be such that every state will connect only to its two immediate neighbors (corresponding to a ±1 value for the counter). In principle, we can consider the problem of controlling such a system (selecting actions to be performed at each state) using a graph model. A graph signal for this problem is a "value function" representing a cost or reward associated with each state (node).

Note that since these system are not deterministic, we do not know *a priori* how likely state transitions will be. Then, it may be possible to observe the system to estimate values for those transitions, and corresponding edge weights. However, a major challenge comes from the fact that, in many such systems, the number of possible states is very large. Graph-based formulations have been proposed as a way to address state explosion, as they provide systematic ways to simplify the graphs [5]. Moreover, value functions are often estimated costs or rewards that do not vary significantly from one node to a neighbor, and thus can be viewed as smooth graph signals.

Box 1.6   Finite state machines

- **Nodes**: one per state
- **Edges**: between states for which a transition is possible
- **Degree distribution**: highly irregular
- **Signal**: some metric to quantify a given state
- **Scale**: potentially very large
- **Application**: reinforcement learning, system adaptation

1.4.5   Social Networks

In an online social network each node corresponds to a user, which may in turn correspond to a person or some other entity. Two users are either connected by an edge or not. Thus, there is no natural notion of distance between nodes, edge weights can take only the values zero or one, and these graphs are generally unweighted. In some cases the graph is undirected, if both users have to agree to "friend" each other (see Box 1.7). In other cases, one user may "follow" another user without necessarily being followed back (see Box 1.8). In this case, the graph is directed. Online social networks are communication and information platforms for their users, and a source of valuable information for the companies owning them. We can view any information available on the network as a signal. In some cases, a numerical representation is obvious (e.g., age), while in other cases, such as preferences about a specific topic, it can be constructed in different ways (e.g., it can be assigned to ± 1 depending on whether users prefer cats or dogs, and set to zero if no information is available).

Box 1.7   Online social network – undirected

- **Nodes**: one per user
- **Edges**: between users and their friends
- **Degree distribution**: highly irregular
- **Signal**: different information associated with each user, such as age
- **Scale**: millions to billions of nodes
- **Application**: information mining

In an undirected social network graph (Box 1.7) the connections in the graph may be used to predict information. For example, it may be desirable to poll some users to gather opinions, but impractical to try to poll all users. Then, if users who are connected are more likely to have similar opinions, it may be possible to "sample" carefully (polling only some users) and to then interpolate (infer what connected users may think on the issues in question given the response of their friends). Note that the connections established in a social network do not necessarily entail similar preferences between users. Prediction accuracy will depend on the specific information (signal). Prediction of political preferences probably does not have much in common with prediction of food preferences.

In a directed social graph (Box 1.8), edges link a user to all his or her followers. These graphs can be particularly irregular. Some users may have millions of followers, while others have a handful, something known as a power-law distribution. The graph connectivity has been used to estimate to what degree some users "influence" others. As an example, a graph signal associated with each user could be the number of messages forwarded by followers (e.g., "re-tweets"). Such a signal could be used as a measure of influence. In this case, as in others, more than one graph can be associated with the data. For example, one could consider a graph connecting hashtags and users who have tweets or re-tweets with those hashtags.

Box 1.8   Online social network – directed

- **Nodes**: one per user
- **Edges**: from user to other users he/she follows, and from followers to a user
- **Degree distribution**: highly irregular
- **Signal**: information associated with each user, e.g., geographical location
- **Scale**: millions to billions
- **Application**: identifying influencers

1.5   Mathematical Models of Graphs

In this book we mostly focus on scenarios where signals on a *given* graph are processed. Thus while we will consider what is possible for some specific *classes* of graphs, e.g., bipartite graphs, we will mostly view graphs as being deterministic, rather than random.

Probabilistic graph models   In the context of the broad field of *network science* there has been a significant amount of work done in developing probabilistic models of classes of graphs [6]. These models can be used to derive estimates of various graph properties that are valid for those specific graph classes. These models are based on defining the probability that any two nodes will be connected. They are used in order to develop closed form expressions for node degree distributions (the probability that a node has a certain number of neighbors). We briefly describe some of the most popular among these models to illustrate the basic concepts and link specific models to some of

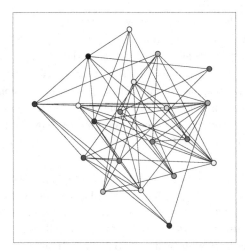

Figure 1.13 Erdős–Rényi graph. These graphs are created by taking every pair of points and connecting them on the basis of a given probability, for the particular graph. Thus the probability gives the proportion of node pairs that are connected. Since the same probability is associated with all pairs of nodes, all nodes will have the same expected degree. Thus, statistically all nodes are the same, even though for a given realization the nodes do not have the same degree (i.e., the graph is not regular).

Code Example 1.2 Code used to generate Figure 1.13. This code takes as a parameter the number of nodes (20), and the probability that any two nodes are connected (0.3)

```
1  erdos20 = grasp_erdos_renyi(20, 0.3, 'directed', 0);
2  h = axes;
3  signal = 255 * rand(1, 20);
4  grasp_show_graph(h, erdos20,...
5                   'node_values', signal, ...
6                   'color_map', 'gray',...
7                   'value_scale', [0 255]);
```

the examples of graphs discussed so far. Details on how these graphs are generated are given in Section B.2.

The question of how well a specific model fits graphs observed in practice is important, but not one we consider in this book. Probabilistic graph models are useful to evaluate specific algorithms (e.g., graph signal sampling) as they allow us to report results averaged over multiple realizations of a certain type of graph. We can use probabilistic graph models to quantify consistency of performance (variance across realizations) or to identify worst case scenarios in terms of model parameters (rather than specific deterministic graphs).

Erdős–Rényi graphs   Erdős–Rényi graphs (Figure 1.13) are defined by a single parameter, the probability that any two nodes are connected. Code Example 1.2 shows the example code used to obtain the graph of Figure 1.13. While this mathematical model allows results about connectivity and degree distribution to be derived, these models may not always capture accurately the structure of many real-world networks. Since every edge has the same probability of being included, all nodes exhibit the same (statistical) behavior, which is rarely the case in practice.

Small-world graphs   An alternative model that addresses some limitations of Erdős–Rényi graphs is the Watts–Strogatz or small-world graph (Figure 1.14). These models

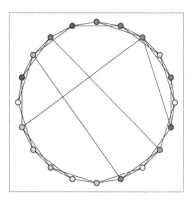

Figure 1.14 Watts–Strogatz or small-world graph. A graph based on this model has the property of being nearly regular (in this example most of the nodes along the circle connect with their immediate neighbors) but also showing random connections across longer distances. This type of graph is designed to capture the behavior of social networks where most connections are local, but a few can go far away, making the average distance between any two nodes in the graph relatively low. This property is popularly known as "six degrees of separation," where any two people are at most six hops apart from each other.

start with a regular graph, such as the cycle graph with N nodes (Figure 1.9). Then, connections are added while ensuring that there are minimal changes in regularity. In the initial model of Figure 1.9, each node has exactly two neighbors. Additional links are included so that each node connects to its k closest neighbors in the original cycle. Finally, with some low probability, edges are added between any two nodes. These added edges provide the small-world property, where it is possible to find relatively short paths between any two nodes.

Barabasi–Albert graphs   Barabasi–Albert or scale-free graphs (Figure 1.15) are de-signed to capture properties observed in social networks (e.g., Twitter) where a few users have orders of magnitude more followers than others (i.e., with some low prob-ability some nodes can have a very high degree). Starting with a connected network, nodes are added so that they connect to a subset of existing nodes, with the probability of connecting to a given node being a function of the degree of the node. Thus, as nodes are added to the network, high-degree nodes are likely to increase their degree even further.

Graph frequencies   The example in Figure 1.15 helps to illustrate the idea of graph frequency. This graph has 20 nodes and so we will be able to create a graph signal representation with 20 elementary frequencies; the corresponding eigenvectors are de-picted on the left. The definition of these frequencies will be the subject of Chapter 2 and Chapter 3, while the visualization will be described in Section 3.2.6. For now, recall from Figure 1.7 that a signal can be represented as a graph signal and as a 1D signal. This visualization builds on this idea by using an **embedding** (Section 7.5.2) that maps graph nodes into 1D. With this embedding each graph signal can be represented as a 1D signal, as shown on the left of Figure 1.15.

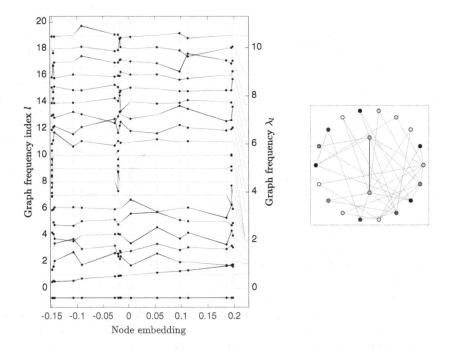

Figure 1.15 Barabasi–Albert or scale-free graph with 20 nodes (right). This type of graph model is meant to reflect the behavior of some real networks where the degree distribution can be modeled by an exponential, that is, it is possible to have nodes with high degree. In this example, the maximum degree is 9, while all nodes have degree at least 2. The graph signal in this example is randomly generated. The plot on the left represents elementary signals associated with the graph frequencies; the vertical displacement of a node represents the value of the signal at the node. Each set of 20 displacements of the 20 nodes constitutes a graph signal representing an elementary graph frequency (an eigenvector of the fundamental graph operator). The left scale (l) and the right scale (λ_l) are the frequency indices and frequency values, respectively, while the lines on the right link the evenly spread elementary graph frequency signals to their corresponding graph frequency (λ_l). This visualization is discussed in Section 3.2.6.

1.6 Roadmap and Approach

The main goal in this book is to provide a basic and intuitive understanding of how to extend to graph signals tools that are standard for regular signals. Chapter 2 introduces basic graph concepts and focuses on graph signal processing in the node domain. Then Chapter 3 introduces graph signal frequencies, building on the tools of spectral graph theory. With these definitions of frequency we can approach classical problems in signal processing such as sampling (Chapter 4) and signal representations (Chapter 5, and also approach more formally the problem of selecting a graph for a specific application (Chapter 6). We then turn our focus to applications (Chapter 7), with examples in several areas including machine learning, sensor networks and imaging. We also provide an overview of basic linear algebra concepts (Appendix A) and introduce in detail the GraSP Matlab toolbox for graph signal processing (Appendix B).

One of the challenges in introducing graph signal processing concepts is that often they can be applied to arbitrary graphs, but the same tools may exhibit very different behaviors for each type of graph. In short, the tools are the same but their behavior is very different. We aim to keep discussions general, while also pointing out these differences in behavior through examples. Thus, there is an important caveat. A specific tool, say a graph filtering method, may be applicable to any graph within a given class (e.g., it can be used for any undirected graph) but the results may not be meaningful for some specific graphs within the class (e.g., for random graphs).

Chapter at a Glance

The main goal of this chapter was to motivate the development of graph signal processing tools. We started by introducing some basic graph concepts and explaining what a graph signal is. We then provided some intuition for how a notion of signal variation can be the basis for a definition of graph signal frequency. Next, we gave a series of examples of graphs arising in many different applications, in order to illustrate their broad use and how significantly they can differ in each application. We showed how graphs can be associated with classical signal processing problems (path and grid graphs), and then considered graphs associated with physical networks, information networks, social networks or those arising in machine learning applications. Finally, we provided examples of random graph models, which can be used to generate graphs with given statistical properties.

Further Reading

The study of graphs has been an inspiration for the development of mathematical tools to solve many problems of interest, such as finding the shortest path between two nodes. Spectral graph theory [7, 8] makes use of the eigenstructure of algebraic representations of graphs to estimate graph structure. As an example, we will see that eigenvectors corresponding to low variation can be used to find approximate solutions to the min-cut problem (i.e., finding the partition of a graph into two subgraphs of equal size such that the number of connections or the weights of edges across subgraphs is minimized).

Graph signal processing takes methods in spectral graph theory as its starting point and develops extensions with the goal of providing tools to analyze and represent graph signals. The interest in GSP can be seen as a natural consequence of our increased ability to monitor and sense the environment (e.g., via wireless sensor networks), create complex networked systems (e.g., smart grids or the Internet) and apply optimization to solve large-scale resource allocation problems (e.g., logistical networks). In many of these problems, our goal is not only to understand the graph topology but also to use the graph topology to understand the graph signals (sensor measurements, Internet traffic etc.). As a field, GSP is at a relatively early stage, and one challenge in writing this book has been to decide how much of the recently published work on this topic should be included. Overview papers such as [9, 10, 11] provide perspectives on the topic, while a recently published monograph [12, 13, 14] gives an in-depth treatment of many of the same topics covered in this book.

GraSP **Examples**

Appendix B provides code examples related to specific topics covered in the book. At the end of each chapter, we provide pointers to examples that might be relevant and could be used as the basis for exercises. As a starting point, install GraSP (Section B.1). In GraSP each graph is represented by a data structure. Random graphs (Section B.2.1) can be generated for the models described in Section 1.5. Using such data structures makes it easy to implement a series of common tasks, such as plotting the graph (Section B.2.1) and associated graph signals (Section B.2.5). While random graph models are useful for developing a better understanding of graph properties, in general it may be necessary to import an existing graph (Section B.2.6). In some cases it may also be useful to build the graph directly with Matlab code (Section B.2.7).

Problems

1.1 *Path graph*

Define a length $N = 16$ path graph as in Box 1.1, with a set of positive edge weights a_{ij} between consecutive nodes. Use the variations TV_t and TV_g defined in (1.2) and (1.3), respectively.

a. Give a definition of the term "node" and a choice of a_{ij} for which the two variations are the same.

b. For the variation (1.2), assuming the points are equally spaced, find a signal x such that: (i) $\sum_k |x(k)| = N$, (ii) $\sum_k x(k) = 0$ and (iii) TV_t is minimal.

c. Assuming that $k_9 - k_8 = 1 - \epsilon$, $\epsilon < 1$, and the remaining intervals remain the same, find the new signal with minimal variation under the conditions in part **a**. Compare the two solutions and discuss.

d. What happens as ϵ in part **c** increases? Relate the change in edge weights to the change in the signal.

1.2 *Images as graphs*

Define a graph associated with an image (Box 1.2), with each pixel connected to its four neighbors (except at image boundaries) and with all edge weights equal to 1. For connected pixels i and j, denote as $\delta_{ij} = a_{ij}|x(i) - x(j)|$ the term in the summation of (1.3).

a. Write Matlab code to load an image and produce another image where each pixel corresponds to one of the δ_{ij} values (i.e., convert each edge in the original graph into a pixel in the resulting image). Discuss how the pixels in the output image depend on the characteristics of the pixels in the original image (e.g., by comparing smooth regions to texture regions in the original image).

b. Repeat part **a** but using, for the same pairs of connected edges, a new weight that depends on pixel intensities:

$$a_{ij} = \exp \frac{-(x(i) - x(j))^2}{\sigma^2}.$$

How do the results change? How does the choice of σ affect the output images?

1.3 *Erdős–Rényi graphs*
Use Code Example 1.2 to generate Erdős–Rényi graphs with different probabilities p. Notice that for lower probabilities it is more likely that the graph will become disconnected.

a. Write a Matlab code to determine whether the graph is disconnected.

b. Use simulations to generate an empirical estimate for the probability of a graph being disconnected (for a given probability p).

1.4 *Watts–Strogatz graphs*
Repeat Problem 1.3 for the Watts–Strogatz model of Section B.2.1.

1.5 *Exploring social network graphs*
This problem can be completed with data from any social network, either directed (Box 1.8) or undirected (Box 1.7). If the number of connections involved is large, use sampling to simplify the process, i.e., select a random subset of neighboring nodes and use only information from that subset.

a. Data collection: for your chosen social network, select one user, *Alice* (this could be your account or another account for which information is accessible) and record the following information: (i) the number of connections *Alice* has, (ii) the number of connections of each of *Alice*'s one-hop neighbors and (iii) the number of shared connections between *Alice* and her connections (if *Alice* and *Bob* are connected, how many people connect to both *Alice* and *Bob*).

b. Discussion: this is an open ended question with the goal of describing the structure of these social network graphs. Here are example questions that could be considered. Is there a lot of variability in (ii) across your neighbors? What are the maximum and minimum variabilities? How do you interpret this variability? What do the shared connections of (iii) tell you about the existence of communities?

1.6 *Random geometric graphs*
Generate a random geometric graph (see Section B.2.3) and note that graph nodes can be interpreted as randomly located sensors in space. Denoting by geom the generated graph, geom.layout gives the position of the nodes in space, while geom.A is the adjacency matrix (to be introduced in Chapter 2). In this symmetric matrix, the entry a_{ij} is equal to the inverse of the distance between node i and node j, $1/d(i, j)$. Typically a threshold is applied so that a_{ij} is set to zero if $1/d(i, j)$ is small. Thus, you can assume that nodes for which $a_{ij} = 0$ are not connected.

a. Write Matlab code to compute (1.3) for geom and use it to find the variation for a random vector **x**.

b. Most sensor networks observe the data generated independently of the sensors. To simulate this, generate a high-resolution random image using Matlab and assume this image represents a regular sampling of the same region as that represented by geom. Then, create a graph signal **x** by assigning to each node location in geom.layout the intensity of the closest pixel to that location in the image you generated.

c. Perform simulations to show how the variation computed in part **a** changes when the images used to generate the graph signal in part **b** have different properties (e.g., smooth images as against noisy images). Discuss.

d. In part **c** we considered a fixed set of sensors and changed the images used to generate the graph signal **x**. Now do the opposite: fix the image to generate graph signals, but compare the variation obtained when different random geometric locations for the sensors are selected. Is the variation consistent across multiple graph realizations? Discuss.

2 Node Domain Processing

Before developing a frequency representation for graph signals (Chapter 3), in this chapter we introduce tools to process graph signals in the node (or vertex) domain. Node-centric processing is a natural choice for graph signals, allowing us to process data at each node on the basis of information from its neighbors. In particular, this will be a useful tool to infer local information within a large graph or to split processing across multiple processors. This chapter will continue building on the idea of variation introduced in Chapter 1. We start by introducing several basic definitions. Next, we discuss more formally the notion of graph locality (Section 2.2) and introduce algebraic representations of graphs (Section 2.3) and how these can be used for processing by introducing graph filters (Section 2.4). Finally, we develop a more formal understanding of graph signals based on a chosen fundamental graph operator (Section 2.5).

2.1 Basic Definitions

In this book we focus on **simple** graphs[1] with at most one edge between any two vertices, but we study both undirected and directed graphs.

DEFINITION 2.1 (GRAPH) A graph $\mathcal{G}(\mathcal{V}, \mathcal{E})$ is defined by a set of nodes $\mathcal{V} = \{v_1, v_2, \ldots, v_N\}$ and a set of edges $\mathcal{E} = \{e_1, e_2, \ldots, e_M\}$. A directed edge ij in $\mathcal{E}$ goes from node v_j to node v_i (see below for a discussion of this convention).

If edge ij can take any real positive weight a_{ij} then the graph is **weighted**. **Unweighted** graphs have edges with weights all equal to 1. In both cases, if the edge ij does not exist then $a_{ij} = 0$. If ji exists whenever ij exists and $a_{ij} = a_{ji}$ then the graph is **undirected**. Otherwise the graph is a **directed** graph (also called a digraph): ij and ji may both exist, but in general $a_{ij} \neq a_{ji}$.

Directed graph convention   For directed graphs we adopt the same convention as in [15]: a_{ij} is non-zero if there exists an edge from v_j to v_i, which will be denoted by an arrow from j to i (see Figure 1.2). We make this choice so that arrows in the graph are consistent with the flow of information as reflected by the corresponding matrix operations (see Section 2.3). If $a_{ij} \neq 0$, but $a_{ji} = 0$, then observations at v_i will directly depend on observations at v_j, but not the other way around. In the example of Figure 1.2,

---

[1] Hypergraphs are defined as graphs where there can be multiple edges between any two nodes.

the traffic at node B directly depends on the traffic at node A, but the link between the traffic at A and the traffic at B is not direct. Therefore, on the graph there is an arrow from A to B but not from B to A.

Paths and cycles   Paths and cycles (Section 1.3) are important graphs that allow us to make a link with conventional signal processing. A path is a simple graph where two nodes are adjacent if and only if they are consecutive in a list (see Figure 1.8 for an example). A cycle is a simple graph with an equal number of nodes and edges, and where two nodes are adjacent if and only if they appear consecutively along a circle. Alternatively, we can think of a cycle graph as a path graph with N nodes, to which an edge from v_N to v_1 has been added (see Figure 1.9).

Complement graph   Given an unweighted graph $\mathcal{G}$, its complement $\mathcal{G}^c$ is a graph with the same nodes but with connections only between nodes that were not connected in the original graph; this means that a signal with low variation in $\mathcal{G}$ will have high variation in $\mathcal{G}^c$ (see Chapter 3).

> DEFINITION 2.2   (COMPLEMENT)   The **complement** of an unweighted graph $\mathcal{G}$ without self-loops is a graph $\mathcal{G}^c$ with same node set $\mathcal{V}$ but with edge set $\mathcal{E}^c$, the complement of $\mathcal{E}$ in the set of possible edges, i.e., an edge $ij \in \mathcal{E}^c$ if and only if $ij \notin \mathcal{E}$. The set of all possible edges corresponds to all the edges in a **complete** graph.

Self-loops   In some cases we use graphs with self-loops or node weights, i.e., edges going from v_i to v_i. Edge weights quantify the similarity between nodes. Since nodes should be maximally similar to themselves, self-loops can be interpreted by considering their weight *relative to the weight of the other edges* (see Section 3.2.2).

Node neighborhood   We will define locality in a graph (Section 2.2.2) on the basis of how the nodes are connected. To do so we define the neighborhood of a node.

> DEFINITION 2.3   (NODE NEIGHBORHOOD)   In an undirected graph, if v_i and v_j are endpoints of an edge, we say these nodes are one-hop neighbors or simply neighbors. We denote by $\mathcal{N}(i)$ the **set of neighbors** of v_i and, by the definition of an undirected graph, if $v_j \in \mathcal{N}(i)$ then $v_i \in \mathcal{N}(j)$.
>   In a directed graph, we define **in-neighbors** and **out-neighbors**. If ij is a directed edge from v_j to v_i then v_i is an out-neighbor of v_j, while v_j is an in-neighbor of v_i.

In Figure 1.2, A is an in-neighbor of B, while B is an out-neighbor of A.

Node degree, regularity and sparsity   Graphs can be characterized by the number of neighbors that each node has (for an unweighted undirected graph this is the same as the node degree), by how much this quantity changes between nodes (the regularity of the graph), and by how it relates to the number of nodes (the density or sparsity).

> **DEFINITION 2.4** (NODE DEGREE) The **degree** d_i of node v_i is the total weight of the edges connecting to this node. For an undirected graph this is
>
> $$d_i = \sum_j a_{ij}.$$
>
> For unweighted undirected graphs, d_i is the number of neighbors of node i. For a directed graph we can define **in-degrees** and **out-degrees**:
>
> $$d_i^{\text{in}} = \sum_{j=1}^{N} a_{ij}, \quad d_i^{\text{out}} = \sum_{j=1}^{N} a_{ji},$$
>
> which correspond to the weights of all the edges that end or that start at i, respectively.

In a directed graph, a node i with $d_i^{\text{in}} = 0$ and $d_i^{\text{out}} \neq 0$ is a **source**, while $d_i^{\text{in}} \neq 0$ and $d_i^{\text{out}} = 0$ corresponds to a **sink**.

A graph is **regular** if all its nodes have the same degree. The cycle graph in Figure 1.9 is exactly regular, with all nodes having degree 2, while the path graph in Figure 1.8 is nearly regular (two of its nodes have degree 1, the others have degree 2). Likewise the grid graph Figure 1.10 is not regular; most nodes have degree 4, while the nodes along the sides and at the corners have degrees 3 and 2, respectively. None of the other graphs from Chapter 1 is exactly regular. A graph with N nodes is **dense** if the number of edges for most nodes is close to N, and **sparse** if the number of edges for any node is much smaller than N.

Most graphs encountered in practice are not exactly regular. In this book, we will often use the terms regularity and sparsity qualitatively. For example, we may say that a graph is more regular than another if it shows less variation in its degree distribution. Similarly, if a graph is "nearly regular," its corresponding frequency definitions are more likely to resemble those used for conventional signals.

Subgraphs Subgraphs of a given graph contain a subset of nodes and edges and are important in the context of graph signal processing.

> **DEFINITION 2.5** (SUBGRAPH) Given a graph $\mathcal{G}(\mathcal{V}, \mathcal{E})$ a subgraph $\mathcal{G}_1(\mathcal{V}_1, \mathcal{E}_1)$ of $\mathcal{G}$ is such that $\mathcal{V}_1 \subset \mathcal{V}$ and $\mathcal{E}_1 \subset \mathcal{E}$, that is, if $a, b \in \mathcal{V}_1$ and $ab \in \mathcal{E}_1$ then we must have that $ab \in \mathcal{E}$.

There are several scenarios where using a subgraph can be preferable to using the original graph. A subgraph that contains all the nodes in the original graph but only some of the edges leads to lower-complexity graph operations, and may have favorable properties (e.g., it may be bipartite). A graph with fewer nodes and edges can be used to replace the original graph signal with a lower-resolution approximation, as part of a multiresolution representation. This can be particularly useful for datasets such as 3D point clouds, where the nodes can number in the millions.

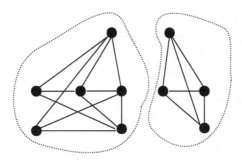

Figure 2.1 This graph consists of two components that are each connected. The corresponding subgraphs are not connected with each other.

Graph signals   We have defined graph signals as data associated with a graph, e.g., measurements in a sensor network or information associated with users in an online social network (see examples in Chapter 1). More formally:

> DEFINITION 2.6   (GRAPH SIGNAL)   A graph signal **x** is a real vector, $\mathbf{x} \in \mathbb{R}^N$, where the entry $x(v)$, $\forall v \in \mathcal{V}$, is the real scalar associated with node v. Unless otherwise stated we will consider only real graph signals, but extensions to complex and vector valued signals are also possible.

Recall that a graph signal, **x**, can only be interpreted with respect to a specific graph (see Section 1.1.2). Therefore **signal smoothness is related to graph locality** (see Section 2.2), which depends on the graph topology. In particular, the same signal **x** will have different interpretations on a graph $\mathcal{G}(\mathcal{V}, \mathcal{E})$ and on a subgraph $\mathcal{G}_1(\mathcal{V}, \mathcal{E}_1)$ of $\mathcal{G}$, where some edges are no longer included.

Box 2.1   Node indexing does not affect graph signal processing

The indexing of nodes on a graph is not important, as all the relevant information is captured by the connections between nodes (i.e., the set of edges). If v_i and v_j are connected and their indices are changed to v_i' and v_j', respectively, then the new edge set $\mathcal{E}'$ will contain $v_i' v_j'$. Likewise, given a matrix representation for a graph (as described in Section 2.3), changing the indices corresponds to a permutation of the rows and columns of the matrix.

Connected graphs   An *undirected* graph is **connected** if any node can be reached through a path from any other node. For *directed* graphs we consider two definitions. A graph is **strongly connected** if there is at least one directed path from any node to any other node; the path can have multiple hops. Note that graphs containing sinks or sources cannot be strongly connected, since no node is reachable from a sink, and a source cannot be reached from any node. A directed graph is **weakly connected** if we can build a connected undirected graph with the same nodes and where each edge connects two nodes if at least one directed edge existed between those two nodes in the original graph. Note that a strongly connected graph is always weakly connected, but the reverse is not true.

Connected components   From a graph signal processing perspective, a graph with two connected components (i.e., two components that are internally connected but not connected to each other, as in Figure 2.1) can be viewed as two separate or independent graphs.

Box 2.2   A graph with two connected components is equivalent to two independent graphs

Without loss of generality we can assume that all graphs are connected. This is so because, in practice, a graph with two separately connected components that have no edges linking the components can be treated as two independent graphs. To see why, note that if a graph is disconnected it can be divided into two sets of nodes $\mathcal{V}_1$ and $\mathcal{V}_2$, with no edges connecting those two sets, as shown in the example of Figure 2.1. Since our definitions of variation and smoothness are based on signal changes *across edges*, it is clear that differences between values in $\mathcal{V}_1$ and $\mathcal{V}_2$ do not affect variation. Examples of disconnected graphs are as follows:

- two sets of sensors, $\mathcal{V}_1$, $\mathcal{V}_2$, that are sufficiently far apart that data measured by sensors in $\mathcal{V}_1$ is not correlated with measurements in $\mathcal{V}_2$;
- a social network, where no user in $\mathcal{V}_1$ has a connection to any user in $\mathcal{V}_2$;
- graphs learned from data under probabilistic models (see Section 6.3 and Box 6.4) where the lack of connection between nodes can be interpreted as conditional independence.

If a directed graph is not weakly connected then each of the connected components can be considered as a separate graph, following the arguments laid out in Box 2.2. However, if a graph is weakly connected this does not guarantee that a meaningful interpretation of graph signals exists. For example, the signal value of a source node is not affected by any other node, since there are no incoming edges, while the signal at a sink node has no effect on any other nodes, since there are no outgoing edges.

Keeping in mind that strong connectedness is indeed a much stricter condition, which may not be easily met by graphs of interest, the question of interpreting graph signals and their frequencies for the directed-graph case remains open. Indeed, in Section 2.5.4 and Example 2.9 we will see how graphs that are not strongly connected can lead to graph operators for which some important properties do not hold.

2.2   Locality

We motivate the importance of "local" processing (Section 2.2.1), define locality in the context of graphs (Section 2.2.2), and introduce elementary local operations (Section 2.2.3) that will serve as the basis for more complex local operations. Efficient distributed processing typically requires partitioning a graph so that each set of nodes is processed separately. We discuss two graph partitioning problems. *Identifying clusters and cuts* (Section 2.2.4) aims at grouping together nodes that are close. In contrast, *sam-*

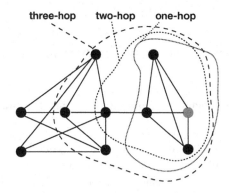

Figure 2.2 Example of locality on a graph. Each set contains neighbors of the gray node on the right. The one-hop, two-hop and three-hop neighborhoods of that node are shown.

pling and coloring (Section 2.2.5) assigns different labels to neighboring nodes, so that no connected nodes share the same label.

2.2.1   Importance of Locality

Localized methods are widely used for processing conventional signals such as audio, images or video. In audio coding, consecutive samples are grouped into **windows** and processed separately, so that the coding parameters adapt to local audio content, leading to increased coding efficiency. Similarly, most image and video coding systems divide images into non-overlapping **blocks**, so that each block (e.g., 8×8 pixels) is compressed separately, adapting to local image properties. Local processing is also useful for graph signals, since there are differences in behavior across the graph (e.g., separate areas monitored by a sensor network or different social groups in a social network).

A further benefit of local processing is computational efficiency. For audio processing, we can segment the signal into non-overlapping windows and encode the signal "on the fly," without having to wait for all of it to be available. Local operations make it easy to parallelize the processing: a large image can be divided into several sub-images, each of which can be assigned to a different processor. Similarly, a large graph can be split into subgraphs so that processing can be accomplished via distributed computations across multiple processors.

2.2.2   What Do We Mean by Local Graph Processing?

The notion of proximity can be generalized by extending Definition 2.3, corresponding to a one-hop neighborhood, and introducing the idea of a k-hop neighborhood.

> DEFINITION 2.7   (k-HOP NEIGHBORHOOD)   Given a node v_i, its k-hop neighborhood, $\mathcal{N}_k(i)$, is the set of all nodes that are part of a path with no more than k edges starting or ending at v_i.

In a k-hop neighborhood (Figure 2.2), for small enough k, "local" processing is possible, involving only a node and its closest neighbors. For example, in a social network, a two-hop neighborhood represents an extended circle of contacts, those we know directly and those to whom we could be introduced via one of our direct contacts.

When is a k-hop neighborhood truly local?   To put locality in perspective, we can ask how small k should be for processing on a k-hop neighborhood to be considered local. This can be done by comparing the number of k-hop neighbors $|\mathcal{N}_k(i)|$ of node i with the size of the graph. If $|\mathcal{N}_k(i)| \ll N$ then processing in a k-hop neighborhood of i can be considered to be local. Alternative methods to quantify locality require defining the distance between nodes.

For an undirected and unweighted graph, the geodesic distance between two nodes is the length, in number of hops, of the shortest path between the two nodes. This idea can be extended to weighted graphs by assuming that the edge weights are a measure of node similarity; distance is then obtained as the reciprocal of similarity.[2] Thus, we can define the geodesic distance between any two nodes i and j in a connected graph as follows.

DEFINITION 2.8   (GEODESIC DISTANCE)   If i, j are connected, $a_{ij} \neq 0$, define distance as $d_{ij} = 1/a_{ij}$, with greater similarity implying shorter distance. If i, j are not directly connected, let $p_{ij} = \{i, j_1, j_2, \ldots, j_k\}$ be a path connecting i and j. Then the geodesic distance between i and j is the minimum over all possible paths:

$$d(i, j) = \min_{p(i,j)} \left(\frac{1}{a_{ij_1}} + \frac{1}{a_{j_1 j_2}} + \cdots + \frac{1}{a_{j_k j}} \right).$$

In an unweighted graph $a_{ij} \in \{0, 1\}$, so that $d(i, j)$ is indeed the length (in number of hops) of the shortest path between i and j.

Note that geodesic distances are dependent on the graph structure. First, as illustrated in Example 1.3 the geodesic distance may not be symmetric, if the graph is directed. Second, the geodesic distance changes if the graph structure changes (e.g., the edge weights change). Thus if any of the edge weights along $p(i, j)$ does change, the distance following that path will change and $p(i, j)$ may no longer be the shortest path.

DEFINITION 2.9   (RADIUS AND DIAMETER OF A GRAPH)   For an arbitrary node v_i in a connected graph $\mathcal{G}$, define $D_i = \max_j d_{ij}$, the distance from v_i to the node farthest from v_i. In an unweighted and undirected graph, D_i is the maximum number of hops that can be taken away from v_i. The diameter and radius of the graph, $d(\mathcal{G})$ and $r(\mathcal{G})$, respectively, are:

$$d(\mathcal{G}) = \max_i D_i \quad \text{and} \quad r(\mathcal{G}) = \min_i D_i.$$

Intuitively, we expect $d(\mathcal{G})$ and $r(\mathcal{G})$ to be related to the graph's regularity. For instance, in the regular cycle graph of Figure 1.9, it is easy to see that $r(\mathcal{G}) = d(\mathcal{G}) = 4$.

---

[2] In this setting, similarity is given and distance is derived from it, but in other cases the distance between nodes is already well defined (e.g., when the nodes correspond to sensors deployed in the environment).

For an unweighted graph, radius and diameter are defined in terms of numbers of hops, so that we can define locality based on the following remark.

Remark 2.1   A k-hop localized node domain operation can only be considered truly "local" if k is significantly smaller than $r(\mathcal{G})$.

As an example, consider carrying out processing on a k-hop neighborhood for $k \geq r(\mathcal{G})$. This would imply that, for at least some nodes in $\mathcal{G}$, a k-hop operation would include **all** the nodes, which can hardly be described as local.

For a weighted graph it may be sufficient to compute the radius or diameter of the corresponding unweighted graph, i.e., the graph with the same topology as the original weighted graph but where all non-zero edge weights have been set to 1. But it is also important to distinguish between locality based strictly on connectivity and the localization implied by similarity. For example, let j be a node in a weighted graph that is exactly k hops away from i. If its similarity, $1/d_{ij}$, is very small, then its effect on a processing operation at i will be low in general. Thus, if all nodes that are k hops away have low similarity, we can view a k-hop operation as being nearly as localized as a $(k-1)$-hop operation. The topic of localization will be studied in more detail in Section 5.1.

2.2.3   Node-Centric Local Processing

Graph processing can be viewed as a series of *node-centric* operations. Consider a simple one-hop averaging operation:

$$y(i) = \frac{1}{d_i} \sum_{j \in \mathcal{N}(i)} a_{ij} x(j), \tag{2.1}$$

where d_i is the degree of node i and $x(i)$, $y(i)$ denote the ith entries of $\mathbf{x}$ and $\mathbf{y}$ respectively. This operation takes an input signal $\mathbf{x}$ and produces an output $\mathbf{y}$, where the ith entry is the average value of the neighbors of node i.

Computing the output $y(i)$ requires access to data from only the one-hop neighbors of i, $\mathcal{N}(i)$, which shows that this computation is *local* on the graph. Moreover, (2.1) is a simple forward operation, where the values in $\mathbf{x}$ are not modified, so that the outputs $y(i)$ can be computed in *any order*, without affecting the result. For example, we can compute $y(i)$ for $i \in 1, \dots, N$ in increasing index order. Since labels associated with nodes are arbitrary (Box 2.1), this ordering (node 1, then 2, 3 etc.) does not necessarily have advantages for processing. This is in contrast with path or grid graphs (see Figure 1.8 or Figure 1.10), for which some natural ordering of nodes exists, e.g., sequential traversal, passing all nodes, from one end-point to the other end-point for a path graph. The following example illustrates further how computation efficiency is closely related to graph structure.

Example 2.1   For the computation of (2.1) assume that the graph is too large to store the weights a_{ij} and $\mathbf{x}$, $\mathbf{y}$ in the processor memory. Then, the input signal $\mathbf{x}$ has to be

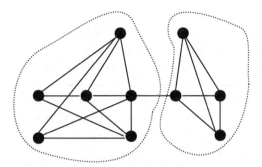

Figure 2.3 Example of clustering. Loosely speaking, for any node in a cluster, the number of connections within the cluster is much greater than connections to nodes outside the cluster. The graph cut is the set of edges which if removed would disconnect the two clusters. In this case only one edge forms the cut.

divided into two smaller signals $\mathbf{x}_1$ and $\mathbf{x}_2$, associated with a partitioning of V into two sets of nodes, V_1 and V_2, such that $V_1 \cup V_2 = V$ and $V_1 \cap V_2 = \emptyset$. Then $\mathbf{x}_1$ and $\mathbf{x}_2$, along with information about the neigborhoods and weights of nodes in each set, are assigned to different processors that can operate in parallel. Assuming that the two subsets are required to have approximately the same size, what would be a good choice for V_1 and V_2?

Solution
Assume that V_1 and V_2 are chosen and the corresponding $\mathbf{x}_1$ and $\mathbf{x}_2$ have been loaded into the respective processors. The processing of $\mathbf{x}_1$ can be completed for all nodes in V_1, *except* for those nodes in V_1 having a neighbor in V_2. If $j \in V_2$ and $j \in N(i)$ for $i \in V_1$ then the two processors have to exchange data to complete the computation. Thus, to reduce this communication overhead, we should choose V_1 and V_2 so as to minimize the number of edges between V_1 and V_2.

A general version of the solution of Example 2.1 requires dividing the original graph into multiple subgraphs, while minimizing the number of edges between nodes assigned to different subgraphs. The problem of dividing a graph into subgraphs in such a way that the connections across subgraphs (graph cuts) meet some conditions is studied next.

2.2.4 Clustering and Graph Cuts

A favorable grouping for distributed processing (Example 2.1) leads to subgraphs that are not strongly connected to each other, as illustrated in Figure 2.3. If V_1 and V_2 are the respective node sets in the two subgraphs, then the nodes in V_1 will have most of their neighbors in V_1, and we can say that the nodes in V_1 (and V_2) form a **cluster**.

More formally, assume that we wish to identify $L \geq 2$ clusters. A specific solution can be described by assigning a label $l \in \{1, \ldots, L\}$ to each node. Denote by $l(i)$ the label assigned to node i. Because the idea of clustering is to group neighbors together, it is natural to define the elementary **cost** of the labeling needed between two nodes as

$$c_{ij} = a_{ij} \, I(l(i) \neq l(j)),$$

where a_{ij} is the edge weight and $\mathcal{I}(l(i) \neq l(j))$ is an indicator function: $\mathcal{I}(l(i) \neq l(j)) = 1$ if $l(i) \neq l(j)$ (i and j are in different clusters) and $\mathcal{I}(l(i) \neq l(j)) = 0$ if $l(i) = l(j)$ (i and j are in the same cluster). The contribution to the overall cost is zero if two nodes are not connected ($a_{ij} = 0$). This allows us to define the clustering problem.

DEFINITION 2.10 (CLUSTERING) For $L \geq 2$, the L-class clustering problem consists of selecting labels $l(i) \in \{1, \ldots, L\}$ for each node i in a graph in such a way as to minimize the cost,

$$C = \sum_{i \sim j} c_{ij} = \sum_{i \sim j} a_{ij} \mathcal{I}(l(i) \neq l(j)), \tag{2.2}$$

where the summation is over all pairs of connected nodes (denoted by $i \sim j$). Variations of this problem may require the size of the clusters to be similar.

This is an NP-complete problem for which spectral techniques (Section 7.5.2) can provide efficient approximations. Clearly, the cost (2.2) can be made zero for a graph with exactly L connected components: in that case we assign each label to one of the connected components and there will be no connections between nodes having different labels. But, as discussed in Box 2.2, when we have multiple connected components we need to consider each of them as distinct graphs.

Graph cuts The problem of clustering can be related to the problem of finding a graph cut, where instead of grouping nodes we select edges.

DEFINITION 2.11 (GRAPH CUT) A graph cut is a set of edges $\mathcal{E}_c \subset \mathcal{E}$ in a connected graph $\mathcal{G}$ such that: (i) $\mathcal{G}$ is disconnected if all edges in $\mathcal{E}_c$ are removed and (ii) after removing all edges of any set $\mathcal{E}'$ such that $\mathcal{E}' \subset \mathcal{E}_c$, $\mathcal{G}$ remains connected. A graph cut is minimal, for given sizes of the resulting connected components, if it has the minimum number of edges, in an unweighted graph, or the minimum edge weights in a weighted graph.

Finding a **minimum graph cut** divides the graph into two sets of nodes, where the weights of edges connecting nodes in the two sets have minimal weight. We can see that the optimal solution to the problem of Definition 2.10 in the case $L = 2$ would correspond to a minimum cut as well. We will explore further the idea of clustering in Section 7.5.2, where we will consider how it connects with our definitions of graph frequencies. Similarly, finding a **maximum cut** also divides the graph into two sets of nodes, but this time the edge weights connecting those two sets are maximized. While a minimum cut can be used for clustering, a maximum cut is useful for coloring.

2.2.5 Graph Coloring

Graph coloring is a classical problem in graph theory, with interesting connections to graph signal processing. A color is a label associated with a set of nodes. A valid **graph coloring** is an assignment of colors to nodes such that no two connected nodes in the

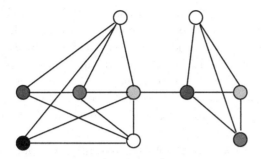

Figure 2.4 Example of coloring. The goal is to assign a color to each node in such a way that no two connected nodes have the same color. In this example, the graph has 10 nodes and is 5-colorable. Each color is represented by a different node shading.

graph are assigned the same color.[3] Note that clustering and coloring seek opposite goals: in clustering we try to minimize differences in labels between connected nodes, while a valid coloring requires labels to be different for connected nodes. A trivial coloring for a graph with N nodes is to assign N different colors, one to each node. Therefore the main challenge in graph coloring is to find the minimum number of colors, also called the **chromatic number**, for which a coloring exists. A k-colorable graph can be defined as follows.

> **DEFINITION 2.12** (k-COLORABLE GRAPH) A graph that is k-**colorable** or k-partite can be described using k sets of nodes, $\mathcal{V}_1, \mathcal{V}_2, \ldots, \mathcal{V}_k$, where $\mathcal{V}_i \cap \mathcal{V}_j = \emptyset$ and $\mathcal{V}_1 \cup \mathcal{V}_2 \cup \cdots \cup \mathcal{V}_k = \mathcal{V}$ for $i \neq j$ and where all edges e are such that $e = v_{k_i} v_{k_j}$, given that v_{k_i} and v_{k_j} are in $\mathcal{V}_i$ and $\mathcal{V}_j$, respectively, for $i \neq j$.

Figure 2.4 shows a coloring (with five colors) for the graph of Figure 2.3. Such bipartite graphs ($k = 2$) are of particular interest for graph signal processing and indeed for conventional signal processing as well (path graphs are always bipartite, but cycle graphs are only bipartite if they have an even number of nodes). The definition of a bipartite graph follows directly from Definition 2.12.

> **DEFINITION 2.13** (BIPARTITE GRAPH) A **bipartite** or 2-colorable graph has two sets of nodes, $\mathcal{V}_1, \mathcal{V}_2$, where $\mathcal{V}_1 \cap \mathcal{V}_2 = \emptyset$ and $\mathcal{V}_1 \cup \mathcal{V}_2 = \mathcal{V}$ and any edge e can be written as $e = v_1 v_2$ or $e = v_2 v_1$, where $v_1 \in \mathcal{V}_1$ and $v_2 \in \mathcal{V}_2$.

Note that the four nodes forming the right-hand cluster in Figure 2.4 are all connected with each other, i.e., they form a **clique**. Because of this, four different colors are needed for those four nodes, so the chromatic number is 4. Similarly a complete graph with N nodes will have chromatic number N. More generally, since coloring has to avoid assigning the same color to two neighbors, the number of colors for a graph with N nodes will depend on the number of edges.

Trees Path graphs are bipartite and do not have any cycles. More generally, graphs that have no cycles are trees. Any node on a tree can be chosen as the **root** and assigned

---

[3] The term graph coloring originates from the problem of assigning colors to countries on a map, where for clarity two countries sharing a border should not be assigned the same color.

to the first **level**. Nodes connected to the root are its **descendants**, or **children**, and belong to the second level, and a node reached through a path with K edges is at level $K + 1$. A node at any level with no descendants is called a **leaf** node. Trees where every node has the same number of descendants are called **regular**, and if it is possible to choose a root node such that all leaf nodes are at the same level then the tree is called **balanced**. Note that every tree is bipartite, but every bipartite graph is not a tree, since bipartite graphs can have cycles; however, the cycles can only have even length.

Approximate k-partition   From previous observations it follows that after removing edges from a graph, the chromatic number of the modified graph can only be the same or smaller. If a k-colorable graph for some prespecified k is needed (e.g., a bipartite graph as in Section 5.6.2) the graph can be modified (keeping all nodes, but removing some edges) to achieve the desired chromatic number. Clearly, we can always reach this goal by removing a sufficiently large number of edges, but this can lead to a graph that is very different from the original one. Thus, the more relevant (and challenging) problem is to find the best approximation, i.e., the graph having the desired chromatic number, while requiring the removal of the least number of edges or the removal of edges with the least total weight.

To illustrate this point, compare the cases of maximum and minimum cuts, which both divide a graph into two subgraphs and produce a bipartite approximation (where only edges that form the cut are kept, and all other edges are removed). However, the maximum cut would lead to the better approximation, since it selects edges with maximum weight across the two sets, and therefore the sum of edge weights within each set will be minimized. Since those are the edges to be removed in the bipartite approximation, minimizing their weight is a reasonable target. Graph approximations are studied in Chapter 6.

Coloring and signal variation   We can use the concept of coloring to gain some insights about graph signal variation. By definition, two nodes that are assigned the same color cannot be immediate neighbors. Consider the bipartite graph of Figure 2.5, with two different signals (Figure 2.5(a) and (b)) defined on the graph. These two signals are similar: both have four positive and three negative values. Notice that the signal in Figure 2.5(b) assigns the same sign to all nodes corresponding to one color. Since the graph is bipartite, all edges connect nodes with values of different signs and the corresponding signal has high variation. In contrast, there are fewer sign changes in the signal of Figure 2.5(a) leading to overall less signal variation.

Coloring and sampling   In graph signal sampling (Chapter 4), the goal is to select a subset of nodes (the sampling set) such that the signal obtained from these observed nodes can be used to estimate the signal at unobserved nodes. If we are going to select K out of N nodes it is desirable for these to be "spread out" so as to provide local information from all parts of the graph. This suggests that, after coloring a graph, one should sample all nodes with one color (since these are automatically kept separate). Thus, in the example of Figure 2.5 if three nodes are to be selected, the nodes on the

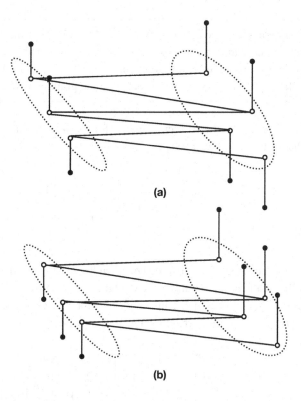

Figure 2.5 Example of bipartite graph with two different graph signals, indicated by the sets of vertical lines. A positive entry is denoted with a vertical line with a black circle above the node. The black circles are below the nodes for negative entries. Notice that both signals have three negative values and four positive values. (a) There is only one **sign change** across the edges of the graph. (b) There are sign changes across all edges.

left (corresponding to one color) would be a good choice: since they are not connected they are less likely to provide redundant information.

2.3 Algebraic Representations of Graphs

As just described, locality can be established on the basis of how nodes are connected. Local linear operations on graphs can be represented more compactly using matrix representations, leading to the definition of graph signal frequencies in Chapter 3.

2.3.1 Adjacency, Incidence and Degree Matrices

For a simple graph $G = (V, \mathcal{E})$ with sets of nodes $V = \{v_1, v_2, \ldots, v_N\}$, and edges $\mathcal{E} = \{e_1, e_2, \ldots, e_M\}$, the adjacency matrix, of size $N \times N$, captures all the connectivity and edge weight information.[4]

---

[4] For directed graphs we follow the edge direction convention described in Section 2.1.

DEFINITION 2.14 (ADJACENCY MATRIX) The adjacency matrix $\mathbf{A}$ is an $N \times N$ matrix where the entry a_{ij} corresponding to the ith row and jth column is equal to the weight of the edge from v_j to v_i, for $i \neq j$. If $a_{ii} = 0$ for all i then the graph has no self-loops, while $a_{ij} = a_{ji}$ indicates that the graph is undirected.

As an example, the adjacency matrix for a four-node unweighted path graph similar to that of Figure 1.8 can be written as

$$
\mathbf{A} = \begin{bmatrix} 0 & 1 & 0 & 0 \\ 1 & 0 & 1 & 0 \\ 0 & 1 & 0 & 1 \\ 0 & 0 & 1 & 0 \end{bmatrix},
$$

labeling the nodes from 1 to 4, left to right. As discussed in Box 2.3, different node labeling leads to a permutation of the adjacency matrix (and of the graph signals) and does not affect processing. Thus, if nodes in the undirected path graph were labeled (left to right) $1, 3, 2, 4$, the corresponding adjacency matrix would be

$$
\mathbf{A}' = \begin{bmatrix} 0 & 0 & 1 & 0 \\ 0 & 0 & 1 & 1 \\ 1 & 1 & 0 & 0 \\ 0 & 1 & 0 & 0 \end{bmatrix} = \begin{bmatrix} 1 & 0 & 0 & 0 \\ 0 & 0 & 1 & 0 \\ 0 & 1 & 0 & 0 \\ 0 & 0 & 0 & 1 \end{bmatrix} \mathbf{A} \begin{bmatrix} 1 & 0 & 0 & 0 \\ 0 & 0 & 1 & 0 \\ 0 & 1 & 0 & 0 \\ 0 & 0 & 0 & 1 \end{bmatrix},
$$

where $\mathbf{A}'$ is obtained by pre- and post-multiplying $\mathbf{A}$ by a permutation matrix representing the change in node indices (nodes 1 and 4 are unchanged, while 2 and 3 are exchanged with each other). We can now restate the ideas of Box 2.1 using permutation matrices.

Box 2.3 Node indexing and permutation

A change in node labeling or indexing can be written as a permutation of $\mathbf{A}$ which leaves unchanged the properties of interest. Let $\mathbf{P}$ be an $N \times N$ **permutation matrix** obtained by reordering the columns of the $N \times N$ identity matrix, $\mathbf{I}$. Given an input signal $\mathbf{x}$, $\mathbf{x}' = \mathbf{P}\mathbf{x}$ is a vector with the same entries as $\mathbf{x}$, but where the entries have been reordered. If the ith column of $\mathbf{I}$ is the jth column of $\mathbf{P}$ then $x'(j) = x(i)$. Then the adjacency matrix $\mathbf{A}'$, given by

$$
\mathbf{A}' = \mathbf{P}^{\mathsf{T}}\mathbf{A}\mathbf{P},
$$

has the same connections as $\mathbf{A}$ but with indices that are modified by permutation. Assume node a is connected to node b in $\mathbf{A}$ and that, after applying permutations, a and b become a' and b', respectively. Then there will be a connection between a' and b' in $\mathbf{A}'$.

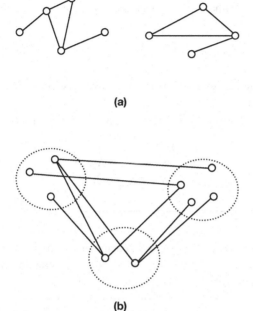

(a)

(b)

Figure 2.6 Examples of graphs. (a) A graph with two connected components leads to a block diagonal matrix with two non-zero blocks. (b) A 3-colorable graph leads to a matrix with three zero blocks along the diagonal.

Directed graphs   For an unweighted directed (left to right) path graph with four nodes, the corresponding adjacency matrix is

$$\mathbf{A} = \begin{bmatrix} 0 & 0 & 0 & 0 \\ 1 & 0 & 0 & 0 \\ 0 & 1 & 0 & 0 \\ 0 & 0 & 1 & 0 \end{bmatrix}, \tag{2.3}$$

so that multiplying a signal by $\mathbf{A}$ corresponds to *shifting* its entries between nodes along the direction of the edge arrows: if the input signal is $\mathbf{x} = [1,0,0,0]^\mathsf{T}$, then $\mathbf{Ax} = [0,1,0,0]^\mathsf{T}$.

Block-structured adjacency matrices   Relabeling can be used to simplify the notation, in particular if a suitable permutation allows us to write adjacency matrices in block form. Two cases of interest are: (i) disconnected graphs, which can be written in block diagonal structure (one block for each connected component) and (ii) k-colorable graphs, which have identically zero diagonal blocks, corresponding to the nodes in each color. These two cases are illustrated in the following example.

Example 2.2   Write down the block form of the adjacency matrices corresponding to the graphs of Figure 2.6(a) and (b).

Solution

The graph in Figure 2.6(a) has two connected components, so its adjacency matrix $\mathbf{A}$ can be written as

$$\mathbf{A} = \begin{bmatrix} \mathbf{A}_1 & \mathbf{0} \\ \mathbf{0} & \mathbf{A}_2 \end{bmatrix}$$

where $\mathbf{A}_1$ and $\mathbf{A}_2$ are adjacency matrices that can have different sizes and the remaining two entries in $\mathbf{A}$ are all-zero matrices.

The graph in Figure 2.6(b) is 3-colorable, and its adjacency matrix can be written as

$$\mathbf{A} = \begin{bmatrix} \mathbf{0} & \mathbf{A}_{12} & \mathbf{A}_{13} \\ \mathbf{A}_{21} & \mathbf{0} & \mathbf{A}_{23} \\ \mathbf{A}_{31} & \mathbf{A}_{32} & \mathbf{0} \end{bmatrix}$$

Incidence matrix   In the adjacency matrix each column (or row) represents all connections of the corresponding node. However, in the **incidence matrix** each column represents one edge, and each row corresponds to a node.

DEFINITION 2.15   (INCIDENCE MATRIX – DIRECTED GRAPH)   The incidence matrix $\mathbf{B}$ of a graph with N nodes and M directed edges is a rectangular $N \times M$ matrix. If the kth edge is $e_k = v_i v_j$, from j to i with weight a_{ij}, then the kth column of $\mathbf{B}$, $\mathbf{b}_k$, has only two non-zero entries, $b_{jk} = -\sqrt{a_{ij}}$ and $b_{ik} = \sqrt{a_{ij}}$. Note that by construction each column adds to zero:

$$\mathbf{1}^{\mathsf{T}}\mathbf{b}_k = 0 \quad \text{so that} \quad \mathbf{1}^{\mathsf{T}}\mathbf{B} = \mathbf{0} \quad \text{and} \quad \mathbf{B}^{\mathsf{T}}\mathbf{1} = \mathbf{0}, \tag{2.4}$$

where $\mathbf{1} = [1, \ldots, 1]^{\mathsf{T}}$ is a vector with all entries equal to 1. Each row i contains the square root of the weights of the edges for which v_i is an end-point, where the sign of an entry is a function of the orientation of the edge.

Thus an incidence matrix $\mathbf{B}$ is such that: (i) each column of $\mathbf{B}$ represents an edge, with two non-zero entries with equal absolute values, the positive one corresponding to the end-point of the edge and (ii) each row of $\mathbf{B}$ represents a node, and non-zero row entries correspond to edges going in or out of that node.

Incidence and signal processing   The incidence matrix of a directed graph allows us to represent the evolution of a graph signal in terms of **flows** between neighboring nodes: if nodes i and j are connected, a quantity flowing through the edge from j to i is added to i and subtracted from j, or shifted from j to i as in (2.3). Denote by $\mathbf{y} \in \mathbb{R}^M$ a vector of flows, where the kth entry of $\mathbf{y}$ represents the flow along the kth edge. Then, multiplying $\mathbf{y}$ by the incidence matrix produces

$$\mathbf{B}\mathbf{y} = \sum_{k=1}^{M} y_k \mathbf{b}_k, \tag{2.5}$$

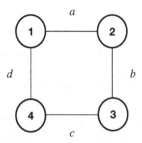

Figure 2.7 Graph for Example 2.3.

where y_k is the kth entry of $\mathbf{y}$ and $\mathbf{b}_k$ is the kth column of $\mathbf{B}$. Thus, if the kth edge goes from node j to node i then $y_k \sqrt{a_{ij}}$ will be added to i and $y_k \sqrt{a_{ij}}$ will be subtracted from j, corresponding to the kth vector in the summation in (2.5). If the original graph signal is $\mathbf{x}$ and a flow $\mathbf{y}$ is applied along the edges, then the resulting graph signal $\mathbf{x}'$ is

$$\mathbf{x}' = \mathbf{x} + \mathbf{B}\mathbf{y} \tag{2.6}$$

so that whatever is subtracted from node j is added to node i and, using (2.4), the sums of the entries of $\mathbf{x}'$ and $\mathbf{x}$ are the same:

$$\mathbf{1}^{\mathsf{T}}\mathbf{x}' = \mathbf{1}^{\mathsf{T}}\mathbf{x}.$$

Note that an undirected graph with M edges can be viewed as a directed graph where edges in both directions have equal weights, i.e., $a_{ji} = a_{ij}$, and we can define the corresponding incidence matrix as an $N \times 2M$ matrix where each edge in the graph appears twice, once in each direction. A more compact definition of the incidence matrix for the undirected case includes each edge just once.

DEFINITION 2.16 (INCIDENCE MATRIX – UNDIRECTED GRAPH) The incidence matrix $\mathbf{B}$ can also be defined as a rectangular $N \times M$ matrix where row i contains the square root of the weights of edges for which v_i is an end-point. If edge k connects i and j with weight a_{ij} then $b_{ik} = \sqrt{a_{ij}}$ and $b_{jk} = -\sqrt{a_{ij}}$, where the sign can be chosen arbitrarily as long as one of the two entries is negative and the other is positive.

Example 2.3 Write down the incidence matrix $\mathbf{B}$ for the undirected graph of Figure 2.7.

Solution
The label that we associate with each edge is not important, and, from Definition 2.16, we can choose arbitrary signs as long as each column sums to zero. Then we can write

$$\mathbf{B} = \begin{bmatrix} \sqrt{a} & 0 & 0 & \sqrt{d} \\ -\sqrt{a} & \sqrt{b} & 0 & 0 \\ 0 & -\sqrt{b} & \sqrt{c} & 0 \\ 0 & 0 & -\sqrt{c} & -\sqrt{d} \end{bmatrix}.$$

Degree matrix   Given a graph with adjacency matrix $\mathbf{A}$ we can define the degree matrix, which will allow us to characterize the graph's regularity.

DEFINITION 2.17   (DEGREE MATRIX)   The degree matrix $\mathbf{D}$ of an undirected graph is an $N \times N$ diagonal matrix such that each diagonal term represents the degree (number of edges or sum of edge weights) for the corresponding node, i.e., $d_{ii} = \sum_j a_{ij}$, where a_{ij} is the i, j entry of $\mathbf{A}$. If the graph is directed then we can define in-degree and out-degree matrices by summing only the weights of the incoming and outgoing edges, respectively.

The degree matrix for an undirected graph can be written more compactly as

$$\mathbf{D} = \mathrm{diag}(\mathbf{A1}),$$

where $\mathrm{diag}(\mathbf{x})$ is a diagonal matrix with the entries $\mathbf{x}$ along the diagonal. For directed graphs we can obtain the *in-degree* and *out-degree* matrices in a similar way:

$$\mathbf{D}_{\mathrm{out}} = \mathrm{diag}(\mathbf{1}^{\mathsf{T}}\mathbf{A}) \quad \text{and} \quad \mathbf{D}_{\mathrm{in}} = \mathrm{diag}(\mathbf{A1}),$$

since the ith row represents the weights of edges ending in node i (Definition 2.4).

As was the case for the incidence matrix, the adjacency and degree matrices define operations on graph signals. That is, we can multiply a graph signal $\mathbf{x}$ by one of those matrices to obtain a new graph signal. Since $\mathbf{D}$ is diagonal, $\mathbf{Dx}$ is simply a scaling of the signal values at each node, while $\mathbf{Ax}$ can be interpreted as a "diffusion operation" where the ith entry of $\mathbf{Ax}$ is obtained as a weighted sum of the entries in $\mathbf{x}$ corresponding to neighbors of i on the graph. General node domain operations constructed with these matrices and the graph Laplacian will be studied in Section 2.4.

Matrix computations and large graphs   Since graphs can be large, it is important to note that their algebraic representation is useful conceptually, but also may be practical for the storage of computation. For storage a straightforward representation of an adjacency matrix is as a 2D array where each element of the array corresponds to one edge, or contains a zero if no edge exists. As an alternative, an *adjacency list* representation stores for each node a linked list of all neighboring nodes so that total required storage increases with the number of edges $|E|$, which can be significantly smaller than N^2 [16]. This approach is much more efficient for very large and sparse graphs. Also note that software packages such as `Matlab` include special representations for sparse matrices.

2.3.2   Graph Laplacians

We introduce several types of graph Laplacians for undirected graphs, which will be used to develop the concepts of graph frequency in Chapter 3.

DEFINITION 2.18 (COMBINATORIAL GRAPH LAPLACIAN MATRIX) The combinatorial graph Laplacian matrix $\mathbf{L}$ of an undirected graph with adjacency and degree matrices $\mathbf{A}$ and $\mathbf{D}$ is the $N \times N$ matrix

$$\mathbf{L} = \mathbf{D} - \mathbf{A}. \tag{2.7}$$

Notice that the Laplacian does not include any additional information that was not already available, since $\mathbf{D}$ is given once $\mathbf{A}$ is known. We can get some insight on how to interpret $\mathbf{L}$ by considering the following example.

Example 2.4 Compute $\mathbf{BB}^\mathsf{T}$ for the incidence matrix of Example 2.3 and express it in terms of $\mathbf{A}$.

Solution
We can easily compute

$$\mathbf{BB}^\mathsf{T} = \begin{bmatrix} a+d & -a & 0 & -d \\ -a & a+b & -b & 0 \\ 0 & -b & b+c & -c \\ -d & 0 & -c & c+d \end{bmatrix}.$$

Recall that

$$\mathbf{A} = \begin{bmatrix} 0 & a & 0 & d \\ a & 0 & b & 0 \\ 0 & b & 0 & c \\ d & 0 & c & 0 \end{bmatrix} \text{ and } \mathbf{D} = \begin{bmatrix} a+d & 0 & 0 & 0 \\ 0 & a+b & 0 & 0 \\ 0 & 0 & b+c & 0 \\ 0 & 0 & 0 & c+d \end{bmatrix}$$

and therefore $\mathbf{BB}^\mathsf{T} = \mathbf{D} - \mathbf{A} = \mathbf{L}$.

The result of Example 2.4 holds in general.

PROPOSITION 2.1 For any undirected graph, with incidence matrix $\mathbf{B}$, we have

$$\mathbf{L} = \mathbf{D} - \mathbf{A} = \mathbf{BB}^\mathsf{T}. \tag{2.8}$$

Proof Let $\mathbf{b}_k$ be the kth column of $\mathbf{B}$ which has non-zero entries $-\sqrt{a_{ij}}$ and $\sqrt{a_{ij}}$, corresponding to rows i and j, respectively. Then we can write $\mathbf{BB}^\mathsf{T}$ as a sum of rank-1 matrices:

$$\mathbf{BB}^\mathsf{T} = \sum_k \mathbf{b}_k \mathbf{b}_k^\mathsf{T} \text{ with } \mathbf{L}_k = \mathbf{b}_k \mathbf{b}_k^\mathsf{T} = \begin{bmatrix} & \vdots & & \vdots & \\ \cdots & a_{ij} & \cdots & -a_{ij} & \cdots \\ & \vdots & & \vdots & \\ \cdots & -a_{ij} & \cdots & a_{ij} & \cdots \\ & \vdots & & \vdots & \end{bmatrix}, \tag{2.9}$$

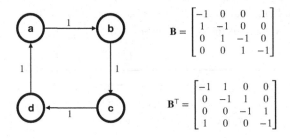

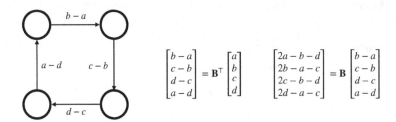

Figure 2.8 Interpretation of incidence matrix **B**.

where only the non-zero terms of $\mathbf{L}_k$ are shown. Therefore the sum of all the $\mathbf{L}_k$ has diagonal terms equal to the degree of each node and (since there is only one edge between nodes i and j) the off-diagonal terms correspond to the negative of the adjacency matrix, from which the result follows.   □

With $\mathbf{L}_k = \mathbf{b}_k \mathbf{b}_k^\mathsf{T}$ and denoting $\mathbf{x}_k = \mathbf{L}_k \mathbf{x}$ we can write

$$\mathbf{L}\mathbf{x} = \left(\sum_k \mathbf{L}_k\right)\mathbf{x} = \sum_k \mathbf{x}_k, \tag{2.10}$$

where each entry of $\mathbf{x}_k$ can be written as follows:

$$x_k(i) = a_{ij}(x(i) - x(j)), \quad x_k(j) = a_{ij}(x(j) - x(i)) \quad \text{and} \quad x_k(n) = 0, \forall n \neq i, j, \tag{2.11}$$

where we see that $x_k(i) + x_k(j) = 0$ and thus the sum of the entries of $\mathbf{x}_k$ is 0 ($\mathbf{1}^\mathsf{T}\mathbf{x}_k = 0$). Thus, adding $\mathbf{x}_k = \mathbf{L}_k \mathbf{x}$ can be viewed as the result of a flow along the kth edge, where the amount added to node i is the amount subtracted from j and we have

$$\mathbf{1}^\mathsf{T}(\mathbf{x} + \mathbf{L}\mathbf{x}) = \mathbf{1}^\mathsf{T}\mathbf{x}. \tag{2.12}$$

This interpretation is valid for both directed and undirected graphs: the operation $\mathbf{B}\mathbf{B}^\mathsf{T}$ first creates a flow $\mathbf{y}$ from $\mathbf{x}$ ($\mathbf{y} = \mathbf{B}^\mathsf{T}\mathbf{x}$) and then maps it back to produce an output $\mathbf{x}' = \mathbf{B}\mathbf{y}$. See Figure 2.8 for a directed graph example.

2.3.3 Normalized Graph Operators

Consider a weighted graph with adjacency matrix $\mathbf{A}$ and denote the output of the corresponding one-hop operation by $\mathbf{y} = \mathbf{A}\mathbf{x}$. Then the new signal value at node i is

$$y(i) = \sum_{j \in \mathcal{N}(i)} a_{ij} x(j), \tag{2.13}$$

where node i contributes $a_{ij}x(i)$ to node j, while node j contributes $a_{ij}x(j)$ to node i. The ith column of $\mathbf{A}$ provides the weights of all outgoing edges from node i, while the jth row includes the weights of all incoming edges for node j. The one-hop operation (2.13) needs to be modified (normalized) in order to provide a suitable model for some specific systems. We consider two important scenarios where normalization is required.

Graph normalization strategies In a **physical distribution network**, the signal at each node represents some physical quantity (e.g., goods being exchanged between nodes) so that the amounts "sent" through all links, i.e., the **outflows**, have to sum up to the original quantity (i.e., each of the items can be sent through only one of the links). Thus the signal at a given node has to be "distributed" across links. In this case, the total outflow of node i, with initial value $x(i)$, can at most be $x(i)$. Since by definition the sum of all outgoing weights is $d(i)$, where $d(i)$ is the degree of node i, normalized outflows are achieved by computing

$$\mathbf{y} = \mathbf{A}\mathbf{D}^{-1}\mathbf{x},$$

where each column of $\mathbf{A}\mathbf{D}^{-1}$ adds to 1.

Conversely, in a **consensus network**, a node receiving inputs from multiple nodes, i.e., **inflows**, may not simply add them. Instead, it could "summarize" them by computing a consensus or average. This can be achieved with using the one-hop average operation of (2.1), which is written in matrix form as

$$\mathbf{y} = \mathbf{D}^{-1}\mathbf{A}\mathbf{x},$$

where each row of $\mathbf{D}^{-1}\mathbf{A}$ adds to 1.

Normalized Laplacians The same normalizations can be applied when the Laplacian $\mathbf{L} = \mathbf{B}\mathbf{B}^\mathsf{T}$ is chosen as the one-hop operator, leading to one-hop operators $\mathbf{L}\mathbf{D}^{-1}$ and $\mathbf{D}^{-1}\mathbf{L}$ for the physical and consensus networks, respectively. These two normalizations can be seen as column or row normalization of the rank-1 matrix $\mathbf{L}_k$ in (2.9), leading to

$$
\mathbf{L}_k \mathbf{D}^{-1} = \begin{bmatrix} & \vdots & & \vdots & \\ \cdots & \dfrac{a_{ij}}{d_i} & \cdots & \dfrac{-a_{ij}}{d_j} & \cdots \\ & \vdots & & \vdots & \\ \cdots & \dfrac{-a_{ij}}{d_i} & \cdots & \dfrac{a_{ij}}{d_j} & \cdots \\ & \vdots & & \vdots & \end{bmatrix}, \quad
\mathbf{D}^{-1}\mathbf{L}_k = \begin{bmatrix} & \vdots & & \vdots & \\ \cdots & \dfrac{a_{ij}}{d_i} & \cdots & \dfrac{-a_{ij}}{d_i} & \cdots \\ & \vdots & & \vdots & \\ \cdots & \dfrac{-a_{ij}}{d_j} & \cdots & \dfrac{a_{ij}}{d_j} & \cdots \\ & \vdots & & \vdots & \end{bmatrix}.
$$

$$\tag{2.14}$$

The second normalization in (2.14) leads to

$$\mathbf{D}^{-1}\mathbf{BB}^{\mathsf{T}} = \mathbf{D}^{-1}\mathbf{L} = \mathbf{D}^{-1}(\mathbf{D} - \mathbf{A}) = \mathbf{I} - \mathbf{D}^{-1}\mathbf{A},$$

which we define as the random walk Laplacian:

> **DEFINITION 2.19   (RANDOM WALK GRAPH LAPLACIAN MATRIX)**   The random walk graph Laplacian matrix $\mathcal{T}$ is an $N \times N$ matrix
>
> $$\mathcal{T} = \mathbf{I} - \mathbf{D}^{-1}\mathbf{A}. \tag{2.15}$$

Note that these two forms of normalization lead to matrices that are not symmetric, as can be observed in (2.14).

The rows of $Q = \mathbf{D}^{-1}\mathbf{A}$ and the columns of $\mathcal{P} = \mathbf{AD}^{-1}$ respectively add to 1, showing that they are row and column **stochastic matrices**, but the same is not true in general for the columns of Q and the rows of $\mathcal{P}$.

Symmetric normalized Laplacian   An alternative approach, which provides normalization while preserving the symmetry in $\mathbf{L}$, is the symmetric normalized Laplacian, where the columns of $\mathbf{B}^{\mathsf{T}}$ and the rows of $\mathbf{B}$ are normalized by the same factor $\mathbf{D}^{-1/2}$.

> **DEFINITION 2.20   (SYMMETRIC NORMALIZED GRAPH LAPLACIAN MATRIX)**   The symmetric normalized graph Laplacian matrix $\mathcal{L}$ is an $N \times N$ matrix
>
> $$\mathcal{L} = \mathbf{D}^{-1/2}\mathbf{L}\mathbf{D}^{-1/2} = \mathbf{I} - \mathbf{D}^{-1/2}\mathbf{A}\mathbf{D}^{-1/2}. \tag{2.16}$$

Note that since we can write $\mathcal{L} = \mathbf{I} - \mathbf{D}^{-1/2}\mathbf{A}\mathbf{D}^{-1/2}$, $\mathcal{L}$ would appear to have the form of a graph Laplacian with degree matrix $\mathbf{I}$ and normalized adjacency matrix $\mathbf{D}^{-1/2}\mathbf{A}\mathbf{D}^{-1/2}$. But in fact the degree matrix for this normalized graph does not equal $\mathbf{I}$, that is, in general,

$$\mathbf{D}^{-1/2}\mathbf{A}\mathbf{D}^{-1/2}\mathbf{1} \neq \mathbf{1},$$

and thus, unless all nodes have the same degree, $\mathcal{L}\mathbf{1} \neq \mathbf{0}$, so that nodes in the normalized graph do not have degree equal to 1. Also note that

$$(\mathcal{L})_{i,j} = -a_{ij}\frac{1}{\sqrt{d_i}\,\sqrt{d_j}}, \quad i \neq j, \tag{2.17}$$

where a_{ij} is the original weight between i and j, and d_i, d_j are the degrees of i and j. We compare $\mathcal{T}$ and $\mathcal{L}$ in the following example.

Example 2.5   Find $\mathcal{T}$ and $\mathcal{L}$ for

$$\mathbf{L} = \begin{bmatrix} 2 & -1 & 0 & -1 \\ -1 & 3 & -2 & 0 \\ 0 & -2 & 4 & -2 \\ -1 & 0 & -2 & 3 \end{bmatrix}.$$

Solution

We have

$$
\mathcal{T} = \begin{bmatrix} 1 & -\frac{1}{2} & 0 & -\frac{1}{2} \\ -\frac{1}{3} & 1 & -\frac{2}{3} & 0 \\ 0 & -\frac{1}{2} & 1 & -\frac{1}{2} \\ -\frac{1}{3} & 0 & -\frac{2}{3} & 1 \end{bmatrix} \quad \text{and} \quad \mathcal{L} = \begin{bmatrix} 1 & -\frac{1}{\sqrt{6}} & 0 & -\frac{1}{\sqrt{6}} \\ -\frac{1}{\sqrt{6}} & 1 & -\frac{1}{\sqrt{3}} & 0 \\ 0 & -\frac{1}{\sqrt{3}} & 1 & -\frac{1}{\sqrt{3}} \\ -\frac{1}{\sqrt{6}} & 0 & -\frac{1}{\sqrt{3}} & 1 \end{bmatrix},
$$

where we can see that $\mathcal{T}$ is not symmetric but has rows that add to zero, while $\mathcal{L}$ is symmetric but its rows do not add up to zero. Thus, while we can create a graph with edge weights obtained with the normalization in Definition 2.20, $\mathcal{L}$ is not the combinatorial Laplacian of that graph.

Matrix representations of graphs are summarized in Table 2.2. In what follows we study how these matrices are used to process graph signals.

2.4 Node Domain Graph Filters

From Definition 2.6, a graph signal is a vector $\mathbf{x} \in \mathbb{R}^N$. In analogy with conventional signal processing, a **linear graph filter** is a linear operator (i.e., a transformation) that can be applied to any signal $\mathbf{x}$ to obtain an output $\mathbf{y} \in \mathbb{R}^N$:

$$
\mathbf{y} = \mathbf{H}\mathbf{x}, \tag{2.18}
$$

where $\mathbf{H}$ is an $N \times N$ matrix. We starting by defining filtering operations based on the one-hop graph operators of Section 2.3.

2.4.1 Simple One-Hop Filters and Their Interpretation

Adjacency matrix   Letting $\mathbf{A}$ be the adjacency matrix of an undirected graph, the ith entry of the filter output $\mathbf{y} = \mathbf{A}\mathbf{x}$, i.e., the output at node i, is

$$
y(i) = \sum_{j \in \mathcal{N}(i)} a_{ij}x(j), \tag{2.19}
$$

where the sum is over all the neighbors of node i. Therefore we can view $\mathbf{A}$ as a one-hop operator, since the output at any one node depends only on the graph signal value at the neighboring nodes. We can modify the operation in (2.19) and replace the sum of the neighboring values by their average, leading to:

$$
y'(i) = \frac{1}{d_i} \sum_{j \in \mathcal{N}(i)} a_{ij}x(j), \tag{2.20}
$$

where $d_i = |\mathcal{N}(i)|$ if the graph is unweighted. More compactly,

$$
\mathbf{y}' = \mathbf{D}^{-1}\mathbf{A}\mathbf{x} = Q\mathbf{x},
$$

so that the random walk matrix, $Q = \mathbf{D}^{-1}\mathbf{A}$, can be seen as a one-hop operator that replaces the signal entry at node i by the average of its neighbors' values.

Random walk Laplacian   The random walk Laplacian of Definition 2.19, $\mathcal{T} = \mathbf{I} - Q$, can be interpreted as subtracting from the signal at node i the average of its one-hop neighborhood:

$$y''(i) = x(i) - \frac{1}{d_i} \sum_{j \in \mathcal{N}_i} a_{ij} x(j) = x(i) - y'(i), \tag{2.21}$$

where the second term in the equation, corresponding to $Q\mathbf{x}$, computes a weighted average in which those nodes that are the most similar (with the largest a_{ij}) are given more weight. Thus, $y''(i)$ can be viewed as a **prediction error**, i.e., if $x(i)$ is close to the neighborhood average then $y''(i)$ will be close to zero. In Chapter 3 we will develop further the link between the properties of graph signals and the output of operators such as $\mathcal{T}$. Signals for which $\mathcal{T}\mathbf{x}$ is close to zero will be deemed smooth or low frequency, while signals for which $|y''(i)|$ is larger will be considered non-smooth or high frequency.

Weighted and unweighted prediction errors   If we think of (2.21) as computing a prediction error, it is interesting to note that this expression does not provide any other information about how reliable the prediction is. In particular this prediction error does not depend on the number of neighbors or their weights. Example 2.6 illustrates why this lack of information may be problematic.

Example 2.6   Consider an unweighted and undirected graph and recall that $d_i = |\mathcal{N}(i)|$ (see Definition 2.3). Let two nodes i and j be such that (i) $|\mathcal{N}(i)| = 1$, $|\mathcal{N}(j)| = k > 1$, (ii) $x(i) = x(j) = 2$ and (iii) $\forall l \in \mathcal{N}(i) \cup \mathcal{N}(j)$ we have $x(l) = 1$. Use (2.21) to compute $y''(i)$ and $y''(j)$. For a smooth signal the prediction error (2.21) will be small. Let $\mathbf{x}$ be a noisy version of a signal $\mathbf{x}_0$ which is not observed but is assumed to be smooth. We would like to determine whether the observed values at nodes i and j are equally "reliable." That is, if we take $x(i)$ and $x(j)$ as estimates of $x_0(i)$ and $x_0(j)$, respectively, do we have the same confidence on both estimates? Discuss how the reliability of $x(j)$ may depend on k.

Solution
From (2.21) we can see easily that $y''(i) = y''(j) = 1$. Thus the prediction error is the same in both cases. However, if we assume that the nodes in $\mathcal{N}(i)$ and $\mathcal{N}(j)$ can provide predictions for the signals at i and j, respectively, we might perhaps expect a more reliable estimate when the neighborhood has more nodes.

Nevertheless, since v_i has only one neighbor, while v_j has k, a prediction error of 1 may be considered to be worse for j, since the number of neighbors of v_j is greater. Mathematically, we can define a weighted error $z(i) = |\mathcal{N}(i)| y''(i)$, which then leads to

$$z(i) = 1, \quad z(j) = k,$$

which makes it explicit that the error is actually worse for node j.

As a concrete illustration of the ideas of Example 2.6, consider temperature measurements, where nodes represent sensors and each sensor is connected on a graph to other sensors that are within a certain distance. Then, if v_i and its single neighbor have different measurements, it is possible that the temperature changes relatively quickly in the environment of v_i. On the other hand, if all neighbors of v_j have the same temperature as each other but different from that of v_i, it may be that the measurement $x(j)$ is incorrect and there is a problem with the corresponding sensor. Thus we expect the prediction error to be more informative (e.g., as an indication of a sensor malfunction) if more nodes are used to compute the prediction (i.e., the node has more neighbors).

Combinatorial Laplacian   On the basis of Example 2.6 we define a **weighted prediction error** $z(i)$, which gives more weight to errors corresponding to nodes with more neighbors (or, more generally, with higher degrees):

$$z(i) = d_i \left(x(i) - \frac{1}{d_i} \sum_{j \in \mathcal{N}(i)} a_{ij} x(j) \right) \qquad (2.22)$$

which can be written in matrix form as

$$\mathbf{z} = \mathbf{D}\mathcal{T}\mathbf{x} = \mathbf{D}(\mathbf{I} - \mathbf{D}^{-1}\mathbf{A})\mathbf{x} = (\mathbf{D} - \mathbf{A})\mathbf{x} = \mathbf{L}\mathbf{x}. \qquad (2.23)$$

Thus, the combinatorial graph Laplacian, $\mathbf{L}$ and the random walk Laplacian, $\mathcal{T}$, are one-hop prediction operators using different weights.

Symmetric normalized Laplacian   Using the edge weights for the symmetric normalized Laplacian of (2.17) we can write $\mathbf{z}' = \mathcal{L}\mathbf{x}$ as a prediction error at node i:

$$z'(i) = x(i) - \sum_{j \in \mathcal{N}(i)} \frac{a_{ij}}{\sqrt{d_i}\sqrt{d_j}} x(j), \qquad (2.24)$$

where, similarly to (2.21), the prediction error does not depend on the size of the neighborhood, but where the weights have been normalized. Also, unlike in (2.21), the **prediction weights** do not add to 1.

2.4.2 Polynomials of One-Hop Operators

From this point forward, denote by $\mathbf{Z}$ a generic one-hop operator, which could be one among those considered so far ($\mathbf{A}, \mathbf{D}^{-1}\mathbf{A}, \mathcal{T} = \mathbf{I} - \mathbf{D}^{-1}\mathbf{A}, \mathcal{L}$ and $\mathbf{L}$). We now make a key observation that will help us interpret more complex operations based on a chosen $\mathbf{Z}$.

Remark 2.2   If $\mathbf{Z}$ is a one-hop graph operator then $\mathbf{Z}^2$ is a two-hop graph operator and, more generally, $\mathbf{Z}^k$ is a k-hop graph operator.

Proof Let δ_i be a signal that is zero everywhere except at node i. By definition, one-hop operators are such that $\mathbf{Z}\delta_i$ can only have non-zero entries at node i and at nodes $j \in \mathcal{N}(i)$. Therefore we will be able to find some scalars α_j such that

$$\mathbf{Z}\delta_i = \sum_{j \in \mathcal{N}(i)} \alpha_j \delta_j,$$

and then, by linearity, we can write

$$\mathbf{Z}^2\delta_i = \mathbf{Z} \sum_{j \in \mathcal{N}(i)} \alpha_j \delta_j = \sum_{j \in \mathcal{N}(i)} \alpha_j \mathbf{Z}\delta_j,$$

which only has non-zero values in the one-hop neighbors of $j \in \mathcal{N}(i)$; by definition these constitute the two-hop neighborhood of i, $\mathcal{N}_2(i)$. The same argument can be applied iteratively to show that $\mathbf{Z}^k\delta_i$ has non-zero values only in $\mathcal{N}_k(i)$. Then, using superposition, it is easy to see that the output of $\mathbf{Z}^2\mathbf{x}$ at node i depends only on the values at the nodes j in the two-hop neighborhood of i. A value at node $k \notin \mathcal{N}_2(i)$ will not affect the output $\mathbf{Z}^2\mathbf{x}$ at node i.   □

More generally, from Remark 2.2 it follows that $p(\mathbf{Z})$, an arbitrary polynomial of $\mathbf{Z}$ of degree K, will produce localized outputs. That is, $p(\mathbf{Z})\delta_i$ will be identically zero for nodes that do not belong to $\mathcal{N}_K(i)$. Note that in our initial definition of a graph filtering operation in (2.18) we simply required the operator $\mathbf{H}$ to be linear, so that any arbitrary $N \times N$ matrix could have been chosen for $\mathbf{H}$. Instead, we now restrict ourselves to filters that can be written as polynomials of $\mathbf{Z}$.

DEFINITION 2.21   (GRAPH FILTERS)   For a given one-hop operator $\mathbf{Z}$, a **graph filter** $\mathbf{H}$ is an $N \times N$ matrix that can be written as a polynomial $p(\mathbf{Z})$ of $\mathbf{Z}$:

$$\mathbf{H} = p(\mathbf{Z}) = \sum_{k=0}^{K} a_k \mathbf{Z}^k, \tag{2.25}$$

where $\mathbf{Z}^0 = \mathbf{I}$ and the scalars a_k are the coefficients of the polynomial.

With this definition, the degree of the polynomial defines the localization of the operator, with the caveat that K-hop localization may lead to a global operation if the radius or the diameter of the graph is small (see Section 2.2.2).

Polynomial graph filters commute   While matrix multiplication is not commutative in general, matrices that are polynomials of a given $\mathbf{Z}$ do commute. This property is analogous to that of shift invariance in conventional signal processing (Box 2.4).

THEOREM 2.1   (COMMUTATIVITY OF POLYNOMIAL OPERATORS)   Let $p(\mathbf{Z})$ be an arbitrary polynomial of $\mathbf{Z}$; then $p(\mathbf{Z})\mathbf{Z} = \mathbf{Z}\,p(\mathbf{Z})$, so that $\mathbf{Z}$ and $p(\mathbf{Z})$ commute, and, for any polynomial $q(\mathbf{Z})$

$$p(\mathbf{Z})\,q(\mathbf{Z}) = q(\mathbf{Z})\,p(\mathbf{Z}).$$

Proof Let $p(\mathbf{Z}) = \sum_{i=0}^{p-1} \alpha_i \mathbf{Z}^i$; then

$$\mathbf{Z}\, p(\mathbf{Z}) = \sum_{i=0}^{p-1} \alpha_i \mathbf{Z}^{i+1} = \left(\sum_{i=0}^{p-1} \alpha_i \mathbf{Z}^i \right) \mathbf{Z}$$

$$= p(\mathbf{Z})\, \mathbf{Z}$$

and the commutativity of two polynomials follows directly from this result. □

Box 2.4   Graph operators versus graph shift operators

Focusing on only those graph filters that can be written in the polynomial form of Definition 2.21 is fundamental to the development of GSP. This also leads to strong analogies with conventional signal processing [15]. Specifically, in a linear shift in-variant system, we can define an elementary shift or delay operator, z^{-1}, and the z-transforms of a filter $H(z)$ and of an input signal, $X(z)$. Then, the output $Y(z)$ of a linear shift-invariant system can be written as [17]

$$Y(z) = H(z)X(z).$$

The shift-invariance property for conventional discrete-time linear systems is expressed as

$$\underbrace{H(z)\left(z^{-1}X(z)\right)}_{(A)} = \underbrace{z^{-1}\left(H(z)X(z)\right)}_{(B)}$$

which states that output of the system when a shift is applied to the input (A) can also be obtained by shifting the output of the original, non-delayed, system (B). Notice that this implies that

$$H(z)z^{-1} = z^{-1}H(z),$$

which clearly shows the analogy with Theorem 2.1, where $p(\mathbf{Z})$ and $\mathbf{Z}$ commute. On the basis of this analogy, $\mathbf{Z}$ is often called the **graph shift operator**, and the property of Theorem 2.1 is described as **shift invariance** [15]. The correspondence between $\mathbf{Z}$ and z is indeed exact when we consider a directed graph cycle [15].

While this analogy is interesting and insightful, in this book we do not use the term "graph shift operator" to describe $\mathbf{Z}$ and instead call $\mathbf{Z}$ the **fundamental graph operator, one-hop graph operator** or just **graph operator** for short. This is simply a difference in names, and in all other respects our definition of $\mathbf{Z}$ is the same as that commonly used in the graph signal processing literature [15, 11].

We choose not to use the term "shift" to avoid potential confusion, given the differences that exist between $\mathbf{Z}$ and z. In particular, for a time-based signal $X(z)$, we can always reverse the delay, i.e., $X(z) = z^{-1}zX(z)$, while in general the graph operators $\mathbf{Z}$ are rarely invertible. Thus if we compute $\mathbf{y} = \mathbf{Z}\mathbf{x}$, we cannot guarantee that this operation can be reversed.

2.5   Graph Operators and Invariant Subspaces

In this section, we develop tools to characterize the output of a graph operator (Definition 2.21) applied to an arbitrary graph signal. We work with polynomials of $\mathbf{Z}$, a one-hop fundamental graph operator (Box 2.4) and make use of the basic linear algebra concepts reviewed in Appendix A. This section addresses two fundamental questions:

- For a graph with N nodes, with $\mathbf{Z}$ of size $N \times N$, how many **degrees of freedom**[5] are there to construct filters in the form of Definition 2.21? Essentially, the answer to this question will tell us what is the maximum degree of a polynomial that has the form of Definition 2.21. We will see that, when we are constructing polynomial operators, the degree can be no greater than N, and is often less than N.
- Can we express $\mathbf{Zx}$ as a function of the outputs $\mathbf{Zu}_i$ for a series of elementary signals $\mathbf{u}_i$? The answer is obviously yes, since $\mathbf{Z}$ is a linear operator and the response can be expressed as a linear combination of responses to elementary signals forming a basis for the space. We can construct **basis sets obtained from invariant subspaces** (where the signals in each subspace E_i are invariant under multiplication by $\mathbf{Z}$: $\mathbf{Zu}_i \in E_i$ if $\mathbf{u}_i \in E_i$). When $\mathbf{Z}$ is diagonalizable each of these subspaces corresponds to one of the N linearly independent eigenvectors.

While the answer to these questions builds on elementary linear algebra concepts (refer to Appendix A for a review), we introduce them step by step so as to highlight their interpretation in the context of graph signal filtering.

2.5.1   Minimal Polynomial of a Vector

As a starting point, consider an arbitrary vector $\mathbf{x} \in \mathbb{R}^N$ and apply the operator $\mathbf{Z}$ successively to this vector. This will produce a series of vectors,

$$\mathbf{x}, \mathbf{Zx}, \mathbf{Z}^2\mathbf{x}, \ldots, \mathbf{Z}^{p-1}\mathbf{x}.$$

Are these p vectors linearly independent? The answer depends on both $\mathbf{x}$ and p. First, note that if $p > N$ then these vectors must be linearly dependent: in a space of dimension N, any set of more than N vectors must be linearly dependent.[6]

Next, for a given $\mathbf{x}$, find the smallest p such that $\mathbf{x}, \mathbf{Zx}, \mathbf{Z}^2\mathbf{x}, \ldots, \mathbf{Z}^{p-1}\mathbf{x}, \mathbf{Z}^p\mathbf{x}$ are linearly dependent. For this p, by the definition of linear dependence, we can find $a_0, a_1, \ldots, a_{p-1}$ such that

$$\mathbf{Z}^p\mathbf{x} = \sum_{k=0}^{p-1} a_k\mathbf{Z}^k\mathbf{x}, \tag{2.26}$$

where as before $\mathbf{Z}^0 = \mathbf{I}$. Therefore,

$$p_x(\mathbf{Z})\mathbf{x} = \left(-\sum_{k=0}^{p-1} a_k\mathbf{Z}^k + \mathbf{Z}^p \right)\mathbf{x} = \mathbf{0}, \tag{2.27}$$

---

[5] By degrees of freedom we mean the number of different parameters that can be selected.

[6] The **span** of $\mathbf{x}, \mathbf{Zx}, \mathbf{Z}^2\mathbf{x}, \ldots, \mathbf{Z}^p\mathbf{x}$ is the order-p Krylov subspace generated by $\mathbf{Z}$ and $\mathbf{x}$.

where $p_x(\mathbf{Z}) = -\sum_{k=0}^{p-1} a_k \mathbf{Z}^k + \mathbf{Z}^p$ is the minimal polynomial of $\mathbf{x}$.

DEFINITION 2.22 (MINIMAL POLYNOMIAL OF A VECTOR) For a given non-zero graph signal $\mathbf{x}$ and a graph operator $\mathbf{Z}$, the minimal polynomial $p_x(\mathbf{Z})$ of $\mathbf{x}$ is the lowest-degree non-trivial polynomial of $\mathbf{Z}$ such that

$$p_x(\mathbf{Z})\mathbf{x} = \mathbf{0}.$$

The minimal polynomial $p_x(\mathbf{Z})$ allows us to simplify operations on $\mathbf{x}$. Any $p(\mathbf{Z})$ can be written as

$$p(\mathbf{Z}) = q(\mathbf{Z})p_x(\mathbf{Z}) + r(\mathbf{Z}) \tag{2.28}$$

where $q(\mathbf{Z})$ and $r(\mathbf{Z})$ are the quotient and residue polynomials, respectively, and $r(\mathbf{Z})$ has degree less than that of $p_x(\mathbf{Z})$. The polynomials $q(\mathbf{Z})$ and $r(\mathbf{Z})$ can be obtained using long division as shown below in Example 2.7. Then, by the definition of $p_x(\mathbf{Z})$, we have

$$p(\mathbf{Z})\mathbf{x} = q(\mathbf{Z})p_x(\mathbf{Z})\mathbf{x} + r(\mathbf{Z})\mathbf{x} = r(\mathbf{Z})\mathbf{x}.$$

Example 2.7 Let $\mathbf{x}$ be a vector with minimal polynomial $p_x(\mathbf{Z}) = \mathbf{Z}^2 + 2\mathbf{Z} + \mathbf{I}$. Find $q(\mathbf{Z})$ and $r(\mathbf{Z})$ such that $p(\mathbf{Z}) = \mathbf{Z}^4$ can be written as in (2.28): $p(\mathbf{Z}) = q(\mathbf{Z})p_x(\mathbf{Z}) + r(\mathbf{Z})$.

Solution
We can express $\mathbf{Z}^4\mathbf{x}$ as a function of $p_x(\mathbf{Z})$ using long division. In the first step we approximate $p(\mathbf{Z})$ by $\mathbf{Z}^2 p_x(\mathbf{Z})$ to cancel out the highest-degree term in $p(\mathbf{Z})$ and compute the resulting residue,

$$r_1(\mathbf{Z}) = \mathbf{Z}^4 - \mathbf{Z}^2(\mathbf{Z}^2 + 2\mathbf{Z} + \mathbf{I}) = -2\mathbf{Z}^3 - \mathbf{Z}^2;$$

then we approximate $r_1(\mathbf{Z})$ to find

$$r_2(\mathbf{Z}) = (-2\mathbf{Z}^3 - \mathbf{Z}^2) + 2\mathbf{Z}(\mathbf{Z}^2 + 2\mathbf{Z} + \mathbf{I}) = 3\mathbf{Z}^2 + 2\mathbf{Z}$$

and finally

$$r(\mathbf{Z}) = (3\mathbf{Z}^2 + 2\mathbf{Z}) - 3\mathbf{I}(\mathbf{Z}^2 + 2\mathbf{Z} + \mathbf{I}) = -4\mathbf{Z} - 3\mathbf{I}$$

so that we have

$$\mathbf{Z}^4 = (\mathbf{Z}^2 - 2\mathbf{Z} + 3\mathbf{I})(\mathbf{Z}^2 + 2\mathbf{Z} + \mathbf{I}) - 4\mathbf{Z} - 3\mathbf{I},$$

where $q(\mathbf{Z}) = \mathbf{Z}^2 - 2\mathbf{Z} + 3\mathbf{I}$. Given that $p_x(\mathbf{Z})$ is the minimal polynomial of $\mathbf{x}$ we have

$$\mathbf{Z}^4\mathbf{x} = (\mathbf{Z}^2 - 2\mathbf{Z} + 3\mathbf{I})p_x(\mathbf{Z})\mathbf{x} - (4\mathbf{Z} + 3\mathbf{I})\mathbf{x} = -(4\mathbf{Z} + 3\mathbf{I})\mathbf{x}.$$

2.5.2   Eigenvectors and Eigenvalues

Clearly, the lowest degree of $p_x(\mathbf{Z})$ for a non-zero vector $\mathbf{x}$ is $p = 1$. Let $\mathbf{u}$ be a vector having minimal polynomial, $p_u(\mathbf{Z})$, of degree 1. Then, from (2.26), $\mathbf{u}$ and $\mathbf{Zu}$ are linearly dependent and so there exists a scalar λ such that $\mathbf{Zu} = \lambda\mathbf{u}$. Thus $\mathbf{u}$ is an eigenvector of $\mathbf{Z}$, with λ the corresponding eigenvalue.

DEFINITION 2.23   (EIGENVECTORS AND EIGENVALUES)   Let $\mathbf{Z}$ be a square matrix. A vector $\mathbf{u}$ and a scalar λ are an eigenvector and an eigenvalue of $\mathbf{Z}$, respectively, if we have

$$\mathbf{Zu} = \lambda\mathbf{u}.$$

An eigenvalue and a corresponding eigenvector form an eigenpair $(\lambda, \mathbf{u})$.

If $\mathbf{u}$ is an eigenvector, the minimal polynomial of $\mathbf{Z}$ with respect to $\mathbf{u}$ is

$$p_u(\mathbf{Z}) = \mathbf{Z} - \lambda\mathbf{I} \tag{2.29}$$

and multiplication by $\mathbf{Z}$ simply scales $\mathbf{u}$. By linearity, any vector along the direction $\mathbf{u}$ is scaled in the same way, so that $\mathbf{Z}(\alpha\mathbf{u}) = \alpha\mathbf{Zu} = \lambda(\alpha\mathbf{u})$. Then $E_u = \text{span}(\mathbf{u})$ is called an **eigenspace** associated with $\mathbf{Z}$ (see Definition A.1 and Definition A.2). Given $|\lambda|$, the magnitude of the eigenvalue λ, if $|\lambda| > 1$ then all vectors in E_u are amplified by the transformation, while if $|\lambda| < 1$ they are attenuated. When $\lambda = 0$, $\mathbf{Zu} = \mathbf{0}$ and $\mathbf{u}$ belongs to the null space of the transformation $\mathbf{Z}$.

Characteristic polynomial   By definition of $p_u(\mathbf{Z})$ in (2.29) we have that $p_u(\mathbf{Z})\mathbf{u} = \mathbf{0}$ and, since $\mathbf{u}$ is assumed to be non-zero, we will need to have $\det(\mathbf{Z} - \lambda\mathbf{I}) = 0$, so that the null space of $\mathbf{Z} - \lambda\mathbf{I}$, $\mathcal{N}(\mathbf{Z} - \lambda\mathbf{I})$, contains the non-zero vector $\mathbf{u}$. To find the eigenvectors $\mathbf{u}$ we need to find all the eigenvalues λ, scalars such that $\mathbf{Z} - \lambda\mathbf{I}$ is a singular matrix, i.e.,

$$\det(\lambda\mathbf{I} - \mathbf{Z}) = 0.$$

This determinant can be written out as a polynomial of the variable λ, leading to the characteristic polynomial of $\mathbf{Z}$:

$$p_c(\lambda) = \lambda^N + c_{N-1}\lambda^{N-1} + \cdots + c_1\lambda + c_0\lambda_0, \tag{2.30}$$

where the c_i are known values that depend on $\mathbf{Z}$ and result from the determinant computation. By the fundamental theorem of algebra, $p_c(\lambda)$ will have N roots in $\mathbb{C}$, the space of complex numbers. Given these roots (real or complex, simple or multiple), we can write

$$p_c(\lambda) = \prod_i (\lambda - \lambda_i)^{k_i}, \tag{2.31}$$

where the roots of this polynomial, $\lambda_1, \lambda_2, \ldots$, are the eigenvalues of $\mathbf{Z}$. From the same theorem, calling k_i the **algebraic multiplicity** of eigenvalue λ_i, we have that

$$\sum_i k_i = N.$$

Invariant subspaces   As noted earlier, each subspace E_u corresponding to an eigenvector $\mathbf{u}$ is **invariant** under $\mathbf{Z}$. This idea can be easily extended, so that any subspace of $\mathbb{R}^N$ (or $\mathbb{C}^N$) having a basis formed by a set of eigenvectors of $\mathbf{Z}$ is also invariant. More generally, for any non-zero vector $\mathbf{x}$ we can construct an invariant subspace E_x.

PROPOSITION 2.2   (INVARIANT SUBSPACE FOR $\mathbf{x}$)   For a given vector $\mathbf{x}$ (not necessarily an eigenvector of $\mathbf{Z}$), with minimal polynomial $p_x(\mathbf{Z})$ of degree p, a subspace E_x, defined as

$$E_x = \text{span}(\mathbf{x}, \mathbf{Z}\mathbf{x}, \mathbf{Z}^2\mathbf{x}, \ldots, \mathbf{Z}^{p-1}\mathbf{x})$$

contains $\mathbf{x}$ and is invariant under multiplication by $\mathbf{Z}$, that is,

$$\forall \mathbf{y} \in E_x, \quad \mathbf{Z}\mathbf{y} \in E_x,$$

where we emphasize that $\mathbf{Z}\mathbf{y}$ is simply required to be in E_x and in general $\mathbf{Z}\mathbf{y} \neq \alpha\mathbf{y}$.

Proof   By the definition of the span of a set of vectors (Definition A.2), any $\mathbf{y} \in E_x$ can be written as

$$\mathbf{y} = a_0\mathbf{x} + a_1\mathbf{Z}\mathbf{x} + \cdots + a_{p-1}\mathbf{Z}^{p-1}\mathbf{x}, \quad \text{so that} \quad \mathbf{Z}\mathbf{y} = a_0\mathbf{Z}\mathbf{x} + a_1\mathbf{Z}^2\mathbf{x} + \cdots + a_{p-1}\mathbf{Z}^p\mathbf{x}.$$

However, from (2.26), if $p_x(\mathbf{Z})$ has degree p then $\mathbf{x}, \mathbf{Z}\mathbf{x}, \mathbf{Z}^2\mathbf{x}, \ldots, \mathbf{Z}^{p-1}\mathbf{x}, \mathbf{Z}^p\mathbf{x}$ are linearly dependent and $a_{p-1}\mathbf{Z}^p\mathbf{x}$ can be written as a function of $\mathbf{x}, \mathbf{Z}\mathbf{x}, \mathbf{Z}^2\mathbf{x}, \ldots, \mathbf{Z}^{p-1}\mathbf{x}$, so that $\mathbf{Z}\mathbf{y} \in E_x$, which proves the space is invariant.   □

Since E_x is invariant under multiplication by $\mathbf{Z}$ the following properties hold for any $\mathbf{y} \in E_x$ and any polynomial $p(\mathbf{Z})$: (i) E_x is invariant under multiplication by $p(\mathbf{Z})$, i.e., $p(\mathbf{Z})\mathbf{y} \in E_x$ and (ii) $p(\mathbf{Z})\mathbf{y}$ can be simplified as in Example 2.7 given that[7] $p_x(\mathbf{Z})\mathbf{y} = \mathbf{0}$.

Complex eigenvalues   Assume $\mathbf{Z}$ is real and μ is a complex root of its characteristic polynomial with multiplicity 1. Since $\mathbf{Z}$ is real, the characteristic polynomial has real coefficients, so that both μ and μ^* are eigenvalues. If $\mathbf{Z}\mathbf{u} = \mu\mathbf{u}$, by conjugating both sides we have $(\mathbf{Z}\mathbf{u})^* = \mathbf{Z}\mathbf{u}^* = (\mu)^*\mathbf{u}^*$, which shows that $(\mu, \mathbf{u})$ and $(\mu^*, \mathbf{u}^*)$ are both eigenpairs of $\mathbf{Z}$. Thus we have a subspace $E_\mu = \text{span}(\mathbf{u}, \mathbf{u}^*) \subset \mathbb{C}^N$ that is invariant under multiplication by $\mathbf{Z}$, since the basis vectors are eigenvectors of $\mathbf{Z}$.

Note that for vectors in $\text{span}(\mathbf{u})$ and $\text{span}(\mathbf{u}^*)$ the minimal polynomials are $\mathbf{Z} - \mu\mathbf{I}$ and $\mathbf{Z} - \mu^*\mathbf{I}$, respectively. From this we can see that, for any $\mathbf{z} \in E_\mu$, $p_\mu(\mathbf{Z})\mathbf{z} = \mathbf{0}$ with

$$p_\mu(\mathbf{Z}) = (\mathbf{Z} - \mu\mathbf{I})(\mathbf{Z} - \mu^*\mathbf{I}) = \mathbf{Z}^2 - (\mu + \mu^*)\mathbf{Z} + |\mu|^2\mathbf{I},$$

where $p_\mu(\mathbf{Z})$ is a real polynomial. Thus, we can construct a basis $(\mathbf{z}, \mathbf{Z}\mathbf{z})$ for E_μ by choosing $\mathbf{z} \in E_\mu$ such that $\mathbf{z} \notin \text{span}(\mathbf{u}) \cup \text{span}(\mathbf{u}^*)$, where we see that the basis vectors are no longer eigenvectors. Also, from the definition of $p_\mu(\mathbf{Z})$

$$\mathbf{Z}^2\mathbf{v} = (\mu + \mu^*)\mathbf{Z}\mathbf{v} - |\mu|^2\mathbf{v}.$$

If we are processing real-valued signals with real-valued polynomials of $\mathbf{Z}$, we can also define a real basis for the space of real vectors in E_μ ($E_\mu \cap \mathbb{R}^N$). This is shown in Box 2.5.

[7]   Note that $p_x(\mathbf{Z})\mathbf{y} = \mathbf{0}$ but $p_y(\mathbf{Z}) \neq p_x(\mathbf{Z})$.

Box 2.5   Real invariant spaces for complex eigenvalues

From the above discussion, $(\mathbf{x}, \mathbf{Zx})$ is a real basis for $E_\mu \cap \mathbb{R}^N$ for any non-zero and real $\mathbf{x} \in E_\mu$. Alternative real-valued basis choices are possible. If $\mathbf{x} \in E_\mu$, $\mathbf{x} = \alpha\mathbf{u}+\beta\mathbf{u}^*$ and $\mathbf{x}$ is real we have that $\mathbf{x}^* = \mathbf{x}$ and therefore

$$\alpha\mathbf{u} + \beta\mathbf{u}^* = (\alpha\mathbf{u} + \beta\mathbf{u}^*)^* = \beta^*\mathbf{u} + \alpha^*\mathbf{u}^*,$$

so that $\alpha = \beta^*$. Thus, any real vector $\mathbf{x}$ in E_μ can be written as

$$\mathbf{x} = (\alpha_1 + j\alpha_2)\mathbf{u} + (\alpha_1 - j\alpha_2)\mathbf{u}^* = \alpha_1(\mathbf{u} + \mathbf{u}^*) + \alpha_2(j(\mathbf{u} - \mathbf{u}^*)),$$

where both α_1 and α_2 are real and $\mathbf{v}_1 = \mathbf{u}+\mathbf{u}^*$ and $\mathbf{v}_2 = j(\mathbf{u}-\mathbf{u}^*)$ are real vectors, that form a basis for $E_\mu \cap \mathbb{R}^N$ (where $\cap$ denotes the intersection). Since $\mathbf{u} = \frac{1}{2}(\mathbf{v}_1 - j\mathbf{v}_2)$ and $\mathbf{u}^* = \frac{1}{2}(\mathbf{v}_1 + j\mathbf{v}_2)$, we have

$$\mathbf{Zv}_1 = \mu\mathbf{u} + \mu^*\mathbf{u}^* = \frac{1}{2}\mu(\mathbf{v}_1 - j\mathbf{v}_2) + \mu^*\frac{1}{2}(\mathbf{v}_1 + j\mathbf{v}_2) = \mathrm{Re}(\mu)\mathbf{v}_1 + \mathrm{Im}(\mu)\mathbf{v}_2$$

and similarly

$$\mathbf{Zv}_2 = j(\mu\mathbf{u} - \mu^*\mathbf{u}^*) = \frac{1}{2}\mu(j\mathbf{v}_1 + \mathbf{v}_2) - \mu^*\frac{1}{2}(j\mathbf{v}_1 - \mathbf{v}_2) = \mathrm{Im}(\mu)\mathbf{v}_1 + \mathrm{Re}(\mu)\mathbf{v}_2.$$

Then for any real vector in the space $E_\mu \cap \mathbb{R}^N$ we will have

$$\mathbf{Zx} = \mathbf{Z}(\alpha_1\mathbf{v}_1 + \alpha_2\mathbf{v}_2) = (\alpha_1\mathrm{Re}(\mu) + \alpha_2\mathrm{Im}(\mu))\mathbf{v}_1 + (\alpha_1\mathrm{Im}(\mu) + \alpha_2\mathrm{Re}(\mu))\mathbf{v}_2,$$

showing explicitly that $E_\mu \cap \mathbb{R}^N$ is invariant under multiplication by $\mathbf{Z}$.

2.5.3   Minimal Polynomial of Z and Invariant Subspaces

So far we have defined minimal polynomials of individual vectors $\mathbf{x}$ and of eigenspaces. Now we study the minimal polynomial for *all* vectors in $\mathbb{R}^N$ (or $\mathbb{C}^N$). This will allow us to characterize all polynomial filtering operations for graph signals. We start by defining the Schur decomposition, which allows us to develop a series of invariant subspaces such as those introduced in Proposition 2.2 but with the added advantage that they lead to a representation for any signal in the space.

THEOREM 2.2   (SCHUR DECOMPOSITION)   Any $N \times N$ (complex) matrix $\mathbf{Z}$ can be written as

$$\mathbf{Z} = \mathbf{U}^H\mathbf{TU}, \tag{2.32}$$

where $\mathbf{U}$ is unitary, $\mathbf{U}^H\mathbf{U} = \mathbf{UU}^H = \mathbf{I}$ (the superscript H denotes the Hermitian conjugate) and $\mathbf{T}$ is upper triangular, with diagonal elements that are the eigenvalues of $\mathbf{Z}$. A decomposition can be obtained for any ordering of the eigenvalues, leading to different $\mathbf{U}$ and $\mathbf{T}$ in each case.

Proof Let $\lambda_1, \lambda_2, \ldots, \lambda_N$ be an ordering of the eigenvalues of $\mathbf{Z}$ (note that eigenvalues of multiplicity greater than 1 are repeated). Let $\mathbf{u}_1$ be an eigenvector corresponding to λ_1, $\mathbf{u}_1^H \mathbf{u}_1 = 1$. Choose a unitary matrix using $\mathbf{u}_1$ as its first column so that $\mathbf{U} = (\mathbf{u}_1 \ \mathbf{U}_2)$, then

$$\mathbf{U}^H \mathbf{Z} \mathbf{U} = \begin{bmatrix} \mathbf{u}_1^H \\ \mathbf{U}_2^H \end{bmatrix} \mathbf{Z} \begin{bmatrix} \mathbf{u}_1 \ \mathbf{U}_2 \end{bmatrix} = \begin{bmatrix} \mathbf{u}_1^H \mathbf{Z} \mathbf{u}_1 & \mathbf{u}_1^H \mathbf{Z} \mathbf{U}_2 \\ \mathbf{U}_2^H \mathbf{Z} \mathbf{u}_1 & \mathbf{U}_2^H \mathbf{Z} \mathbf{U}_2 \end{bmatrix} = \begin{bmatrix} \lambda_1 & \mathbf{u}_1^H \mathbf{Z} \mathbf{U}_2 \\ \mathbf{0} & \mathbf{U}_2^H \mathbf{Z} \mathbf{U}_2 \end{bmatrix},$$

where the last equality uses the facts that $\mathbf{u}_1$ is an eigenvector, so that $\mathbf{u}_1^H \mathbf{Z} \mathbf{u}_1 = \lambda_1 \mathbf{u}_1^H \mathbf{u}_1 = \lambda_1$, and that $\mathbf{U}_2^H \mathbf{u}_1 = \mathbf{0}$ because $\mathbf{U}$ is unitary. Next denote $\mathbf{Z}_2 = \mathbf{U}_2^H \mathbf{Z} \mathbf{U}_2$ and observe that $\mathbf{Z}$ and $\mathbf{U}^H \mathbf{Z} \mathbf{U}$ have the same eigenvalues. From this it follows that $\mathbf{Z}_2$ must have eigenvalues $\lambda_2, \lambda_3, \ldots, \lambda_N$, so that we can apply recursively the approach above (define $\mathbf{u}_2$ as an eigenvector, find a unitary matrix, etc.) until we obtain the decomposition (2.32). □

Let $\mathbf{U}$ be a unitary matrix given by the Schur decomposition of $\mathbf{Z}$ in (2.32). This decomposition is not unique: a different $\mathbf{U}$ is obtained for each possible ordering of the eigenvalues[8] of $\mathbf{Z}$. Denote by $\mathbf{u}_i$ the ith column of $\mathbf{U}$ and let $E_i = \text{span}(\mathbf{u}_i)$. Then, because $\mathbf{U}$ is unitary, we have that $\mathbb{C}^N = \bigoplus_{i=1}^{N} E_i$, i.e., $\mathbb{C}^N$ is the direct sum of the E_i subspaces and the following property holds.

PROPOSITION 2.3   (INVARIANT SUBSPACES FROM SCHUR DECOMPOSITION) The subspaces $F_k = \bigoplus_{i=1}^{k} E_i, k = 1, \ldots, N$ are invariant under $\mathbf{Z}$, that is, if $\mathbf{x} \in F_k$ then $\mathbf{Z}\mathbf{x} \in F_k$.

Proof Choose an arbitrary $\mathbf{x} \in F_k$. By the definition of F_k that means that $\mathbf{x} = \mathbf{U}^H \mathbf{y}$, where $\mathbf{y}$ is such that its last $N - k$ entries are zero. Then

$$\mathbf{Z}\mathbf{x} = \mathbf{U}^H \mathbf{T} \mathbf{U} \mathbf{U}^H \mathbf{y} = \mathbf{U}^H \mathbf{T} \mathbf{y} = \mathbf{U}^H \mathbf{y}',$$

where the last $N - k$ entries of $\mathbf{y}'$ are also zero, because $\mathbf{T}$ is upper triangular, and thus $\mathbf{U}^H \mathbf{y}' \in F_k$. □

As in Proposition 2.2, the invariance in Proposition 2.3 is not as strong as that associated with eigensubspaces (i.e., there is no scalar α such that $\mathbf{Z}\mathbf{x} = \alpha \mathbf{x}$), but it is important because it applies to any square $\mathbf{Z}$, including a $\mathbf{Z}$ lacking a complete set of eigenvectors.

The Schur decomposition of Theorem 2.2 can be used to identify minimal polynomials for all vectors in the space. First, we write a matrix polynomial for an arbitrary $N \times N$ matrix, $\mathbf{X}$, with the same coefficients as the characteristic polynomial of (2.30):

$$p_c(\mathbf{X}) = \mathbf{X}^N + c_{N-1}\mathbf{X}^{N-1} + \cdots + c_1\mathbf{X} + c_0\mathbf{I}. \tag{2.33}$$

For a given $\mathbf{X}$, $p_c(\mathbf{X})$ is also an $N \times N$ matrix. The Cayley–Hamilton theorem states that $\mathbf{Z}$ is a root of the characteristic polynomial.

---

[8] In addition, for a given ordering of the eigenvalues, the representation may not be unique if there are eigenvalues of multiplicity greater than 1.

> THEOREM 2.3   (CAYLEY–HAMILTON) Let $p_c(\lambda)$ be the characteristic polynomial of the matrix $\mathbf{Z}$. For an arbitrary square matrix $\mathbf{X}$ let the corresponding matrix polynomial be $p_c(\mathbf{X})$, as defined in (2.33). Then, we have that $\mathbf{Z}$ is a root of $p_c(\mathbf{X})$, i.e., for **any** $\mathbf{x} \in \mathbb{R}^N$,
>
> $$p_c(\mathbf{Z})\mathbf{x} = \mathbf{0}$$
>
> so that
>
> $$p_c(\mathbf{Z}) = \mathbf{0}.$$

Proof From Theorem 2.2 we can write $\mathbf{Z} = \mathbf{U}^H\mathbf{T}\mathbf{U}$ with $\mathbf{U}$ unitary and $\mathbf{T}$ upper triangular with λ_i along the diagonals. Since $\mathbf{T}$ and $\mathbf{Z}$ are similar they will have the same characteristic polynomial. To show that $\mathbf{Z}$ is a root of $p_c(\mathbf{X})$, we need to show that $p_c(\mathbf{Z})\mathbf{x} = \mathbf{0}$ for any $\mathbf{x}$. Since

$$p_c(\mathbf{Z})\mathbf{x} = \mathbf{U}^H p_c(\mathbf{T})\mathbf{U}\mathbf{x},$$

all we need to do is show that $P_c(\mathbf{T}) = \mathbf{0}$. Recall that we can write

$$p_c(\mathbf{T}) = \prod_i (\mathbf{T} - \lambda_i\mathbf{I})^{k_i} \tag{2.34}$$

and note that the entries in the first column of $\mathbf{T} - \lambda_1\mathbf{I}$ are all zero, so that the first column of the product of matrices in (2.34) is $\mathbf{0}$. Then, the product of the first two matrices in $p_c(\mathbf{T})$ is

$$\begin{bmatrix} 0 & t_{21} & \cdots \\ 0 & \lambda_2 - \lambda_1 & \cdots \\ 0 & 0 & \\ \vdots & \vdots & \end{bmatrix} \begin{bmatrix} 0 & t_{21} & \cdots \\ 0 & \lambda_2 - \lambda_2 & \cdots \\ 0 & 0 & \\ \vdots & \vdots & \end{bmatrix} = \begin{bmatrix} 0 & t_{21} & \cdots \\ 0 & \lambda_2 - \lambda_1 & \cdots \\ 0 & 0 & \\ \vdots & \vdots & \end{bmatrix} \begin{bmatrix} 0 & t_{21} & \cdots \\ 0 & 0 & \cdots \\ 0 & 0 & \\ \vdots & \vdots & \end{bmatrix}$$

which shows that the second column is $\mathbf{0}$. Applying the same reasoning for successive columns shows that they are all $\mathbf{0}$ and thus $p_c(\mathbf{T}) = \mathbf{0}$ and $p_c(\mathbf{Z}) = \mathbf{0}$. $\square$

As a consequence of Theorem 2.3 we can factor $p(\mathbf{Z})$ as follows:

$$p_c(\mathbf{Z}) = \prod_i (\mathbf{Z} - \lambda_i\mathbf{I})^{k_i} = \mathbf{0}, \tag{2.35}$$

where $k_i \geq 1$ is the multiplicity of the root λ_i of $p_c(\lambda)$. Note that Theorem 2.3 implies that there exist $c_0, c_1, \ldots, c_{N-1}$ such that for *any* $\mathbf{x}$

$$\mathbf{Z}^N\mathbf{x} = -\left(\sum_{i=0}^{N-1} c_i\mathbf{Z}^i\right)\mathbf{x}. \tag{2.36}$$

Thus, we can write $\mathbf{Z}^N\mathbf{x}$ as a linear combination of N vectors, $\mathbf{x}, \mathbf{Z}\mathbf{x}, \ldots, \mathbf{Z}^{N-1}\mathbf{x}$. From Theorem 2.3 $p_c(\mathbf{Z})\mathbf{x} = \mathbf{0}$ for *any* $\mathbf{x}$. Then we can ask what is the lowest degree (minimal) polynomial for a *specific* $\mathbf{x}$, $p_x(\mathbf{Z})$. The degree of the minimal polynomial $p_x(\mathbf{Z})$ depends on $\mathbf{x}$ and in general it can be lower than the degree of p_c. For example, if $\mathbf{u}$ is an eigenvector of $\mathbf{Z}$ with eigenvalue λ then we will have $p_c(\mathbf{Z})\mathbf{u} = \mathbf{0}$ and $(\mathbf{Z} - \lambda\mathbf{I})\mathbf{u} = \mathbf{0}$, so that $p_u(\mathbf{Z}) = (\mathbf{Z} - \lambda\mathbf{I})$ is the minimal polynomial for $\mathbf{u}$. This leads us to ask the following question: what is the minimal-degree polynomial $p_{\min}(\mathbf{Z})$, for which $p_{\min}(\mathbf{Z})\mathbf{x} = \mathbf{0}$ for *all* $\mathbf{x}$?

Minimal polynomial The minimal polynomial of $\mathbf{Z}$, $p_{\min}(\cdot)$, is the polynomial of minimal degree for which $\mathbf{Z}$ is a root. In other words, for any $\mathbf{x} \in \mathbb{R}^N$ we have that

$$p_{\min}(\mathbf{Z})\mathbf{x} = \left(\mathbf{Z}^p + \sum_{i=0}^{p-1} a_i \mathbf{Z}^i\right)\mathbf{x} = \mathbf{0}.$$

The importance of the minimal polynomial for node domain processing (graph signal filtering) is that it determines the maximum degree of any polynomial operator on $\mathbf{Z}$ (see Box 2.6 below). The minimal polynomial is closely related to the characteristic polynomial.

PROPOSITION 2.4 (ROOTS OF MINIMAL POLYNOMIAL) The polynomial $p_{\min}(\mathbf{Z})$ divides $p_c(\mathbf{Z})$ without residue and therefore can be factored as follows:

$$p_{\min}(\mathbf{Z}) = \prod_i (\mathbf{Z} - \lambda_i \mathbf{I})^{p_i}, \qquad (2.37)$$

where $p_i \le k_i$ and $\sum p_i = p \le N$. For each eigenvalue λ_i, k_i is the **algebraic multiplicity** of the eigenvalue.

Proof By definition of the minimal and characteristic polynomials we must have that $p_{\min}(\mathbf{Z}) = \mathbf{0}$ and $p_c(\mathbf{Z}) = \mathbf{0}$. Since $p_{\min}(\mathbf{X})$ corresponds to the minimal polynomial of $\mathbf{Z}$, $p_{\min}(\lambda)$, its degree has to be $p \le N$, because $p_c(\lambda)$, and thus $p_c(\mathbf{X})$, has degree N by construction. Then, we can use long division as in Example 2.7 to express the characteristic polynomial as a function of the minimal polynomial and a remainder:

$$p_c(\mathbf{X}) = q(\mathbf{X})p_{\min}(\mathbf{X}) + r(\mathbf{X}),$$

where the degree of $r(\mathbf{X})$ has to be less than p. Then, by Theorem 2.3,

$$\mathbf{0} = p_c(\mathbf{Z}) = q(\mathbf{Z})p_{\min}(\mathbf{Z}) + r(\mathbf{Z});$$

but, given that $p_{\min}(\mathbf{Z}) = \mathbf{0}$ and is the minimal-degree polynomial, the polynomial $r(\mathbf{Z})$ must be zero otherwise there would be a non-zero polynomial of lower degree than $p_{\min}$ for which $r(\mathbf{Z}) = \mathbf{0}$. Therefore, we can write

$$p_c(\mathbf{Z}) = q(\mathbf{Z})p_{\min}(\mathbf{Z}),$$

which means that $p_{\min}(\mathbf{Z})$ has the same roots as $p_c(\mathbf{Z})$, and therefore (2.37) follows. □

The main implication of the existence of $p_{\min}$ is that some polynomial operations on the graph can be simplified, which allows us to answer the first question we posed at the beginning of this section: the number of degrees of freedom is determined by the degree of the minimal polynomial.

> **Box 2.6   Maximum-degree polynomials for a given graph**
>
> Let $p(\mathbf{Z})$ be an arbitrary polynomial of $\mathbf{Z}$ of degree greater than that of $p_{\min}(\mathbf{Z})$. Then we can write
>
> $$p(\mathbf{Z}) = q(\mathbf{Z})p_{\min}(\mathbf{Z}) + r(\mathbf{Z}),$$
>
> where $q(\mathbf{Z})$, the quotient, and $r(\mathbf{Z})$, the remainder, are polynomials of $\mathbf{Z}$ that can be obtained by long division (see Example 2.7). Then for any input signal $\mathbf{x}$ it is easy to see that
>
> $$p(\mathbf{Z})\mathbf{x} = (q(\mathbf{Z})p_{\min}(\mathbf{Z}) + r(\mathbf{Z}))\mathbf{x} = r(\mathbf{Z})\mathbf{x}, \qquad (2.38)$$
>
> since $p_{\min}(\mathbf{Z}) = \mathbf{0}$. An important consequence of this observation is that the locality of polynomial operations on the graph depends on both the graph's radius and diameter (as discussed in Section 2.4.2) and also on the algebraic characteristics of the one-hop operator (i.e., the degree of its minimal polynomial).

Given $p_c(\mathbf{Z})$, in order to identify $p_{\min}(\mathbf{Z})$ all we need to do is find the powers $p_i \leq k_i$ associated with each of the factors in (2.37) that guarantee that $p_{\min}(\mathbf{Z})\,\mathbf{x} = \mathbf{0}$ for all $\mathbf{x}$. Recall that polynomials of $\mathbf{Z}$ commute (Theorem 2.1) and thus we can write the factors in (2.37) in any order. Note that if the factor $(\mathbf{Z} - \lambda_i \mathbf{I})^{p_i}$ is present in (2.37), any vector $\mathbf{x} \in \mathcal{N}((\mathbf{Z} - \lambda_i \mathbf{I})^{p_i})$, the null space of $(\mathbf{Z} - \lambda_i \mathbf{I})^{p_i}$, will be such that $p(\mathbf{Z})\mathbf{x} = \mathbf{0}$ for any polynomial $p(\mathbf{Z})$ including a term $(\mathbf{Z} - \lambda_i \mathbf{I})^{p_i}$. Thus, the problem of finding the minimal polynomial $p_{\min}(\mathbf{Z})$ is the problem of finding p_i such that we can construct a basis for $\mathbb{C}^N$ where each basis vector is in one of the null spaces $\mathcal{N}((\mathbf{Z} - \lambda_i \mathbf{I})^{p_i})$. The choice of p_i thus depends on the dimension of $\mathcal{N}((\mathbf{Z} - \lambda_i \mathbf{I})^{p_i})$. We consider this next.

2.5.4   Algebraic and Geometric Multiplicities and Minimal Polynomials

We have seen already that the algebraic multiplicity k_i of eigenvalue λ_i is the power of the corresponding factor in the characteristic polynomial of (2.35), i.e., $(\mathbf{Z} - \lambda_i \mathbf{I})^{k_i}$. Since λ_i is an eigenvalue there must be at least one non-zero vector (an eigenvector) in $\mathcal{N}(\mathbf{Z} - \lambda_i \mathbf{I})$. The **geometric multiplicity** m_i of λ_i is the **dimension** of $\mathcal{N}(\mathbf{Z} - \lambda_i \mathbf{I})$, i.e., the maximum size of a set of linearly independent vectors in $\mathcal{N}(\mathbf{Z} - \lambda_i \mathbf{I})$.

Consider a specific case where $(\mathbf{Z} - \lambda_i \mathbf{I})^{k_i}$ is a term in the characteristic polynomial and $(\mathbf{Z} - \lambda_i \mathbf{I})^{p_i}$ is the corresponding term in the minimal polynomial. Denote by E_i the subspace associated with eigenvalue λ_i. As discussed above our goal is to represent any $\mathbf{x}$ as follows:

$$\mathbf{x} = \sum_i \mathbf{x}_i$$

where $\mathbf{x}_i \in E_i = \mathcal{N}((\mathbf{Z} - \lambda_i \mathbf{I})^{p_i})$. For each of the following two cases, depending on whether k_i and m_i are equal or not, we discuss how p_i can be obtained.

Case 1 $m_i = k_i, k_i \geq 1$

In this case $E_i = \mathcal{N}(\mathbf{Z} - \lambda_i\mathbf{I})$ has dimension k_i and we are able to express all vectors in E_i in terms of a basis comprising $m_i = k_i$ linearly independent eigenvectors, so that $E_i = \text{span}(\mathbf{u}_{i,1}, \mathbf{u}_{i,2}, \ldots, \mathbf{u}_{i,m_i})$, where each $\mathbf{u}_{i,j}$ is one of the eigenvectors associated with the eigenvalue λ_i. Thus, for any $\mathbf{x} \in E_i$ we have

$$\mathbf{x} = \sum_j \alpha_j \mathbf{u}_{i,j}$$

and as a consequence

$$(\mathbf{Z} - \lambda_i\mathbf{I})\mathbf{x} = \mathbf{0}, \quad \forall \mathbf{x} \in E_i,$$

so that $\mathbf{Z} - \lambda_i\mathbf{I}$ is the minimal polynomial for vectors in E_i, and $p_i = 1$. Moreover

$$\mathbf{Z}\mathbf{x} = \lambda_i \sum_j \alpha_j \mathbf{u}_{i,j} \in E_i,$$

which shows that E_i is **invariant** under multiplication by any polynomial of $\mathbf{Z}$. Denote by $\mathbf{U}_i$ the $N \times m_i$ matrix of the linearly independent eigenvectors that form a basis for E_i and let $\tilde{\mathbf{U}}_i$ be $N \times m_i$ and such that $\tilde{\mathbf{U}}_i^H \mathbf{U}_i = \mathbf{I}_{m_i}$, where $\mathbf{I}_{m_i}$ is the identity matrix of size $m_i \times m_i$. For any $\mathbf{x} \in E_i$ there is a vector $\mathbf{a} = [\alpha_1, \ldots, \alpha_{m_i}]^T$ such that:[9]

$$\mathbf{x} = \mathbf{U}_i\mathbf{a}, \quad \text{where} \quad \mathbf{a} = \tilde{\mathbf{U}}_i^H\mathbf{x},$$

and the graph operator for $\mathbf{x} \in E_i$ can be written as

$$\mathbf{Z}\mathbf{x} = \mathbf{U}_i(\lambda_i\mathbf{I}_{m_i})\tilde{\mathbf{U}}_i^H\mathbf{x}.$$

If in addition $m_i = k_i$ for *all* i, then $\mathbf{Z}$ is **diagonalizable** and can be written as:

$$\mathbf{Z} = \mathbf{U}\mathbf{\Lambda}\mathbf{U}^{-1}$$

where each column of $\mathbf{U}$ is one of N linearly independent eigenvectors and $\mathbf{\Lambda}$ is diagonal, with each diagonal entry an eigenvalue.

Case 2 $m_i < k_i, k_i > 1$

In this scenario, $\mathbf{Z}$ is **defective** or **non-diagonalizable**: m_i, the dimension of $\mathcal{N}(\mathbf{Z}-\lambda_i\mathbf{I})$, is less than k_i, the dimension of $\mathcal{N}((\mathbf{Z} - \lambda_i\mathbf{I})^{k_i})$. Note that $\mathbf{x} \in \mathcal{N}(\mathbf{Z} - \lambda_i\mathbf{I})$ implies that $\mathbf{x} \in \mathcal{N}((\mathbf{Z} - \lambda_i\mathbf{I})^2)$, so that $\mathcal{N}(\mathbf{Z} - \lambda_i\mathbf{I}) \subseteq \mathcal{N}((\mathbf{Z} - \lambda_i\mathbf{I})^2)$.

The term in the minimal polynomial $(\mathbf{Z}-\lambda_i\mathbf{I})^{p_i}$ has to set to zero all the vectors in $E_i = \mathcal{N}((\mathbf{Z}-\lambda_i\mathbf{I})^{k_i})$. Thus we are looking for the minimal p_i such that $\mathcal{N}((\mathbf{Z}-\lambda_i\mathbf{I})^{p_i}) = \mathcal{N}((\mathbf{Z}-\lambda_i\mathbf{I})^{k_i})$. Then $E_i = \text{span}(\mathbf{u}_{i,1}, \mathbf{u}_{i,2}, \ldots, \mathbf{u}_{i,k_i})$, where the linearly independent vectors $\mathbf{u}_{i,j}$ span $\mathcal{N}((\mathbf{Z} - \lambda_i\mathbf{I})^{p_i}) = \mathcal{N}((\mathbf{Z} - \lambda_i\mathbf{I})^{k_i})$, so that

$$(\mathbf{Z} - \lambda_i\mathbf{I})^{k_i}\mathbf{x} = (\mathbf{Z} - \lambda_i\mathbf{I})^{p_i}\mathbf{x} = \mathbf{0}, \quad \forall \mathbf{x} \in E_i. \tag{2.39}$$

Note that in Case 1 the basis for E_i was formed with linearly independent eigenvectors

---

[9] As discussed in Appendix A, $\tilde{\mathbf{U}}_i$ is the dual basis of $\mathbf{U}_i$, which always exists and is unique because the column vectors in $\mathbf{U}_i$ are linearly independent. If $\mathbf{U}$ has columns forming a basis for $\mathbb{R}^N$ then $\tilde{\mathbf{U}} = \mathbf{U}^{-1}$.

corresponding to λ_i. In contrast, in Case 2 the basis vectors cannot all be eigenvectors and the basis for E_i can be constructed in different ways. One approach for basis construction is illustrated in Example 2.8 below.

Example 2.8   Basis for invariant subspaces of a defective matrix   Let $\mathbf{Z}$ be a defective matrix having an eigenvalue λ of algebraic multiplicity $k = 2$ and geometric multiplicity $m = 1$. Find a basis for the space E of all vectors $\mathbf{x}$ such that $(\mathbf{Z} - \lambda\mathbf{I})^2\mathbf{x} = \mathbf{0}$.

Solution
If $\mathbf{u}^{(0)}$ is an eigenvector corresponding to λ, we have that $(\mathbf{Z} - \lambda\mathbf{I})\mathbf{u}^{(0)} = \mathbf{0}$, by definition, and thus $(\mathbf{Z} - \lambda\mathbf{I})^2\mathbf{u}^{(0)} = \mathbf{0}$ so that $\mathbf{u}^{(0)} \in E$. Since the algebraic multiplicity is 2, the invariant space E has dimension 2 but its geometric multiplicity is 1, so that we cannot find two linearly independent eigenvectors. Thus, we need to find a vector $\mathbf{u}^{(1)}$ such that $\mathbf{u}^{(0)}$ and $\mathbf{u}^{(1)}$ are linearly independent and $(\mathbf{Z} - \lambda\mathbf{I})^2\mathbf{u}^{(1)} = \mathbf{0}$. Choose $\mathbf{u}^{(1)}$ such that

$$(\mathbf{Z} - \lambda\mathbf{I})\mathbf{u}^{(1)} = \mathbf{u}^{(0)}. \tag{2.40}$$

For any such $\mathbf{u}^{(1)}$, we have $(\mathbf{Z} - \lambda\mathbf{I})^2\mathbf{u}^{(1)} = (\mathbf{Z} - \lambda\mathbf{I})\mathbf{u}^{(0)} = \mathbf{0}$ and, since $\mathbf{u}^{(0)}$ is non-zero, this means that $\mathbf{u}^{(1)}$ does not belong to $\mathcal{N}(\mathbf{Z} - \lambda\mathbf{I})$ and therefore $\mathbf{u}^{(0)}$ and $\mathbf{u}^{(1)}$ are linearly independent.

Jordan canonical form   The construction sketched in Example 2.8 forms the basis for the Jordan canonical form. For any λ_i of algebraic multiplicity $k_i > 1$, if the geometric multiplicity is $m_i < k_i$ then we can find m_i linearly independent eigenvectors. For each of these eigenvectors we can follow the procedure sketched in Example 2.8 to create a *Jordan chain*. As in Example 2.8, $\mathbf{u}^{(1)}$ can be found by first solving (2.40); the next step is to find $\mathbf{u}^{(2)}$ such that

$$(\mathbf{Z} - \lambda\mathbf{I})\mathbf{u}^{(2)} = \mathbf{u}^{(1)}, \quad (\mathbf{Z} - \lambda\mathbf{I})^2\mathbf{u}^{(2)} = \mathbf{u} \quad \text{and} \quad (\mathbf{Z} - \lambda\mathbf{I})^3\mathbf{u}^{(2)} = \mathbf{0},$$

or, equivalently, such that

$$\mathbf{Z}\mathbf{u}^{(0)} = \lambda\mathbf{u}^{(0)}, \quad \mathbf{Z}\mathbf{u}^{(1)} = \lambda\mathbf{u}^{(1)} + \mathbf{u}^{(0)} \quad \text{and} \quad \mathbf{Z}\mathbf{u}^{(2)} = \lambda\mathbf{u}^{(2)} + \mathbf{u}^{(1)}.$$

For any vector $\mathbf{x} \in \mathrm{span}(\mathbf{u}^{(2)}, \mathbf{u}^{(1)}, \mathbf{u}^{(0)})$, so that $\mathbf{x} = x_2\mathbf{u}^{(2)} + x_1\mathbf{u}^{(1)} + x_0\mathbf{u}^{(0)}$, multiplication by $\mathbf{Z}$ can be written in a compact form:

$$\mathbf{y} = \mathbf{Z}\mathbf{x} = \mathbf{Z}(x_2\mathbf{u}^{(2)} + x_1\mathbf{u}^{(1)} + x_0\mathbf{u}^{(0)}) = \lambda x_2\mathbf{u}^{(2)} + (\lambda x_1 + x_2)\mathbf{u}^{(1)} + (\lambda x_0 + x_1)\mathbf{u}^{(0)},$$

where we can see that $\mathbf{y}$ and $\mathbf{x}$ are in the same subspace (since both are linear combinations of $\mathbf{u}^{(2)}$, $\mathbf{u}^{(1)}$, $\mathbf{u}^{(0)}$) and therefore $\mathrm{span}(\mathbf{u}^{(2)}, \mathbf{u}^{(1)}, \mathbf{u}^{(0)})$ is invariant under multiplication by $\mathbf{Z}$; $\mathbf{y}$ can then be written as

$$\begin{bmatrix} y_0 \\ y_1 \\ y_2 \end{bmatrix} = \begin{bmatrix} \lambda & 1 & 0 \\ 0 & \lambda & 1 \\ 0 & 0 & \lambda \end{bmatrix} \begin{bmatrix} x_0 \\ x_1 \\ x_2 \end{bmatrix}. \tag{2.41}$$

The matrix in (2.41) is an example of a Jordan block. A Jordan chain starts with each of the m_i eigenvectors and p_i will have the length of the longest Jordan chain.

Example 2.9 Defective graph As a concrete example, consider a directed path graph with four nodes, leading to the adjacency matrix

$$\mathbf{A} = \begin{bmatrix} 0 & 1 & 0 & 0 \\ 0 & 0 & 1 & 0 \\ 0 & 0 & 0 & 1 \\ 0 & 0 & 0 & 0 \end{bmatrix}$$

where we note that one node has only one outgoing link, while another only has an incoming link. Prove that $\mathbf{A}$ is defective, find three independent vectors in the invariant subspace and interpret their behavior.

Solution
First note that, since the determinant of an upper triangular matrix is the product of its diagonal terms, we have $\det(\lambda\mathbf{I} - \mathbf{A}) = \lambda^4$, where $\lambda = 0$ is an eigenvalue with multiplicity 4. Then, solving for an eigenvector leads to $u(2) = u(3) = u(4) = 0$, so that $\mathbf{u}_1 = \mathbf{e}_1 = [1,0,0,0]^\mathsf{T}$ is the only eigenvector. Notice that $\mathbf{u}_2 = \mathbf{e}_2$ is such that $\mathbf{A}\mathbf{u}_2 = \mathbf{u}_1$, and therefore $\mathbf{A}^2\mathbf{u}_2 = \mathbf{0}$, so that $\mathbf{u}_2$ is not an eigenvector but belongs to the invariant space with polynomial $\mathbf{A}^4$. Likewise $\mathbf{e}_3$ and $\mathbf{e}_4$ are also in that invariant subspace. For any vector in $\mathbb{R}^4$ we have $\mathbf{A}^4\mathbf{x} = \mathbf{0}$. We can also observe directly that $\mathbf{A}$ is a Jordan block as in (2.41) with diagonal values $\lambda = 0$. In conclusion, since $\lambda = 0$ has algebraic multiplicity 4, but there is only one eigenvector, $\mathbf{e}_1$, corresponding to this eigenvalue, the geometric multiplicity is smaller than the algebraic multiplicity and $\mathbf{A}$ is defective.

Diagonalization and invariance Our goal in this section is to understand graph filters, polynomials of $\mathbf{Z}$, from the perspective of their invariance properties. It is important to emphasize that invariance under multiplication by $\mathbf{Z}$ exists whether or not $\mathbf{Z}$ can be diagonalized.

> *Remark* 2.3 Let eigenvalue λ_i have algebraic multiplicity k_i, and define $E_i = \mathcal{N}\left((\mathbf{Z} - \lambda_i\mathbf{I})^{k_i}\right)$. Then E_i is invariant under multiplication by $\mathbf{Z}$.

Proof By definition $\mathbf{x} \in E_i$ if $(\mathbf{Z} - \lambda_i\mathbf{I})^{k_i}\mathbf{x} = \mathbf{0}$, so that, using the commutativity of polynomials (Theorem 2.1),

$$(\mathbf{Z} - \lambda_i\mathbf{I})^{k_i}\mathbf{Z}\mathbf{x} = \mathbf{Z}(\mathbf{Z} - \lambda_i\mathbf{I})^{k_i}\mathbf{x} = \mathbf{0},$$

which shows that $\mathbf{Z}\mathbf{x} \in E_i$ and proves that the space is invariant. □

Any vector in $\mathbb{R}^N$ can be written as a linear combination of vectors belonging to subspaces $E_1, E_2, \ldots, E_M$, where M is the number of distinct eigenvalues. The only difference between the diagonalizable and defective cases is that in the former we can form a basis for E_i consisting of eigenvectors, while this is not possible for all E_i in the latter case. For a given subspace E_i the power p_i of the term $(\mathbf{Z} - \lambda_i \mathbf{I})^{p_i}$ in the minimal polynomial is the minimal value such that $(\mathbf{Z} - \lambda_i \mathbf{I})^{p_i} \mathbf{x} = \mathbf{0}$ for any $\mathbf{x} \in E_i$. These ideas are summarized in Table 2.1.

Table 2.1 Summary: Diagonalizable and non-diagonalizable operators

	Diagonalizable	Non-diagonalizable
Multiplicity	$m_i = k_i, \forall i$	$m_i < k_i$ for at least one i
$\dim(\mathcal{N}((\mathbf{Z} - \lambda_i \mathbf{I})^{k_i}))$	$\dim(\mathcal{N}(\mathbf{Z} - \lambda_i \mathbf{I}))$	$\dim(\mathcal{N}((\mathbf{Z} - \lambda_i \mathbf{I})^{p_i}))$
Invariant subspaces	N	$M = \sum_i m_i$
$p_c(\mathbf{Z})$	$\prod_i (\mathbf{Z} - \lambda_i \mathbf{I})^{k_i}$	$\prod_i (\mathbf{Z} - \lambda_i \mathbf{I})^{k_i}$
$p_{\min}(\mathbf{Z})$	$\prod_i (\mathbf{Z} - \lambda_i \mathbf{I})$	$\prod_i (\mathbf{Z} - \lambda_i \mathbf{I})^{p_i}$
$\mathbf{U}$	N eigenvectors	$\sum_i m_i$ eigenvectors
Λ	Diagonal	Block diagonal

Schur decomposition for defective matrices   The Jordan canonical form for a defective $\mathbf{Z}$ has well-known numerical problems. An alternative approach is to start from the Schur decomposition (see Theorem 2.2) and obtain a block diagonal form from it. See for example [18, 19] for a general description. The idea of using the Schur decomposition as an alternative to the Jordan form was first proposed in the context of GSP in [20] and also implemented in the GraSP Matlab toolbox [21] described in Appendix B. Denote $\mathbf{U}_i$ as a matrix with k_i columns forming a basis for E_i and let $\tilde{\mathbf{U}}_i$ be $N \times k_i$ and such that $\tilde{\mathbf{U}}_i^H \mathbf{U}_i = \mathbf{I}_{k_i}$, where $\mathbf{I}_{k_i}$ is the identity matrix of size $k_i \times k_i$. In this case we again have, for $\mathbf{x}_i \in E_i$,

$$\mathbf{x}_i = \mathbf{U}_i \mathbf{a}, \tag{2.42}$$

and therefore

$$\mathbf{a} = \tilde{\mathbf{U}}_i^H \mathbf{x}_i \tag{2.43}$$

so that

$$\mathbf{Z}\mathbf{x}_i = \mathbf{U}_i \Lambda_i \tilde{\mathbf{U}}_i^H \mathbf{x}_i \tag{2.44}$$

where the main difference with respect to Case 1 is that Λ_i is not diagonal.

Importance of the defective case   In some cases making use of the original graph is important and so it will be necessary to work with a defective $\mathbf{Z}$. In other cases we may consider changing the graph. Informally, it is always possible to find a diagonalizable matrix "close" to any non-diagonalizable one. Thus, if $\mathbf{Z}$ cannot be be diagonalized, we could look for an alternative, $\bar{\mathbf{Z}}$, that: (i) can be diagonalized, (ii) is close to $\mathbf{Z}$ (e.g., in terms of the Frobenius norm of the difference) and (iii) represents a graph (e.g., $\bar{\mathbf{Z}}$ is a valid adjacency matrix).

If selecting a new $\bar{\mathbf{Z}}$ is an option, several methods can be used. Recalling that symmetric matrices can always be diagonalized, one approach would be to convert the directed graph into an undirected one. For a directed $\mathbf{Z} = \mathbf{A}$, a natural symmetrization would be $\bar{\mathbf{Z}} = \mathbf{A} + \mathbf{A}^{\mathsf{T}}$, with the bibliometric symmetrization $\bar{\mathbf{Z}} = \mathbf{A}\mathbf{A}^{\mathsf{T}} + \mathbf{A}^{\mathsf{T}}\mathbf{A}$ as a possible alternative [5]. Other approaches that preserve the directed nature of the graph can be based on structural criteria, which modify the graph for a specific application. An example of this approach is the teleportation idea in PageRank [4]. Moreover, since some graph structures (such as path sources or path sinks) lead to a defective $\mathbf{Z}$, removing those can lead to graphs with better behavior.

More generally, as discussed in Chapter 6, in many cases the choice of graph is in itself a problem, and thus, when deciding what graph to choose in a particular case, the relevant properties of the corresponding graph operator $\mathbf{Z}$ should be taken into account.

2.5.5 Practical Implications for Graph Filter Design

To conclude this chapter we summarize the key consequences of our study of polynomials $\mathbf{Z}$ for the design of practical graph filters.

Localization depends on polynomial degree and graph topology As discussed in detail in Section 2.2, a degree-k polynomial corresponds to k-hop localized processing around every node in the graph. But the choice of an appropriate parameter k should be a function of prior knowledge about the graph signal of interest (e.g., over what size subgraphs can we expect to see locally similar behavior) as well as properties of the graph topology, such as radius or diameter. It is clear that a polynomial of degree $k_1 < k_2$ will provide more localized processing than one of degree k_2, but just how localized this is depends on the graph properties. This is particularly important in applications where graphs are reduced (see Section 6.1.5) and become denser as the number of nodes is decreased. Thus, if a k-hop filter is appropriate for the original graph, it may not be the right choice for a smaller graph derived from the original one.

Invariant subspaces and graph signal analysis We have described in detail subspaces of vectors that are invariant under multiplication by $\mathbf{Z}$. From a graph signal analysis perspective the corresponding minimal polynomials, which by definition produce a zero output for any signal in that subspace, may be useful. If E is an invariant subspace, and if $P_E(\mathbf{Z})$ is the corresponding minimal polynomial, then $P_E(\mathbf{Z})\mathbf{x} = \mathbf{0}$ for any $\mathbf{x} \in E$. Therefore, when analyzing some arbitrary signals $\mathbf{y} \in \mathbb{R}^N$ we can use $P_E(\mathbf{Z})$ to eliminate any component of such a signal that belongs to E. The signal $P_E(\mathbf{Z})\mathbf{y}$ will contain no information corresponding to subspace E.

In Chapter 3 we will discuss the problem of filter design, with an emphasis on the case where the graph operator $\mathbf{Z}$ can be diagonalized. But, more generally, even for defective $\mathbf{Z}$ the effect of a polynomial $P(\mathbf{Z})$ can be completely characterized by its effect on each invariant subspace. In particular, the dimension of each of these invariant subspaces is important, since for spaces of dimension greater than 1 the choice of basis functions is not unique, as discussed below in Box 2.7.

Box 2.7   Non-uniqueness of bases for invariant subspaces

If an eigenvalue λ_i is not simple, $k_i > 1$, we have a fundamental ambiguity. The invariant subspace E_i has dimension greater than 1 and thus an infinite number of basis functions can be selected for E_i. Since we are characterizing signals by their response to $\mathbf{Z}$ all the vectors in E_i have the same properties and same minimal polynomial $(\mathbf{Z} - \lambda_i \mathbf{I})^{p_i}$. Thus, any bases for vectors in E_i should be equivalent.

As noted in [22], unless there is a standardized way of selecting a basis for an invariant space E_i with $\dim(E_i) > 1$ we may not be able to compare representations of the same signal $\mathbf{x}$ produced with different implementations. The alternative proposed in [22] is to use the expressions (2.42) and (2.43). Letting $\mathbf{x}$ be any vector in $\mathbb{R}^N$ we choose its (non-orthogonal) projection $\mathbf{x}_i$ onto E_i as follows:

$$\mathbf{x}_i = \mathbf{U}_i \tilde{\mathbf{U}}_i^{\mathsf{H}} \mathbf{x}. \tag{2.45}$$

The vector $\mathbf{x}_i$ is independent of the basis chosen to represent E_i. This is not an orthogonal projection: the approximation error is orthogonal to the dual vectors, $\tilde{\mathbf{U}}_i$, rather than the spanning vectors, $\mathbf{U}_i$, as would be the case for an orthogonal projection:

$$\tilde{\mathbf{U}}_i^{\mathsf{H}}(\mathbf{x} - \mathbf{x}_i) = \tilde{\mathbf{U}}_i^{\mathsf{H}}(\mathbf{I} - \mathbf{U}_i \tilde{\mathbf{U}}_i^{\mathsf{H}})\mathbf{x} = (\tilde{\mathbf{U}}_i^{\mathsf{H}} - \tilde{\mathbf{U}}_i^{\mathsf{H}})\mathbf{x} = \mathbf{0}$$

where we have used the fact that $\tilde{\mathbf{U}}_i^{\mathsf{H}} \mathbf{U}_i = \mathbf{0}$, from the definition of biorthogonal bases (Definition A.8), and the conjugate transpose appears because the basis may be complex.

In summary, given invariant subspaces E_i the representation

$$\mathbf{x} = \sum_i \mathbf{x}_i,$$

where $\mathbf{x}_i$ is computed as in (2.45) and there is one term $\mathbf{x}_i$ per subspace E_i, which is unique and eliminates any ambiguity due to the choice of basis vectors for each E_i.

Chapter at a Glance

In this chapter we started by introducing basic definitions associated with graphs and using those to develop an understanding of node domain processing. Since graphs are characterized by one-hop connections between nodes, we explored how these induce a notion of locality, and how it can be extended so that a one-hop connection induces a k-hop neighborhood around a node. A natural follow-up step is to consider neighborhoods around multiple close nodes, leading to the notion of cuts and clustering. We then introduced algebraic representations of graphs and discussed how these can be viewed as one-hop operations (see Table 2.2). This was then generalized to define graph filters as polynomials of an elementary graph operator $\mathbf{Z}$. We showed that the algebraic properties of $\mathbf{Z}$ determine what operations can be performed on a graph. In particular, vectors that are invariant under multiplication by $\mathbf{Z}$ can be used to understand the effect

Table 2.2 Summary of one-hop operators

	$\mathbf{Z}$	$\mathbf{y} = \mathbf{Z}\mathbf{x}$
Adjacency matrix	$\mathbf{A}$	$y(i) = \sum_{j \in \mathcal{N}(i)} a_{ij} x(j)$
Random walk	$Q = \mathbf{D}^{-1}\mathbf{A}$	$y(i) = \frac{1}{d_i} \sum_{j \in \mathcal{N}(i)} a_{ij} x(j)$
Combinatorial Laplacian	$\mathbf{L} = \mathbf{D} - \mathbf{A} = \mathbf{B}\mathbf{B}^{\mathsf{T}}$	$y(i) = d_i x(i) - \sum_{j \in \mathcal{N}(i)} a_{ij} x(j)$
Normalized Laplacian	$\mathcal{L} = \mathbf{I} - \mathbf{D}^{-1/2}\mathbf{A}\mathbf{D}^{-1/2}$	$y(i) = x(i) - \sum_{j \in \mathcal{N}(i)} \frac{a_{ij}}{\sqrt{d_i}\sqrt{d_j}} x(j)$
Random walk Laplacian	$\mathcal{T} = \mathbf{I} - Q = \mathbf{D}^{-1}\mathbf{B}\mathbf{B}^{\mathsf{T}}$	$y(i) = x(i) - \frac{1}{d_i} \sum_{j \in \mathcal{N}(i)} a_{ij} x(j)$

of polynomials of $\mathbf{Z}$ on arbitrary vectors. When $\mathbf{Z}$ is diagonalizable there exists a full set of linear independent invariant vectors (i.e., eigenvectors). When $\mathbf{Z}$ is not diagonalizable, we can instead find invariant subspaces and use bases for those subspaces to create a complete representation for graph signals in the space. In either case every vector in the space can be written as a linear combination of elementary vectors, each belonging to a space invariant under multiplication by $\mathbf{Z}$.

Further Reading

Graph theory is a classical topic, with many textbooks available to cover topics such as graph coloring, graph cuts and so on [23]. Texts more focused on spectral graph theory, such as [8], also provide basic graph definitions and cover the elementary algebraic operators described in this chapter. For the development of graph signal processing from the perspective of a polynomial of an elementary one-hop operator and its connection with conventional signal processing see [15, 10] as well as the general algebraic signal processing framework of [24, 25]. The representation of a space in terms of a basis derived from invariance under multiplication by a matrix $\mathbf{Z}$ is developed in [26]. A detailed discussion of graph Fourier transforms in the context of non-diagonalizable $\mathbf{Z}$ is provided by [22]. The main difference in our presentation with respect to that of [22] is that we show that the Jordan blocks, which we also used, are only one of the possible block diagonal representations that could be chosen. Because of the well-known numerical issues associated with the Jordan form, other block diagonal representations may be preferable.

GraSP Examples

Section B.3 provides several examples of node domain filtering operations to illustrate the different approaches that can be used for a specific filter defined as a function of a fundamental graph operator. A direct approach, where a full matrix is computed and applied to the signal, is shown first. This approach may not be efficient for a large graph, however. Thus, a second example considers implementation via successive applications of the fundamental graph operator, where the filter is a polynomial of the operator.

Problems

2.1   *Coloring*

Finding a graph coloring can be simple. If we have N nodes we can assign a different color to each node and the graph will be N-colorable. The challenge is finding the minimum coloring, i.e., the one with the minimum number of colors.

Find the minimum coloring for the following graphs:

a.   a complete unweighted graph with N nodes;

b.   a star graph with N nodes;

c.   a tree with K levels.

2.2   *Coloring and number of edges*

(True/False) Given an unweighted graph with N nodes and chromatic number C (see Section 2.2.5), if we add an edge to connect two nodes that are not yet connected, then the chromatic number always increases.

2.3   *Trees*

In this problem we consider an unweighted tree with K levels. Discuss whether your answers depend on specific tree properties (balanced or not, minimum and maximum number of children per node, etc.).

a.   Find its diameter and radius.

b.   Is it possible to remove an edge in such a way that the diameter is increased while the graph remains connected?

c.   Is it possible to add one edge to the graph (which then may no longer be a tree) in such a way that the diameter is reduced while the radius remains unchanged?

d.   What is the chromatic number of the tree?

2.4   *Connectedness*

In this problem we consider an unweighted directed graph with N nodes with corresponding adjacency matrix $\mathbf{A}$. Assume that $\mathbf{A}$ is reducible, so that it is possible to find a permutation of $\mathbf{A}$ to put it into block upper triangular form:

$$\mathbf{A} = \begin{bmatrix} \mathbf{A}_1 & \mathbf{A}_{12} \\ \mathbf{0} & \mathbf{A}_2 \end{bmatrix}$$

where $\mathbf{A}_1$ and $\mathbf{A}_2$ are square matrices of sizes $N_1 \times N_1$ and $N_2 \times N_2$ with $N_1 + N_2 = N$.

a.   Prove that the corresponding graph is not strongly connected.

b.   Assuming that the graph is weakly connected, what would be the minimum number of edges to be added in order to make it strongly connected?

c.   Consider the special case where only entries directly above the diagonal are non-zero. What is this graph? Is it weakly connected?

2.5   *Processing signals on a tree*

Consider a balanced binary tree with K levels.

a.   Assuming there are two processors, what is the best way to split the tree for processing into two roughly equal subgraphs?

b. Assuming we are performing a k-hop localized graph filter on this tree, how much data has to be exchanged between the two processors?

2.6 *Normalization*

In Section 2.3.3 we discussed two normalization strategies, one where the goal was to achieve consensus, the other where some physical goods were transported. Here we consider an extension of the latter case, where physical goods are transported but where a percentage of the goods is lost while being stored at a node before being distributed. Assume only a fraction $0 < \alpha < 1$ of the goods at any node is actually transported.

a. Let **L** be the combinatorial Laplacian of the original graph. For a given α propose a normalization matrix **K** as an alternative for the above approach using normalization strategies mentioned above.

b. In the case in part **a**, given an initial signal $\mathbf{x}^{(0)}$ with all positive entries and defining $N_k = \mathbf{1}^\mathsf{T} \mathbf{x}^{(k)}$, find k such that $N_k/N_0 < \epsilon$ for a given $0 < \epsilon < 1$.

c. Repeat parts **a** and **b** for the case where the loss is node-dependent, that is, there is an $0 < \alpha_i < 1$ associated with each node i.

2.7 *Adjacency matrices and permutations*

a. Write the adjacency matrix of an eight-node undirected and unweighted cycle graph with consecutive labeling of the nodes along the cycle.

b. Repeat this for the case where the labels along the cycle are $1, 3, 5, 7, 2, 4, 6, 8$.

c. With the first labeling the adjacency matrix is circulant, that is, each row can be obtained by shifting by one entry the row immediately above (and the first row can be obtained from the eighth). Assuming we add an edge from node 1 to node 4, what other edges should be added so that the resulting adjacency matrix is still circulant?

2.8 *Minimal polynomials of vectors*

Let $\mathbf{x}_1$ and $\mathbf{x}_2$ be two graph signals with respective minimal polynomials $p_1(\mathbf{Z})$ and $p_2(\mathbf{Z})$ having degrees n_1 and n_2. Assume that $\mathbf{Z}$ is $n \times n$ and that all its eigenvalues are simple.

a. What is the minimal polynomial for the vectors in the set $S = \text{span}(\mathbf{x}_1, \mathbf{x}_2)$?

b. Letting $p_{\min}(\mathbf{Z})$ be the minimal polynomial of $\mathbf{Z}$, is it possible for the minimal polynomial of S to be $p_{\min}(\mathbf{Z})$? If not, justify why this is so. If on the other hand it is possible then give an example of two vectors $\mathbf{x}_1$, $\mathbf{x}_2$ for which this is true.

2.9 *Non-diagonalizable* $\mathbf{Z}$

Consider a directed unweighted path graph of length 4.

a. Prove that the the operator $\mathbf{Z} = \mathbf{A}$ is not diagonalizable.

b. What are the invariant subspaces of $\mathbf{Z}$ as defined in **a**?

2.10 *Directed acyclic graphs*

Let **A** be the adjacency matrix of a directed acyclic graph (DAG). Prove that **A** is not diagonalizable.

2.11 *Path sinks and path sources*

In a directed graph G we define a path sink (or source) with n nodes as a structure where a node i_0 is connected to the rest of the graph by only one incoming edge (for a sink) or one outgoing edge (for a source) and where the connection to the rest of the graph is via a directed path connecting nodes $(i_1, i_2, \ldots, i_n)$.

a.   Using a suitable permutation, write the adjacency matrices for two graphs having a single path sink and a single path source, respectively, with both paths having n nodes.

b.   Prove that for $n > 1$ the corresponding adjacency matrices cannot be diagonalized.

2.12   *Nilpotent* **Z**

Let $\mathbf{Z}$ be nilpotent, such that there is a $k > 1$ for which $\mathbf{Z}^k = \mathbf{0}$ while $\mathbf{Z}^l \neq \mathbf{0}$ for $1 \leq l < k$. Prove that $\mathbf{Z}$ cannot be diagonalized.

2.13   *Symmetrization*

Let $\mathbf{A}$ be the adjacency matrix of a directed graph, with the corresponding incidence matrix $\mathbf{B}$, where there are $M = |E|$ edges and $\mathbf{b}_k$ denotes the kth column of $\mathbf{B}$.

a.   Find the incidence matrix of $\mathbf{A} + \mathbf{A}^{\mathsf{T}}$.

b.   Find the incidence matrices for $\mathbf{A}\mathbf{A}^{\mathsf{T}}$, $\mathbf{A}^{\mathsf{T}}\mathbf{A}$ and $\mathbf{A}\mathbf{A}^{\mathsf{T}} + \mathbf{A}^{\mathsf{T}}\mathbf{A}$.

c.   Provide a qualitative comparison of the symmetrizations $\mathbf{A}_1 = \mathbf{A} + \mathbf{A}^{\mathsf{T}}$ and $\mathbf{A}_2 = \mathbf{A}\mathbf{A}^{\mathsf{T}} + \mathbf{A}^{\mathsf{T}}\mathbf{A}$ with the corresponding incidence matrices $\mathbf{B}_1$ and $\mathbf{B}_2$ in terms of their number of edges (sparsity).

3 Graph Signal Frequency – Spectral Graph Theory

In Chapter 1 we first discussed extending the idea of "variation" from time signals to graph signals. Since there is no natural ordering of graph nodes (see Box 2.1), variation has to be defined on the basis of graph topology, i.e., as a function of the differences between the signal values at a node and at its immediate neighbors. Then, a graph signal will have more variation (and thus higher frequency) if there are large differences in signal values between neighboring nodes (see Figure 1.7).

In Chapter 2 we introduced node domain linear filters, built using a fundamental graph operator $\mathbf{Z}$. Because $\mathbf{Z}$ is a *local* (e.g., one-hop) operator, for any signal $\mathbf{x}$, $\|\mathbf{x} - \mathbf{Z}\mathbf{x}\|$, where $\|\cdot\|$ indicates the norm of the vector (Definition A.5), quantifies how much the operator modifies the signal locally. Also in Chapter 2 we introduced subspaces of signals invariant under $\mathbf{Z}$. If $\mathbf{u}$ is an invariant vector (eigenvector), then $\mathbf{Z}\mathbf{u} = \lambda\mathbf{u}$ and thus $\|\mathbf{u} - \mathbf{Z}\mathbf{u}\| = |1 - \lambda|\|\mathbf{u}\|$. If $\mathbf{Z}$ can be diagonalized then the local variation is the same for all signals in each subspace and depends on the specific eigenvalue λ for that subspace.

In this chapter we bring together the ideas of graph signal variation and graph operators with the goal of constructing frequency representations for graph signals. The notions of frequency we introduce here have been studied in the context of spectral graph theory (see for example [7, 8]). Note that in this book we are interested in definitions of frequency that allow us to analyze graph signals, while in spectral graph theory the goal is to use the graph spectrum to analyze the graph structure. The main results presented in this chapter are as follows.

- Any graph signal can be represented as a weighted sum of elementary signals, each corresponding to a graph frequency, which together form the graph Fourier transform (GFT) (Section 3.1). These elementary signals correspond to the basis vectors for the invariant subspaces defined in Chapter 2.
- We describe the GFTs associated with several commonly used graph operators, $\mathbf{Z}$, and their corresponding variation (Section 3.2).
- Using the GFT we revisit graph signal filtering from a frequency domain perspective · (Section 3.3).
- The GFTs of common graph operators are related (Section 3.4).
- The properties of eigenvectors and eigenvalues depend on the graph topology (Section 3.5).

In what follows the choice of operator determines how the invariant vectors $\mathbf{u}$ and their corresponding λ can be interpreted. As shown in Table 2.2, we have considered both smoothing operators, such as Q, and "differencing" operators such as L; their actions can be seen by observing that $Q1 = 1$, while $L1 = 0$. Therefore the interpretation of a $(\lambda, \mathbf{u})$ eigenpair will also depend on the type of operator. For a smoothing operator, a smooth eigenvector $\mathbf{u}$ will have a larger λ, while the opposite will be true for a differencing operator. While we seek to be general, in practice most of the results that follow pertain to (possibly weighted) undirected graphs. Whenever results apply only to specific graphs, we make this explicit.

3.1 Graph Fourier Transforms (GFTs)

3.1.1 From Invariant Subspaces to Graph Signal Representations

In Chapter 2 we introduced node domain processing based on a fundamental graph operator, $\mathbf{Z}$. Linear graph filters were defined as polynomials of $\mathbf{Z}$ and a series of invariant subspaces, E_i, were found for the elementary operation $\mathbf{Z}\mathbf{x}$: any $\mathbf{x}_i \in E_i$ is such that $\mathbf{Z}\mathbf{x}_i \in E_i$.

Bases from invariant subspaces   If $\mathbf{Z}$ is diagonalizable we can find N linearly independent eigenvectors, each associated with an invariant subspace, and we can form a basis for $\mathbb{R}^N$ using these eigenvectors (see Section A.2 for a review of the concept of a basis). Alternatively, if $\mathbf{Z}$ cannot be diagonalized, we have K linearly independent invariant subspaces, such that for any signal $\mathbf{x}$ we can write

$$\mathbf{x} = \mathbf{x}_1 + \mathbf{x}_2 + \cdots + \mathbf{x}_K, \quad \text{with } \mathbf{x}_i \in E_i, \forall i = 1, \ldots, K \tag{3.1}$$

and such that $\sum_i \dim(E_i) = N$. Therefore we can form a basis for vectors in $\mathbb{R}^N$ where each basis vector belongs to one of the invariant subspaces E_i. Expressing an arbitrary signal $\mathbf{x}$ in the invariant basis as in (3.1) makes it easy to write $\mathbf{Z}\mathbf{x}$:

$$\mathbf{Z}\mathbf{x} = \mathbf{Z}\mathbf{x}_1 + \mathbf{Z}\mathbf{x}_2 + \cdots + \mathbf{Z}\mathbf{x}_K = \mathbf{y}_1 + \mathbf{y}_2 + \cdots + \mathbf{y}_K,$$

where, by the definition of invariance, $\mathbf{y}_i \in E_i$ for all i. In other words, when $\mathbf{x}$ is written as a linear combination of basis vectors from invariant subspaces, the product $\mathbf{Z}\mathbf{x}$ can be expressed as a block diagonal operation. If $\mathbf{Z}$ can be diagonalized then all graph filtering operations can also be expressed in diagonal form when signals are represented as a function of the basis vectors derived from the E_i (Section 3.3).

Invariant signals and graph frequency   The shared properties of signals in a given subspace E_i can be expressed in terms of the $\mathbf{Z}$ quadratic form, which will be used later to define graph signal variation.

DEFINITION 3.1   (**Z** QUADRATIC FORM)   For a graph operator **Z** the operator quadratic form $\Delta_Z(\mathbf{x})$ and normalized quadratic form $\bar{\Delta}_Z(\mathbf{x})$ are functions that are associated with any signal $\mathbf{x} \in \mathbb{R}^N$ (that must be non-zero for the normalized case):

$$\Delta_Z(\mathbf{x}) = \mathbf{x}^T \mathbf{Z} \mathbf{x} \quad \text{and} \quad \bar{\Delta}_Z(\mathbf{x}) = \frac{\mathbf{x}^T \mathbf{Z} \mathbf{x}}{\mathbf{x}^T \mathbf{x}}.$$

For real **Z** and **x**, the quadratic form $\Delta_Z(\mathbf{x}) = \langle \mathbf{x}, \mathbf{Z}\mathbf{x} \rangle$ can be interpreted as the *similarity* between **x** and **Zx** (see Section A.3 for a discussion of inner products as a measure of similarity). With the normalization, a large $|\bar{\Delta}_Z(\mathbf{x})|$ indicates that the operator **Z** preserves **x**. We consider the cases of diagonalizable and defective **Z** separately.

3.1.2 Diagonalizable Z

If **Z** is diagonalizable, we can find linearly independent eigenvectors corresponding to eigenvalue λ_i to form a basis for E_i, so that, for any $\mathbf{x}_i \in E_i$,

$$\Delta_Z(\mathbf{x}_i) = \mathbf{x}_i^T \mathbf{Z} \mathbf{x}_i = \lambda_i \mathbf{x}_i^T \mathbf{x}_i \tag{3.2}$$

and all vectors in E_i are eigenvectors and have the same $\bar{\Delta}_Z(\mathbf{x}_i)$:

$$\bar{\Delta}_Z(\mathbf{x}_i) = \frac{\mathbf{x}_i^T \mathbf{Z} \mathbf{x}_i}{\mathbf{x}_i^T \mathbf{x}_i} = \lambda_i. \tag{3.3}$$

In Section 3.2 we show that, for some **Z**, $\bar{\Delta}_Z(\mathbf{x}_i)$ can be interpreted as a graph signal variation. The graph Fourier transform (GFT) can be constructed by combining the bases of all E_i subspaces.

DEFINITION 3.2   (GRAPH FOURIER TRANSFORM – DIAGONALIZABLE **Z**)   Since **Z** is diagonalizable, it has a full set of eigenvectors $\mathbf{u}_i$ that form a basis for $\mathbb{R}^N$. We can construct **U**, the graph Fourier transform, an invertible matrix where each column is one of the eigenvectors, and write

$$\mathbf{Z} = \mathbf{U}\mathbf{\Lambda}\mathbf{U}^{-1}, \tag{3.4}$$

where $\mathbf{\Lambda}$ is the diagonal matrix of eigenvalues. Thus we can write any **x** in terms of the basis vectors (the columns of **U**) representing its graph frequencies:

$$\mathbf{x} = \mathbf{U}\tilde{\mathbf{x}} \quad \text{with} \quad \tilde{\mathbf{x}} = \mathbf{U}^{-1}\mathbf{x}$$

where $\tilde{\mathbf{x}}$ is the GFT of graph signal **x**.

Note that this GFT is only unique if all the eigenvalues have multiplicity 1 (see Box 2.7 and Section 3.1.4). Note also that when the graph is undirected, **Z** is a symmetric matrix and since any real symmetric matrix is guaranteed to have a full set of orthogonal eigenvectors, we have

$$\mathbf{U}^T\mathbf{U} = \mathbf{U}\mathbf{U}^T = \mathbf{I}$$

and the corresponding eigenvalues are all real. This leads to the definition of the GFT for an undirected graph.

DEFINITION 3.3   (GRAPH FOURIER TRANSFORM – UNDIRECTED GRAPH)   Let $\mathbf{Z}$ be the $N \times N$ graph operator of a weighted undirected graph; then we can write

$$\mathbf{Z} = \mathbf{U}\boldsymbol{\Lambda}\mathbf{U}^\mathsf{T} \tag{3.5}$$

where $\mathbf{U}$ is the graph Fourier transform (GFT). Thus we can write any $\mathbf{x}$ in terms of the basis vectors (the columns of $\mathbf{U}$) representing its graph frequencies:

$$\mathbf{x} = \mathbf{U}\tilde{\mathbf{x}} \quad \text{with} \quad \tilde{\mathbf{x}} = \mathbf{U}^\mathsf{T}\mathbf{x}$$

where $\tilde{\mathbf{x}}$ is the GFT of graph signal $\mathbf{x}$.

3.1.3   Defective Z

For a defective (non-diagonalizable) matrix, and for all vectors in E_i, of dimension k_i, multiplication by $\mathbf{Z}$ can be written as in (2.44). Denoting $\mathbf{a} \in \mathbb{R}^{k_i}$ and $\mathbf{U}_i$ an $N \times k_i$ matrix with columns that form a basis for E_i, so that $\mathbf{x}_i = \mathbf{U}_i\mathbf{a}$, we have

$$\Delta_{\mathbf{Z}}(\mathbf{x}_i) = \mathbf{x}_i^\mathsf{T}\mathbf{Z}\mathbf{x}_i = \mathbf{a}^\mathsf{T}\mathbf{U}_i^\mathsf{T}\mathbf{U}_i\boldsymbol{\Lambda}_i\mathbf{a},$$

which cannot be simplified further, because in general the columns of $\mathbf{U}_i$ are not orthogonal. Thus, if the operator $\mathbf{Z}$ cannot be diagonalized, invariant subspaces exist but the vectors in these subspaces are no longer simply scaled eigenvectors. While we can write $\mathbf{Z}$ in terms of bases for each of the invariant subspaces E_i, we no longer have a factorization that produces a diagonal matrix, as in (3.4).

DEFINITION 3.4   (GRAPH FOURIER TRANSFORM (DEFECTIVE $\mathbf{Z}$))   A diagonal form for $\mathbf{Z}$ is no longer available and we have two possible choices.
    We can use the Schur decomposition as in Theorem 2.2:

$$\mathbf{Z} = \mathbf{U}\mathbf{T}\mathbf{U}^\mathsf{H}, \tag{3.6}$$

where $\mathbf{U}$ is a unitary matrix and $\mathbf{T}$ is upper triangular.
    Alternatively, we can write $\mathbf{Z}$ in terms of the **Jordan canonical form**:

$$\mathbf{Z} = \mathbf{V}\boldsymbol{\Lambda}_b\mathbf{V}^{-1}, \tag{3.7}$$

where $\boldsymbol{\Lambda}_b$ is block diagonal and $\mathbf{V}$ is invertible. Both $\mathbf{U}$ and $\mathbf{V}$ form bases for $\mathbb{R}^N$, so we can represent any graph signal in terms of those bases:

$$\mathbf{x} = \mathbf{U}\check{\mathbf{x}}, \quad \text{where} \quad \check{\mathbf{x}} = \mathbf{U}^\mathsf{H}\mathbf{x} \quad \text{or} \quad \mathbf{x} = \mathbf{V}\tilde{\mathbf{x}}, \quad \text{where} \quad \tilde{\mathbf{x}} = \mathbf{V}^{-1}\mathbf{x}.$$

3.1.4   How Many GFTs Are There?

It is important to keep in mind that there is no single GFT. The transform we use depends on multiple choices, as follows.

1 For a given dataset, multiple choices of graphs are often possible (directed or undirected, with different weight choices, etc.). Methods to select or "learn" graphs are
discussed in Chapter 6.

2 The GFT depends on the choice of **Z**. As noted in Section 3.2 and through the rest of
this chapter, this choice has major implications for the properties of the elementary
basis functions.

3 In cases where the operator cannot be diagonalized, or when it can be diagonalized
but some eigenvalues have multiplicity greater than 1, we have multiple choices for
constructing a basis (Box 2.7).

In general, the choice to be made will depend on various factors, of which the most important should be the properties of the signals to be processed: it will be highly desirable
for the elementary bases, or at least some of them, to have a meaningful interpretation
in the application of interest (see Section 3.4.5 for an example).

3.2   Graph Signal Variation and Z Quadratic Forms

In Section 1.1.2, we introduced informally the idea of graph signal variation based on
how different the signal value $x(i)$ at node i is from the signal values $x(j)$ at the nodes in
its neighborhood, $j \in \mathcal{N}(i)$. The fundamental operators of Chapter 2 allow us to quantify
local similarity by computing **Zx** with one-hop graph operators such as **L**, $\mathcal{L}$ and $\mathcal{T}$, as
summarized in Table 2.2. For these operators, if **Zx** is small, the graph signal variation
is low and the corresponding signal is smooth. Since $\Delta_Z(\mathbf{x})$ (Definition 3.1) quantifies
the similarity between **x** and **Zx** we will now see how it can be used as a measure
of *total signal variation*. We start with the undirected graph case, with combinatorial
and normalized Laplacian matrices, and then briefly discuss directed graphs. The same
ideas are developed for smoothing operators, such as Q or **A**, in Section 3.2.5. If **Z** is a
smoothing operator then a large $\Delta_Z(\mathbf{x})$ value indicates that **x** is smooth. As an example,
for an undirected graph and for any **x**, $\Delta_{\mathcal{T}}(\mathbf{x}) = 1 - \Delta_Q(\mathbf{x})$.

3.2.1   Combinatorial Laplacian

Consider first the combinatorial Laplacian operator **L** of Definition 2.18. The Laplacian
quadratic form $\Delta_L(\mathbf{x}) = \mathbf{x}^\mathsf{T} \mathbf{L} \mathbf{x}$ can be interpreted as a measure of signal variation.

Remark 3.1   (LAPLACIAN QUADRATIC FORM)   The Laplacian quadratic form quantifies the total graph signal variation:

$$\Delta_L(\mathbf{x}) = \mathbf{x}^\mathsf{T} \mathbf{L} \mathbf{x} = \sum_{i \sim j} a_{ij} (x(i) - x(j))^2, \tag{3.8}$$

where the summation is over all edges. Each edge contributes to the variation as
a function of its weight, a_{ij}, and of the difference between the signal values at its
end-points, $(x(i) - x(j))^2$.

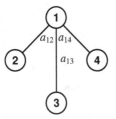

Figure 3.1 Example graph

Proof Note that (2.22), giving the weighted prediction error, $z(i)$, can be rewritten as

$$z(i) = d_i x(i) - \sum_{j \in N_i} a_{ij} x(j) = \sum_{j \in N_i} a_{ij} (x(i) - x(j))$$

where we have used the fact that $d_i = \sum_{j \in N_i} a_{ij}$. Then we can write

$$\mathbf{x}^T \mathbf{L} \mathbf{x} = \sum_{i=1}^{N} \sum_{j \in N(i)} a_{ij} (x(i) - x(j)) x(i). \tag{3.9}$$

Note that for an undirected graph $a_{ij} = a_{ji}$ and thus the term a_{ij} appears twice; grouping those two occurrences we can rewrite (3.9) as

$$\mathbf{x}^T \mathbf{L} \mathbf{x} = \sum_{i,j,i<j} a_{ij} [(x(i) - x(j)) x(i) + (x(j) - x(i)) x(j)],$$

so that

$$\mathbf{x}^T \mathbf{L} \mathbf{x} = \sum_{i,j,i<j} a_{ij} (x(i) - x(j))^2 = \sum_{i \sim j} a_{ij} (x(i) - x(j))^2. \tag{3.10}$$

$\square$

Recall how the graph signal variation was introduced in Section 1.1.2 and observe that the Laplacian quadratic form computes an aggregate measure of the variation across all edges. Then, for a signal $\mathbf{x} \in E_i$, the invariant subspace associated with eigenvalue λ_i, $\mathbf{L}\mathbf{x} = \lambda_i \mathbf{x}$, and thus the graph signals in E_i, will tend to be smooth (with similar values at connected graph nodes) if λ_i is small. This is illustrated by the following example.

Example 3.1   For the four-node graph of Figure 3.1 with adjacency matrix

$$\mathbf{A} = \begin{bmatrix} 0 & a_{12} & a_{13} & a_{14} \\ a_{12} & 0 & 0 & 0 \\ a_{13} & 0 & 0 & 0 \\ a_{14} & 0 & 0 & 0 \end{bmatrix},$$

denote a graph signal $\mathbf{x} = [x(1), x(2), x(3), x(4)]^T$. Compute the Laplacian quadratic form $\mathbf{x}^T \mathbf{L} \mathbf{x}$ for this signal and provide examples of low- and high-variation signals.

Solution
From (3.8):

$$\mathbf{x}^T \mathbf{L} \mathbf{x} = a_{12} (x(1) - x(2))^2 + a_{13} (x(1) - x(3))^2 + a_{14} (x(1) - x(4))^2,$$

where we see that a signal with different values in two nodes contributes to the variation only if the corresponding nodes are connected. Thus, if $x(3) \neq x(4)$ this has no direct effect on variation. As a low-variation signal, we can choose $x(1) = x(2) = x(3) = x(4) = 1$, which will lead to $\mathbf{x}^T\mathbf{L}\mathbf{x} = 0$. Since all the terms in (3.8) are positive (because the weights $a_{ij} \geq 0$), this has to be the minimal possible variation.

For higher variation we can choose signals with values that are different across edges in the graph. To maximize the variation, note that v_1 connects to all the other nodes, while the remaining nodes do not connect with each other. We can choose a signal that maximizes the number of zero crossings (changes in sign between the signal values at the ends of an edge), e.g., selecting $x(1) = 1$ and $x(2) = x(3) = x(4) = -1$, so that $\mathbf{x}^T\mathbf{L}\mathbf{x} = 4(a_{12} + a_{13} + a_{14})$.

3.2.2 Self-Loops and Variation

Assume that a graph includes positive self-loops and let us discuss their effect on the graph signal variation. Adding self-loops leads to a generalized graph Laplacian, $\mathbf{L}_g$:

$$\mathbf{L}_g = \mathbf{C} + \mathbf{D} - \mathbf{A}$$

where $\mathbf{C}$ is the diagonal matrix for the self-loops, with $c_i \geq 0$ denoting the self-loop at node i. By linearity, the variation $\mathbf{x}^T\mathbf{L}_g\mathbf{x}$ is obtained by adding an extra term to (3.8):

$$\mathbf{x}^T\mathbf{L}_g\mathbf{x} = \sum_i c_i x(i)^2 + \sum_{i \sim j} a_{ij}(x(i) - x(j))^2. \tag{3.11}$$

Given the two terms in (3.11), the relative values of self-loops and edge weights determine how much of the variation depends on the node itself and how much depends on the differences with respect to its neighbors. This is illustrated by a simple example.

Example 3.2 A path graph with three nodes has two edges $a_{12} = a_{23} = 1$ and self-loops $c_1 = c_3 = 0$, $c_2 = \epsilon > 0$. Find the gradient of (3.11) as a function of c_2 at the point giving the optimal non-zero solution to (3.10) and discuss.

Solution
For a signal $\mathbf{x}$ the quadratic form for $\mathbf{L}_g$ can be written as

$$\Delta_{\mathbf{L}_g}(\mathbf{x}) = \mathbf{x}^T\mathbf{L}_g\mathbf{x} = (x(1) - x(2))^2 + c_2 x(2)^2 + (x(2) - x(3))^2,$$

and taking derivatives with respect to $x(1), x(2), x(3)$ leads to the gradient:

$$\nabla_{\mathbf{x}}\Delta_{\mathbf{L}_g}(\mathbf{x}) = 2 \begin{bmatrix} x(1) - x(2) \\ (2x(2) - x(1) - x(3)) + \epsilon x(2) \\ (x(3) - x(2)) \end{bmatrix} \quad \text{so that} \quad \nabla_{\mathbf{x}}\Delta_{\mathbf{L}_g}(\mathbf{1}) = \begin{bmatrix} 0 \\ 2\epsilon \\ 0 \end{bmatrix},$$

which shows that the optimal solution will require $x(2)$ to be smaller than that for the combinatorial Laplacian $\mathbf{L}$ ($\mathbf{C} = \mathbf{0}$), which led to minimum variation at $\mathbf{1}$.

As the node weight c_i increases, a larger $x(i)$ will lead to an increase in variation that is independent of the difference between that node and neighboring nodes. Thus, when $c_i = 0$, the minimal variation is achieved when $x(i)$ is equal to $x(j)$ for all $j \in \mathcal{N}(i)$. On the other hand, when c_i grows, $x(i)$ should be increasingly small, which means that minimum variation may occur for $\mathbf{x}$ such that $x(i) \neq x(j)$. The difference between the cases $c_i = 0$ and $c_i \neq 0$ can be illustrated by comparing the basis vectors for the DCT and the asymmetric discrete sine transform (ADST), as shown in Figure 7.2 and Figure 7.3, which correspond to identical path graphs, the only difference being that a self-loop $c_1 = 1$ is associated with the leftmost node of the ADST graph. Notice that for the lowest frequency in ADST, not all node signal values are equal, owing to this added self-loop weight.

3.2.3   GFTs and Variation

Given Remark 3.1 we can now interpret the basis vectors $\mathbf{u}_i$ associated with the operator $\mathbf{L}$, i.e., the GFT of Definition 3.3. First, from (3.8), and for any graph, we have

$$\mathbf{1}^T\mathbf{L}\mathbf{1} = 0 \quad \text{and} \quad \mathbf{f}^T\mathbf{L}\mathbf{f} \geq 0,$$

where the second relation follows from the fact that $\mathbf{L}$ is positive semidefinite. Thus, the constant signal $\mathbf{1}$ achieves *minimal* variation, corresponding to the lowest possible graph frequency. Notice that this is analogous to **regular domain** signals, where a constant signal in time has zero frequency. Next, the computation of successive eigenvectors of $\mathbf{L}$ to build an orthogonal basis for $\mathbb{R}^N$ can be viewed as an optimization problem based on the Rayleigh quotient. Starting from $\mathbf{u}_1 = \frac{1}{\sqrt{N}}\mathbf{1}$ (the normalization factor is chosen so that the vector has norm 1, i.e., $\mathbf{u}_1^T\mathbf{u}_1 = 1$), we find successive vectors $\mathbf{u}_k$ that have minimum variation $\Delta_{\mathbf{L}}(\mathbf{u}_k)$ (minimum frequency) while being orthogonal to all the previously chosen vectors, and also having norm equal to 1. Thus we look for $\mathbf{u}_2$ such that

$$\mathbf{u}_2 = \min_{\mathbf{x}\perp\mathbf{u}_1,\|\mathbf{x}\|=1} \mathbf{x}^T\mathbf{L}\mathbf{x},$$

and, generalizing, $\mathbf{u}_k$ such that

$$\mathbf{u}_k = \min_{\mathbf{x}\perp\mathbf{u}_1,\mathbf{u}_2,\dots,\mathbf{u}_{k-1}\|\mathbf{x}\|=1} \mathbf{x}^T\mathbf{L}\mathbf{x}.$$

By solving this problem iteratively we find the eigenvectors of $\mathbf{L}$, leading to the GFT $\mathbf{U}$ as introduced in Definition 3.3. Now we can use this definition to compute $\Delta_{\mathbf{L}}(\mathbf{x})$ in the frequency domain as follows.

Remark 3.2   (Laplacian quadratic form in frequency domain)   The Laplacian quadratic form $\mathbf{x}^T\mathbf{L}\mathbf{x}$ for $\mathbf{x} = \mathbf{U}\tilde{\mathbf{x}}$ can also be written as

$$\Delta_{\mathbf{L}}(\mathbf{x}) = \sum_k \lambda_k \tilde{x}_k^2, \tag{3.12}$$

where the summation is over all graph frequencies λ_k.

Proof From the definitions of the quadratic form and the GFT:

$$\mathbf{x}^\mathsf{T}\mathbf{L}\mathbf{x} = \mathbf{x}^\mathsf{T}\mathbf{U}\mathbf{\Lambda}\mathbf{U}^\mathsf{T}\mathbf{x} = (\tilde{\mathbf{x}})^\mathsf{T}\mathbf{\Lambda}\tilde{\mathbf{x}} = \sum_k \lambda_k \tilde{x}_k^2.$$

□

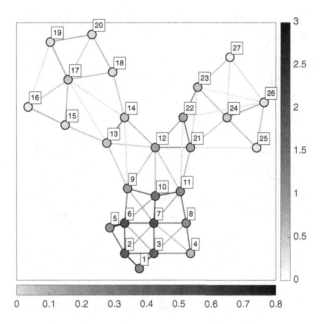

Figure 3.2 Simple weighted graph. The edge weights in this graph decrease as distance (as shown in the figure) increases. Thus, the edges connecting to node 7 have larger weights than those connecting to node 17. The bottom bar shows the mapping between edge weight and edge color, while the bar on the right corresponds to the mapping between node color and node degree. Refer to Section B.2.6 for more details on edge weight visualization.

As an example, selected Laplacian eigenvectors of the graph in Figure 3.2 are shown in Figure 3.3. The eigenvectors show increasing variation on the graph as λ increases. $\lambda = 0$ corresponds by definition to $\mathbf{u}_1 = \mathbf{1}$ and thus minimal variation. The next higher frequency, $\mathbf{u}_2$ ($\lambda_2 = 0.02$), shows roughly the same number of positive and negative values, as do all the other frequency vectors. This is as expected, given that $\mathbf{1}^\mathsf{T}\mathbf{u}_i = 0$ for all $i > 1$, and so the sum of entries for $\mathbf{u}_2$, $\mathbf{u}_3$, etc. has to be equal to 1.

Observe the difference between $\mathbf{u}_2$ ($\lambda_2 = 0.02$) and $\mathbf{u}_{12}$ ($\lambda_{12} = 0.58$) in Figure 3.3. In $\mathbf{u}_2$, the eigenvector corresponding to the second lowest frequency, only those edges near nodes 9 and 12 connect nodes with opposite signs. In contrast, for $\mathbf{u}_{12}$ we see a greater number of edges connecting positive and negative nodes. While this appears to be analogous to the behavior of sinusoids of increasing frequency shown in Figure 1.6,

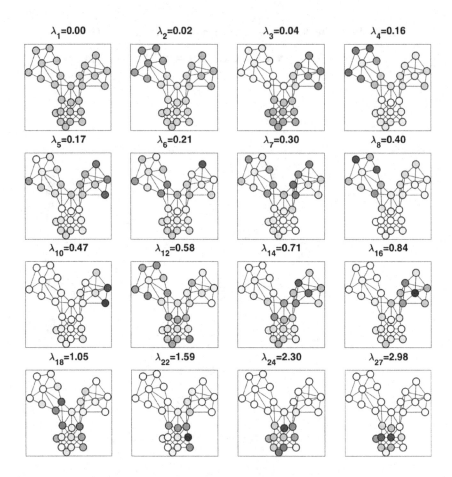

Figure 3.3 Eigenvectors for the graph of Figure 3.2. Each eigenvector is a graph signal, with the shading in each node representing the signal at that node. At the top of each plot is the corresponding eigenvalue with the subscript indicating the eigenvalue's index. Notice that many eigenvectors are highly localized, with most values very close to zero (white in the plot). Since the eigenvectors are normalized, eigenvectors with many nodes close to zero also have nodes with higher values. Note that in this figure only the magnitude of entries is plotted. Thus, darker nodes correspond to large magnitude (positive or negative) entries.

we can also note that the actual entries at higher frequencies are in fact very small in some nodes. For example $\mathbf{u}_{27}$ has entries that are zero in two clusters (around nodes 17 and 24), and achieves high variation ($\lambda_{27} = 2.98$) because several strongly connected nodes have very different values (in the cluster around node 7). Thus, frequencies can be highly localized, so that only a few signal values are non-zero, quite unlike the typical behavior for frequencies in regular domains. The difference between the localization of elementary frequency modes in conventional signal processing and in graph signal processing is discussed in Section 5.1, while other GFT properties are studied in Section 3.4 and Section 3.5.

3.2.4 Symmetric Normalized Laplacian

For $\mathbf{Z} = \mathcal{L}$, the symmetric normalized Laplacian (Definition 2.20), the only change with respect to $\mathbf{L}$ comes from the edge weights.

Remark 3.3   (NORMALIZED LAPLACIAN QUADRATIC FORM)   The normalized Laplacian quadratic form $\mathbf{x}^T \mathcal{L} \mathbf{x}$ can be written as

$$\Delta_{\mathcal{L}}(\mathbf{x}) = \mathbf{x}^T \mathcal{L} \mathbf{x} = \sum_{i \sim j} a_{ij} \left(\frac{x(i)}{\sqrt{d_i}} - \frac{x(j)}{\sqrt{d_j}} \right)^2, \tag{3.13}$$

where as before the summation is over all the edges.

Proof   Recall that $\mathbf{L} = \mathbf{D}^{1/2} \mathcal{L} \mathbf{D}^{1/2}$ and therefore

$$\mathbf{x}^T \mathbf{L} \mathbf{x} = (\mathbf{D}^{1/2} \mathbf{x})^T \mathcal{L} (\mathbf{D}^{1/2} \mathbf{x}),$$

so that

$$\Delta_{\mathcal{L}}(\mathbf{x}) = \mathbf{x}^T \mathcal{L} \mathbf{x} = (\mathbf{D}^{-1/2} \mathbf{x})^T \mathbf{L} (\mathbf{D}^{-1/2} \mathbf{x}),$$

from which we obtain the desired expression using Remark 3.1.   □

From Remark 3.1 and Remark 3.3 we can see that $\Delta_{\mathbf{L}}(\mathbf{x})$ and $\Delta_{\mathcal{L}}(\mathbf{x})$ quantify the variation across *connected* nodes for a given signal. The fact that only pairs of connected nodes in the graph contribute to the variation is again analogous to how we would work with time domain signals, where we would be considering the difference between consecutive samples (i.e., nodes connected in the path graph).

3.2.5 Variation Based on Adjacency Matrices

Quadratic form interpretation   From (2.19) and (2.20), we see that $\mathbf{Ax}$ and $\mathbf{Qx}$ correspond to a one-hop sum and one-hop average, respectively. Therefore, using the interpretation of the inner product as a measure of similarity (Remark A.1) we can view $\mathbf{x}^T \mathbf{Ax}$ and $\mathbf{x}^T \mathbf{Qx}$ as measuring the total similarity between all nodes and their respective neighbors. For $\mathbf{A}$ the measure of similarity

$$\mathbf{x}^T \mathbf{Ax} = \sum_i x(i) \sum_{j \in \mathcal{N}_i} a_{ij} x(j) \tag{3.14}$$

in its normalized version leads to the normalized quadratic form

$$\bar{\Delta}_{\mathbf{A}}(\mathbf{x}) = \frac{\mathbf{x}^T \mathbf{Ax}}{\mathbf{x}^T \mathbf{x}}, \tag{3.15}$$

which we will revisit in Section 3.4.1. When the graph is undirected we can simplify (3.14) by using $a_{ij} = a_{ji}$ to obtain

$$\mathbf{x}^T \mathbf{Ax} = 2 \sum_{i \sim j} a_{ij} x(i) x(j). \tag{3.16}$$

Since the weights a_{ij} are non-negative, it is clear that (3.14) and (3.16) are maximized for vectors having all positive entries. This insight will be formalized as the Perron–Frobenius theorem (Theorem 3.2).

Variation   Notice that in contrast with **L** and $\mathcal{L}$, $\bar{\Delta}_A(\mathbf{x})$ is maximized for a very smooth signal (all signal values positive, no zero crossings), thus having low variation. Operators similar to **L** and $\mathcal{L}$ can be introduced, including for example

$$\mathcal{T} = \mathbf{I} - Q \quad \text{and} \quad \mathcal{M} = \mathbf{I} - \frac{1}{\lambda_1}\mathbf{A},$$

where $\lambda_1 > 0$ is the largest eigenvalue of **A** (see Theorem 3.2) and is a normalization factor to ensure that all eigenvalues of $\mathcal{M}$ are such that $|\lambda_i| \leq 1$. The operators $\mathcal{T}$ and $\mathcal{M}$ have the same eigenvectors as Q and **A**, respectively, but with reverse ordering.

For a directed graph, with associated diagonalizable **Z**, we still have that $\mathbf{x}^\mathsf{T}\mathbf{Z}\mathbf{x} = \lambda_i \mathbf{x}^\mathsf{T}\mathbf{x}$ for any vector in E_i. But while in the undirected case all eigenvalues were real, in this case they may be complex and thus their ordering is not straightforward. More broadly, even in the undirected case, ordering eigenvectors on the basis of their real eigenvalues appears to be a natural choice but could be questionable. For example, as noted in [27], if graphs are constructed for points in a Euclidean space (e.g., points in 2D space) the 1D ordering provided by the eigenvalues does not match the ordering of 2D frequencies obtained using standard multidimensional signal processing methods.

3.2.6   GFT Visualization

Figure 1.7 gave a comparison of a 1D signal and a graph signal with exactly the same values, with the goal of illustrating the challenge of frequency visualization for graph signals. While for the temporal signal of Figure 1.7 we can get a good intuition of its frequency content by observing the variations along the path, this is not so easy for a graph. In a relatively sparse graph example (Figure 3.3) we can observe frequency increases by noticing that sign changes across graph edges become more frequent for larger values of λ. However, it becomes impractical to map larger graphs to a 2D plane (as in Figure 3.3) and increased numbers of nodes and edges make it difficult to observe frequency and other properties (e.g., localization) of the GFT basis vectors.

We briefly present an alternative tool to visualize the GFT vectors, included in the GraSP Matlab toolbox. This approach was proposed in [28] and is described in more detail in Section B.4.1. It is motivated by the relative ease of visualizing the frequency and variation in 1D signals (Figure 1.7) and starts by embedding the nodes of an arbitrary graph into 1D (see Section 7.5.2 for a more detailed discussion of embedding). After this 1D embedding, each node corresponds to a point on the real line and each of the GFT basis vectors is a discrete signal with irregularly spaced samples (corresponding to the position of each node in 1D).

As an example, the random Watts–Strogatz graph of Figure 3.4 results in the GFT shown in Figure 3.5. With the 1D embedding, each node is assigned a real value equal to the entry corresponding to that node in $\mathbf{u}_2$, the second eigenvector of its graph Laplacian.

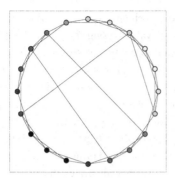

Figure 3.4 Eigenvector for a Watts–Strogatz graph, showing a 2D embedding of the graph where the node tone varies from white to black, corresponding to the leftmost and rightmost values on Figure 3.5, respectively.

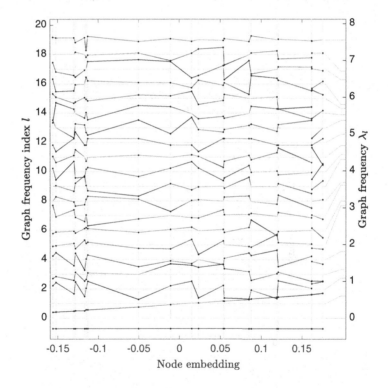

Figure 3.5 Eigenvectors for the Watts–Strogatz graph of Figure 3.4 . The eigenvectors as 1D signals are ordered from the lowest frequency (bottom) to the highest (top).

Thus, node i is assigned the position $u_2(i)$ along the horizontal axis. In Figure 3.5 nodes are mapped to points in the interval $[-1, 1]$ and are grouped into two clusters, where values close to -1 in Figure 3.5 correspond to nodes given a lighter shade in Figure 3.4, while nodes close to $+1$ in Figure 3.5 are darker in Figure 3.4.

The GFT visualization includes two vertical axes. The axis on the left side has evenly spaced integer values corresponding to the eigenvalue index. Since the graph in Figure 3.4 has 20 nodes, there are 20 integer values along the vertical axis in Figure 3.5, each marking the end of a light horizontal line which represents the zero value for the corresponding eigenvector viewed as a 1D signal. For example, $l = 1$ along the vertical axis corresponds to $\mathbf{u}_2$ and we can see that the resulting 1D embedding of $\mathbf{u}_2$ is a monotonically increasing function with negative values on the left and positive on the right. The vertical axis on the right allows us to show the actual eigenvalues, which are not evenly spaced in general. Thus, for any of the integer indices on the left the corresponding light horizontal line ends with short upward or downward segment that shows the exact position of the graph frequency λ. This illustrates the irregular nature of the spacing between eigenvalues. As an example, the largest frequency index in this plot is $l = 19$ (left) and the corresponding eigenvalue is around 7.5 (right). As an additional example, Figure 3.6 depicts the elementary frequencies for the graph of Figure 3.2.

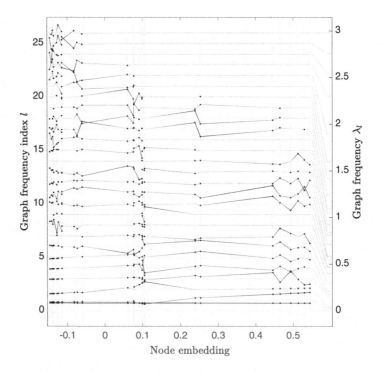

Figure 3.6 Eigenvectors for the graph of Figure 3.2. Some of these have already been plotted in Figure 3.3. Note that the 1D embedding makes it more obvious that the eigenvectors are localized. In particular, nodes that are close in the graph will be close also in the 1D embedding, and we can observe that localization in the node domain translates into localization in 1D. Specifically, the nodes embedded to negative values in this figure correspond to the cluster around node 7 in Figure 3.2. As also seen in Figure 3.3, the highest-frequency eigenvectors are highly localized, with their non-zero values concentrated in nodes within this cluster.

Another important observation is that GFT basis vectors can be highly localized. In Figure 3.3 we can see that values at or near zero can be found, in particular at high frequencies. The graph of Figure 3.2 contains three clusters. In Figure 3.3 we can see that the nodes in the denser cluster (nodes 1 to 11, mapped to negative values by the node embedding) have non-zero values at the highest frequencies (at the top of the figure), while the values in the other cluster are identically zero. This behavior is unlike that seen for frequency representations in conventional signal processing, where there is a well-established trade-off between frequency localization and time localization (see also Section 5.1).

3.3 Graph Filtering – Frequency Domain

For each fundamental operator $\mathbf{Z}$, we have seen how the invariance property associated with its eigenvectors has an interpretation in terms of graph signal variation. Thus, in what follows, eigenvalues are defined as frequencies and eigenvectors are elementary basis vectors associated with each frequency. The transformed signals of (3.2), Definition 3.3 and Definition 3.4 produce frequency domain representations of signals. We now establish the connection between the node domain (Section 2.4) and frequency domain filtering methods.

3.3.1 Frequency Domain Representation of Polynomial Filters

As noted in Section 2.4, a graph signal, $\mathbf{x}$, is simply a vector of dimension N, $\mathbf{x} \in \mathbb{R}^N$, to which we could apply many different transforms to obtain an output $\mathbf{y} = \mathbf{Hx}$. Then, we raised the question of what makes a specific matrix $\mathbf{H}$ a graph filter. In Section 2.4 we considered this problem from a node domain perspective and defined graph filters to be those that could be written as a polynomial of the elementary one-hop operator, $\mathbf{Z}$. We can now interpret this idea from a frequency domain perspective. Assuming that $\mathbf{Z}$ is a diagonalizable operator, with $\mathbf{Z} = \mathbf{U}\boldsymbol{\Lambda}\mathbf{U}^{-1}$, we can show that a graph filter (a polynomial of $\mathbf{Z}$) leads to a diagonal operation in the frequency domain as follows.

Remark 3.4 (GRAPH FILTERS IN FREQUENCY DOMAIN) Any graph filter $\mathbf{H}$, defined as in Definition 2.21, as a polynomial of an operator $\mathbf{Z}$ with GFT $\mathbf{U}$, can be written as

$$\mathbf{H} = h(\mathbf{Z}) = \sum_{i=0}^{k} h_i \mathbf{Z}^i = \mathbf{U}h(\boldsymbol{\Lambda})\mathbf{U}^{-1}, \tag{3.17}$$

where $h(\boldsymbol{\Lambda})$ is a diagonal matrix with jth entry

$$h(\lambda_j) = \sum_{i=0}^{k} h_i \lambda_j^i. \tag{3.18}$$

Proof By the definitions of $\mathbf{Z}$ and $\mathbf{U}$ we have

$$\mathbf{Z} = \mathbf{U}\boldsymbol{\Lambda}\mathbf{U}^{-1} \tag{3.19}$$

with $\mathbf{\Lambda} = \mathrm{diag}\{\lambda_1, \lambda_2, \ldots\}$ the diagonal matrix of eigenvalues of $\mathbf{Z}$. Then

$$\mathbf{Z}^i = \mathbf{U}\mathbf{\Lambda}^i\mathbf{U}^{-1}$$

for all i, where $\mathbf{\Lambda}^i$ is diagonal, and therefore

$$\mathbf{H} = \sum_{i=0}^{k} h_i\mathbf{Z}^i = \sum_{i=0}^{k} h_i\mathbf{U}\mathbf{\Lambda}^i\mathbf{U}^{-1} = \mathbf{U}\left(\sum_{i=0}^{k} h_i\mathbf{\Lambda}^i\right)\mathbf{U}^{-1} \tag{3.20}$$

where denoting

$$h(\mathbf{\Lambda}) = \sum_{i=0}^{k} h_i\mathbf{\Lambda}^i \tag{3.21}$$

we obtain (3.17), and we can also see that $h(\mathbf{\Lambda})$ is diagonal (because it is a sum of diagonal matrices). □

3.3.2   Node and Frequency Domain Filtering Implementations

The graph filtering operation $\mathbf{y} = \mathbf{H}\mathbf{x}$ can be applied to an arbitrary graph signal $\mathbf{x}$ in the frequency domain by first computing the GFT of the signal:

$$\tilde{\mathbf{x}} = \mathbf{U}^{-1}\mathbf{x}, \tag{3.22}$$

applying the (diagonal) frequency representation of the filter operator,

$$\tilde{\mathbf{y}} = h(\mathbf{\Lambda})\tilde{\mathbf{x}},$$

and then obtaining the filtered output in the node domain by computing the inverse GFT,

$$\mathbf{y} = \mathbf{U}\tilde{\mathbf{y}}. \tag{3.23}$$

The steps in (3.22) and (3.23) correspond exactly to those performed for time domain signals filtered using the discrete Fourier transform (DFT). A frequency domain implementation allows us to design "ideal" filters (see Box 3.1 below) with any desired frequency response. However, unlike for the Fourier transform, fast algorithms do not necessarily exist to compute (3.22) and (3.23).

Box 3.1   Ideal filters

An ideal bandpass filter $\mathbf{H}_{k_1,k_2}$ selects a range of frequencies k_1 to k_2:

$$\mathbf{H}_{k_1,k_2} = \mathbf{U}\,\mathrm{diag}(0, 0, \ldots, 0, 1, \ldots, 1, 0, \ldots, 0)\,\mathbf{U}^{-1},$$

where the diagonal matrix has entries equal to 1 for frequencies in the range k_1 to k_2, and zero otherwise. A particular case of interest discussed in Chapter 5 is the ideal half-band low pass filter for $\mathbf{H}_0$, where the first $N/2$ frequencies are selected, and the corresponding half-band high pass filter $\mathbf{H}_1$, which selects the top $N/2$ frequencies. For an undirected graph ($\mathbf{U}^{-1} = \mathbf{U}^{\mathsf{T}}$), we have that

$$\mathbf{x}_0 = \mathbf{H}_0\mathbf{x}, \quad \mathbf{x}_1 = \mathbf{H}_1\mathbf{x}$$

are orthogonal projections (Proposition A.4).

Note that for a given graph there is a finite number of frequencies, N, corresponding to the number of nodes, and thus any diagonal operation in the frequency domain requires setting just N values. An important consequence of Remark 3.4 is that any choice of N such values leads to a polynomial graph filter.

Starting with a polynomial $\mathbf{H} = p(\mathbf{\Lambda})$ as in (3.20), Remark 3.4 shows that:

$$
\mathbf{H} = \mathbf{U}h(\mathbf{\Lambda})\mathbf{U}^{-1} = \mathbf{U}
\begin{bmatrix}
\sum_{i=0}^{k} h_i \lambda_1^i & 0 & \cdots & 0 \\
0 & \sum_{i=0}^{k} h_i \lambda_2^i & \cdots & 0 \\
\vdots & \vdots & \ddots & \vdots \\
0 & 0 & \cdots & \sum_{i=0}^{k} h_i \lambda_N^i
\end{bmatrix}
\mathbf{U}^{-1}
$$

where we can see that the diagonal values of $\mathbf{H}$ can be obtained by finding $h(\lambda)$ at each of the eigenvalues λ_i. Note that if some eigenvalues have multiplicity greater than 1, this means that the corresponding diagonal entries in $h(\mathbf{\Lambda})$ will be repeated. Conversely we can show the following.

Remark 3.5 (DIAGONAL FREQUENCY DOMAIN OPERATORS AND POLYNOMIALS) Let $\mathbf{H}$ be a diagonal operator in the frequency domain:

$$
\mathbf{H} = \mathbf{U}
\begin{bmatrix}
h_1 & 0 & \cdots & 0 \\
0 & h_2 & \cdots & 0 \\
\vdots & \vdots & \ddots & \vdots \\
0 & 0 & \cdots & h_N
\end{bmatrix}
\mathbf{U}^{-1}.
$$

Here the h_i are subject to an additional constraint: if an eigenvalue has multiplicity greater than 1 then its corresponding diagonal entries are all equal, e.g., if λ_k has multiplicity 2, then $\lambda_{k+1} = \lambda_k$ and $h_{k+1} = h_k$. Then we can construct a polynomial $h(\lambda)$ such that $\mathbf{H} = h(\mathbf{Z}) = \mathbf{U}h(\mathbf{\Lambda})\mathbf{U}^{-1}$. The rationale for imposing this constraint is discussed in Box 3.2.

Proof In this case we can use the Lagrange interpolation formula to obtain a polynomial that produces the desired h_k for the distinct eigenvalues λ_k, i.e., define $h(\lambda)$ as follows:

$$
h(\lambda) = \sum_{k=1}^{N'} l_k(\lambda) h_k,
$$

where N' is the number of distinct eigenvalues and

$$
l_k(\lambda) = \prod_{i=1, i \neq k}^{N'} \frac{\lambda - \lambda_i}{\lambda_k - \lambda_i},
$$

where each $l_k(\lambda)$ is a polynomial for which $l_k(\lambda_i) = 0$ for $i \neq k$ and $l_k(\lambda_k) = 1$. Note that these polynomials have maximum degree N'. That is, as illustrated in Box 2.6, if we define a polynomial in the node domain of degree greater than N', the same transformation can be expressed by a polynomial of degree N' or lower. $\square$

Box 3.2   Why assign the same gain to all vectors in an invariant subspace?

Assume that for $\mathbf{Z}$ there is at least one invariant space E_i, corresponding to eigenvalue λ_i, that has dimension greater than 1. There are several reasons why a filter $\mathbf{H}$ should assign the same gain $h(\lambda_i)$ to all vectors in that subspace:

- there is an infinite number of possible basis choices for E_i (Box 2.7);
- $\mathbf{H}$ cannot be written as a polynomial of $\mathbf{Z}$ unless the same value is assigned to all vectors in the subspace;
- the amount of variation (and thus frequency) along any direction in E_i is the same, so it is unclear that associating different gains with each vector in a basis for E_i leads to a meaningful result.

Because of this, in what follows we assume that all filters are designed using diagonal matrices where $h(\lambda_i)$ is the same for all the basis vectors corresponding to E_i.

Least squares approximate designs   The approach used in the proof of Remark 3.5 is mathematically exact but may not lead to numerically stable solutions. In particular, notice that terms of the form $\lambda_k - \lambda_i$ in the denominator of $l_k(\lambda)$ can lead to very large values if two frequencies are very close, which tends to happen in actual graphs. As an alternative, given exact knowledge of all frequencies $\lambda_1, \ldots, \lambda_N$ and for a desired response given as $h(\lambda_i)$ at each frequency λ_i, under the assumption of Box 3.2, we can identify a polynomial filter of some desired degree K by noting that we need to find the coefficients $(h_1, \ldots, h_K)$ such that for every λ_i we have

$$h(\lambda_i) = h_1 \lambda_i + h_2 \lambda_i^2 + \cdots + h_K \lambda_i^K$$

leading to a system with N equations and K unknowns, for which we can find the best least squares solution $(h_1, \ldots, h_K)$:

$$
\begin{bmatrix} h(\lambda_1) \\ h(\lambda_2) \\ \vdots \\ h(\lambda_{N-1}) \\ h(\lambda_N) \end{bmatrix} = \begin{bmatrix} \lambda_1 & \lambda_1^2 & \cdots & \lambda_1^K \\ \lambda_2 & \lambda_2^2 & \cdots & \lambda_2^K \\ \vdots & \vdots & \ddots & \vdots \\ \lambda_{N-1} & \lambda_{N-1}^2 & \cdots & \lambda_{N-1}^K \\ \lambda_N & \lambda_N^2 & \cdots & \lambda_N^K \end{bmatrix} \begin{bmatrix} h_1 \\ h_2 \\ \vdots \\ h_K \end{bmatrix},
$$

so that an approximate polynomial filter can be found for any choice of K.

Autoregressive moving average (ARMA) filtering   In the least squares example, the accuracy of the approximation can be improved only by increasing the polynomial degree. Alternative designs [29] have been proposed where the frequency domain filter response is a rational function,

$$h(\lambda_i) = \frac{p(\lambda_i)}{q(\lambda_i)}$$

where $p(\cdot)$ and $q(\cdot)$ are polynomials of λ. Any rational function can be written as a sum of polynomial fractions with denominators of order 1, and each of these can be computed recursively as follows:

$$\mathbf{y}^{(t+1)} = \alpha \mathbf{Z} \mathbf{y}^{(t)} + \beta \mathbf{x},$$

where $\alpha, \beta \in \mathbb{C}$; defining $r = -\beta/\alpha$ and $p = 1/\alpha$, this iteration converges to

$$h_{p,r}(\lambda) \frac{r}{\lambda - p} \quad \text{for} \quad |p| > |\lambda_{\max}|,$$

where $|\lambda_{\max}|$ is the magnitude of the maximum eigenvalue of $\mathbf{Z}$ [29, 30].

As for conventional signal processing, ARMA designs have the advantage of requiring a smaller number of design parameters. Notice, however, that from Remark 3.5, given that the frequency response of these filters is diagonal, any ARMA filter can be represented by a polynomial (but potentially one of high degree). ARMA filters may be desirable when the filter coefficients are "learned" and a small number of parameters is desirable, e.g., in graph convolutional networks (Section 7.5.4).

Alternative filter designs The Lagrange interpolation method and least squares technique both require exact knowledge of all eigenvalues, which can be quite costly for large graphs. Thus, while frequency domain designs allow maximum control of the filter response, most filter designs used in practice either (i) choose, by construction, $h(\lambda)$ to be a polynomial of desired degree, or (ii) choose an arbitrary $h(\lambda)$ and then find a polynomial approximation (Section 5.5.2). More broadly, the properties of filter designs (numerical stability in general, convergence in the case of ARMA filters, etc.) depend significantly on the distribution of eigenvalues of $\mathbf{Z}$, so that different techniques may be preferable depending on the specific graph properties. The GraSP toolbox has implemented several filter design methods, including the least squares approach described earlier and the Chebyshev polynomial method (Section 5.5.2). These and other filter design techniques and their corresponding implementation are discussed in Section B.4.4 and the box Going Further B.4.2. We refer the interested reader to [30] for additional discussion of the numerical issues and alternative design methods.

Box 3.3 Summary: what is a graph filter?

We summarize the main ideas about graph filtering presented in this section on the frequency domain and in Section 2.4 on the node domain. For a given operator $\mathbf{Z}$:

1 A graph filter $\mathbf{H}$ is a polynomial of $\mathbf{Z}$.
2 Assuming that $\mathbf{Z}$ can be diagonalized, and that $\mathbf{U}$ is its GFT, then a graph filter can be defined as $\mathbf{H} = \mathbf{U}h(\mathbf{\Lambda})\mathbf{U}^{-1}$, where $h(\mathbf{\Lambda})$ is diagonal, and if an eigenvalue has multiplicity greater than 1 then all its corresponding entries in $h(\mathbf{\Lambda})$ are equal.
3 Filter localization can be determined by representing it as a polynomial in $\mathbf{Z}$, keeping in mind that the actual degree of the polynomial may be lower, as in (2.38), and also that the localization is relative to the diameter of the graph.

Note that if $\mathbf{Z}$ can be diagonalized then conditions 1 and 2 above are equivalent.

3.3.3 Interpretations of $\mathbf{Z}^k$

Graph filters are defined as polynomials of the operator $\mathbf{Z}$. From the previous discussion, the filter $\mathbf{H} = \mathbf{Z}$ has a response $h(\lambda_i) = \lambda_i$ for all vectors in subspace E_i. From this it follows that $\mathbf{H} = \mathbf{Z}^k$ will have a response $h(\lambda_i) = \lambda_i^k$, leading to greater attenuation if $|\lambda_i| < 1$ and increased gain otherwise. Also, we discussed how, for a one-hop localized $\mathbf{Z}$, the kth power $\mathbf{Z}^k$ is k-hop localized. We now summarize additional properties of interest for powers of the adjacency matrix.

Powers of the adjacency matrix and the number of paths   For the adjacency matrix $\mathbf{A}$ of an unweighted undirected graph, the i, j entry of $\mathbf{A}^2$ can be written as

$$(\mathbf{A}^2)_{i,j} = \sum_k a_{ik}a_{kj}$$

and $a_{ii} = a_{jj} = 0$ if the graph has no self-loops. Thus, even if i and j are connected we have $a_{ii}a_{ij} = a_{ji}a_{jj} = 0$, so that $(\mathbf{A}^2)_{i,j}$ will have non-zero entries only if there are nodes k that connect to both i and j, i.e., $a_{ik} \neq 0$ and $a_{jk} \neq 0$.

Thus, for an unweighted graph, we have that $(\mathbf{A}^2)_{i,j}$ represents the number of paths of length 2 between i and j. A particular case of interest is $(\mathbf{A}^2)_{i,i}$ which counts the number of paths of length 2 that start and end at i, thus providing the *degree* of node i. By induction, higher powers, $\mathbf{A}^k$, can be used to count the number of paths of length k.

Furthermore, using the fact that

$$\text{trace}(\mathbf{A}^k) = \sum_i \lambda_i^k = c_k$$

we can say that c_k represents the total number of closed paths of length k. It is also easy to see that the average degree is c_2/N and the total number of undirected edges is $c_2/2$ (since every edge is counted twice towards the degree of the two nodes it connects).

Eigenvalues of A and graph topology   It can also be shown (see [31]) that the chromatic number of a graph $c(G)$ is bounded as follows:

$$c(G) \leq 1 + \lambda_1(\mathbf{A}),$$

where $\lambda_1(\mathbf{A})$ is the largest (positive) eigenvalue of $\mathbf{A}$. Note that this is a tighter bound than $c(G) \leq 1 + d_{max}$ because $\lambda_1(\mathbf{A}) \leq d_{max}$. Also, since $\mathbf{A}_{i,j}^k$ is non-zero only if there exists a path of length k from i to j, we can use the properties of powers of the adjacency matrix to determine the diameter of the graph.

THEOREM 3.1   (EIGENVALUES AND DIAMETER)   Assume $\mathbf{A}$ is diagonalizable and has K distinct eigenvalues; then the diameter, d satisfies

$$d < K.$$

Proof   If the diameter is K then there exist two nodes i and j such that i can only be reached from j in K hops. That means that $\mathbf{A}_{ij}^K > 0$ while $\mathbf{A}_{ij}^k = 0$ for $k = 1, \ldots, K - 1$ (otherwise this would not be the shorter path). But because $\mathbf{A}$ has K distinct eigenvalues and is diagonalizable it has a minimal polynomial of degree K, so that $\mathbf{A}^K$ can be written as a function of its lower powers. But then we have a contradiction: $\mathbf{A}_{ij}^K > 0$ cannot be obtained as a linear combination of $\mathbf{A}_{ij}^k = 0$.   □

Note that this result applies to both weighted and unweighted graphs. In the weighted graph case a non-zero entry in $\mathbf{A}_{ij}^k$ indicates the existence of a path with k hops between i and j. The result of Theorem 3.1 is important in that it allows us to formalize further the ideas of Box 2.6: the localization of a graph filter is determined by its degree relative to that of the minimal polynomial of the graph operator. Since the degree of the minimal polynomial bounds the diameter of the graph this confirms that filters whose degree is close to that of the minimal polynomial cannot be viewed as being local.

3.3.4 Graph Filters, Graph Signals and Convolutions

We now discuss the connection between graph filtering and convolution operations in conventional signal processing following the ideas in [15], where the algebraic signal processing framework of [24, 25] is used for graph operators. To do so, we start by establishing that some graph signals correspond to polynomials of the fundamental operator $\mathbf{Z}$, where for some simplicity we assume that $\mathbf{Z}$ is diagonalizable.

Graph signals as polynomials Let $\mathbf{x}$ be a vector whose representation in the frequency domain $\tilde{\mathbf{x}} = \mathbf{U}^{-1}\mathbf{x}$ meets the conditions of Remark 3.5, that is, for any λ_k with multiplicity greater than 1, the corresponding entries of $\tilde{\mathbf{x}}$ are identical. For example, if λ_k has multiplicity 2 then $\tilde{x}(k) = \tilde{x}(k+1)$. Denote by $\mathcal{A}(\mathbf{Z}) \subset \mathbb{C}^N$ the set of all $\mathbf{x}$ having this property. Clearly, $\mathcal{A}(\mathbf{Z})$ is a vector subspace of $\mathbb{C}^N$, since it is closed under addition and multiplication by a scalar (see Definition A.1). But, unless all the eigenvalues of $\mathbf{Z}$ are simple, $\mathcal{A}(\mathbf{Z}) \neq \mathbb{C}^N$. For example, if λ_k has multiplicity 2 and $\tilde{x}(k) \neq \tilde{x}(k+1)$ then $\mathbf{x} \notin \mathcal{A}(\mathbf{Z})$, but of course by definition $\mathbf{x} \in \mathbb{C}^N$. Thus, while we can compute the GFT $\tilde{\mathbf{x}}$ of any $\mathbf{x} \in \mathbb{C}^N$, not every $\mathbf{x}$ belongs to $\mathcal{A}(\mathbf{Z})$.

Multiplication of signals With this definition, there is a one-to-one mapping between $\mathbf{x} \in \mathcal{A}(\mathbf{Z})$ and a matrix $\mathbf{X}$, which from Remark 3.5 corresponds to a polynomial $p_x(\mathbf{Z})$:

$$\mathbf{X} = p_x(\mathbf{Z}) = \mathbf{U}\operatorname{diag}(\tilde{\mathbf{x}})\mathbf{U}^{-1}.$$

This allows us to define multiplication between vectors in $\mathcal{A}(\mathbf{Z})$. For any two vectors $\mathbf{x}, \mathbf{y} \in \mathcal{A}(\mathbf{Z})$, the vector $\mathbf{x} \cdot \mathbf{y}$ is defined as follows:

$$\mathbf{x} \cdot \mathbf{y} = \mathbf{XY} = p_x(\mathbf{Z})p_y(\mathbf{Z}) = \mathbf{U}\operatorname{diag}(\tilde{\mathbf{x}})\operatorname{diag}(\tilde{\mathbf{y}})\mathbf{U}^{-1}, \tag{3.24}$$

where it is clear that $\mathbf{x} \cdot \mathbf{y} \in \mathcal{A}(\mathbf{Z})$ since $\tilde{x}(k)\tilde{y}(k)$ has the properties of Remark 3.5 if $\tilde{x}(k)$ and $\tilde{y}(k)$ have those properties. The multiplication (3.24) is also commutative and, with these properties, $\mathcal{A}(\mathbf{Z})$ is an algebra, i.e., both a vector space and a ring [24, 25, 15]. Notice that while $p_{xy}(\mathbf{Z}) = p_x(\mathbf{Z})p_y(\mathbf{Z})$ can have degree greater than N, the residue $r_{xy}(\mathbf{Z}) = p_{xy}(\mathbf{Z}) \bmod p_{\min}(\mathbf{Z})$ has degree at most the number of unique eigenvalues. Also, for any vector $\mathbf{a}$, $r_{xy}(\mathbf{Z})\mathbf{a} = p_{xy}(\mathbf{Z})\mathbf{a}$ (Box 2.6), and therefore there is a one-to-one correspondence between $\mathcal{A}(\mathbf{Z})$ and the set of polynomials $p(\mathbf{Z})$ modulo $p_{\min}(\mathbf{Z})$.

Polynomial multiplication and convolution   For discrete-time signals, the output $y(n)$ of a linear shift invariant filter with impulse response $h(n)$ when the input is $x(n)$ can be written as a convolution [17]:

$$y(n) = h(n) * x(n) = x(n) * h(n) = \sum_k x(k)h(n-k), \qquad (3.25)$$

which can be viewed as the inner product between the input signal and the shifted and time-reversed impulse response $h(n)$. While by definition convolution involves shifts and time reversals, a graph convolution can be defined by analogy. In the z-transform domain, introduced in Box 2.4 below, the convolution (3.25) can be written as

$$Y(z) = H(z)X(z) = X(z)H(z), \qquad (3.26)$$

where $X(z), H(z), Y(z)$ are polynomials of z. It is common in the GSP literature to describe the multiplication of vectors in $\mathcal{A}(\mathbf{Z})$, defined in (3.24), as a convolution operation. This is based on the analogy between the multiplications (3.26) and (3.24), which both involve polynomials of a fundamental operator. Note that, as is the case for conventional signal processing, two graph filters $h_1(\mathbf{Z})$ and $h_2(\mathbf{Z})$ can be cascaded, resulting in an operator $h_1(\mathbf{Z})h_2(\mathbf{Z})$ or equivalently $h_1(\mathbf{Z})h_2(\mathbf{Z}) \bmod p_{\min}(\mathbf{Z})$. Filter implementations using this definition of convolution are discussed in Section B.4.3.

3.4   GFTs of Common Graph Operators

Up to this point we have described how to use a one-hop operator, $\mathbf{Z}$, to construct general graph filters (as polynomials of $\mathbf{Z}$) that allow us to analyze signals $\mathbf{x}$ in terms of basis vectors representing elementary frequencies (through the GFT $\mathbf{U}$ consisting of the eigenvectors of $\mathbf{Z}$). As has been made clear throughout this book, these concepts are general but specific results depend on the characteristics of each graph *and* the choice of graph operator. In this section we study common graph operators derived from the adjacency matrix, $\mathbf{A}$, their respective GFTs and how these relate to each other.

3.4.1   Adjacency Matrix

We start by focusing on the case $\mathbf{Z} = \mathbf{A}$. The following important theorem is valid for adjacency matrices of both directed and undirected graphs, and can be applied to graph Laplacians as well.

> THEOREM 3.2   (PERRON–FROBENIUS)   If $\mathbf{A}$ is a non-negative matrix (i.e., it has no negative entries), corresponding to a graph adjacency matrix with all positive weights, then λ_1 is a real eigenvalue λ_1 with multiplicity 1 if the graph is connected, and the corresponding eigenvector has all positive entries. Furthermore, any other eigenvalue λ_k is such that $\lambda_1 \geq |\lambda_k|$.
>
> If the graph is undirected, $\mathbf{A}$ is symmetric, all its eigenvalues are real and we have $\lambda_1 \geq \lambda_2 \geq \cdots \geq \lambda_N$, with $\lambda_1 > \lambda_2$ if the graph is connected.

To interpret the action of the operator $\mathbf{A}$, note that the **Rayleigh quotient**

$$R(\mathbf{x}) = \bar{\Delta}_A(\mathbf{x}) = \frac{\mathbf{x}^\mathsf{T} \mathbf{A} \mathbf{x}}{\mathbf{x}^\mathsf{T} \mathbf{x}} \tag{3.27}$$

is maximized for $\mathbf{x} = \alpha \mathbf{u}_1$, where $\mathbf{u}_1$ is an eigenvector of $\mathbf{A}$, so that in this case $R(\mathbf{x}) = \lambda_1$. From Theorem 3.2, all the entries of $\mathbf{u}_1$ will be positive (no zero crossings) if the graph is connected. Furthermore, for undirected graphs $\mathbf{A}$ is symmetric and its eigenvectors can be chosen to be orthogonal. Therefore, for any $\mathbf{u}_i$, $i > 1$, we will have $\mathbf{u}_i^\mathsf{T} \mathbf{u}_1 = 0$, so that $\mathbf{u}_i$ will have both positive and negative entries (zero crossings). For the normalized adjacency matrix $Q = \mathbf{D}^{-1}\mathbf{A}$, the output at node i of $Q\mathbf{x}$ is the average of $\mathbf{x}$ at the nodes in $\mathcal{N}(i)$, as shown in (2.20). Then, since the eigenvector $\mathbf{u}_1$ of Q is maximally similar to $Q\mathbf{u}_1$ (it is the average of the neighbors), we can view $\mathbf{u}_1$ as an invariant signal with maximal smoothness: the value at each node is as close as possible to that of its neighbors. We study normalized adjacency matrices in Section 3.4.3.

Graph reduction and graph frequencies   The following theorem is important for understanding how reducing the graph size (removing nodes) affects the graph frequencies.

THEOREM 3.3   (INTERLACING THEOREM)   Let $\mathbf{A}$ be real and symmetric with eigenvalues (e.g., corresponding to an undirected graph)

$$\lambda_1 \geq \lambda_2 \geq \lambda_3 \geq \cdots \geq \lambda_N;$$

now let $\mathbf{A}'$ be a principal submatrix of $\mathbf{A}$ (correspondingly, there is an induced subgraph, where we have removed a node and all its associated edges). Then the eigenvalues μ_i of $\mathbf{A}'$ are such that

$$\mu_1 \geq \mu_2 \geq \mu_3 \geq \cdots \geq \mu_{N-1}$$

and

$$\lambda_i \geq \mu_i \geq \lambda_{i+1}, \quad \forall i = 1, \ldots, N-1.$$

Intuitively, removing one node and its corresponding edges will lead to eigenvectors with lower overall similarity between $\mathbf{u}_i$ and $\mathbf{A}\mathbf{u}_i$. Thus, given a frequency index i, we have $\lambda_i \geq \mu_i$, i.e., the eigenvector $\mathbf{u}_i$ of $\mathbf{A}$ will have a higher normalized similarity than the corresponding eigenvector for $\mathbf{A}'$. By applying the above interlacing theorem recursively, after removing k nodes we will have that

$$\lambda_i \geq \mu_i^{(k)} \geq \lambda_{i+k},$$

where $\mu_i^{(k)}$ is an eigenvalue of a subgraph of the original graph where k nodes have been removed.

3.4.2   Graph Laplacians of Undirected Graphs

Focusing on undirected graphs, we now study important spectral properties of graph Laplacians, which are widely used as fundamental operators in graph signal processing.

We start with the combinatorial Laplacian (Definition 2.18) and the symmetric normalized Laplacian (Definition 2.20):

$$\mathbf{L} = \mathbf{D} - \mathbf{A} \quad \text{and} \quad \mathcal{L} = \mathbf{D}^{-1/2}\mathbf{L}\mathbf{D}^{-1/2} = \mathbf{I} - \mathbf{D}^{-1/2}\mathbf{A}\mathbf{D}^{-1/2},$$

while the random walk Laplacian (Definition 2.19) will be considered in Section 3.4.3. For undirected graphs, $\mathbf{A}$, $\mathbf{L}$ and $\mathcal{L}$ are all symmetric matrices and thus, as stated in Definition 3.3, can always be diagonalized; their corresponding eigenvalues are real and their eigenvectors are orthogonal.

Properties of first eigenvector   We can apply Theorem 3.2 to understand the properties of the **first eigenvector** of $\mathbf{L}$ or $\mathcal{L}$. First, note that $-\mathbf{L}$ has negative entries only along the diagonal. Therefore, choosing α such that $\alpha > d_{\max}$ we can see that $\alpha\mathbf{I} - \mathbf{L}$ is non-negative and has highest eigenvalue $\alpha - \lambda_1$, corresponding to the eigenvector $\mathbf{u}_1$ of $\mathbf{L}$. From this it follows that for a connected graph the first eigenvectors of $\mathbf{L}$ and $\mathcal{L}$ have non-negative entries.

Indeed, $\mathbf{u}_1 = \mathbf{1}$ is also always an eigenvector of the combinatorial Laplacian, because all the rows of this matrix add to zero; thus $\mathbf{L}\mathbf{1} = \mathbf{0} = 0 \cdot \mathbf{1}$. From this we can conclude that $\mathbf{u}_1' = \mathbf{D}^{1/2}\mathbf{1}$ is the corresponding eigenvector for eigenvalue $\lambda_1 = 0$ in the symmetric normalized Laplacian, $\mathcal{L}$. This can be easily seen because by definition $\mathcal{L} = \mathbf{D}^{-1/2}\mathbf{L}\mathbf{D}^{-1/2}$:

$$\mathcal{L}\mathbf{D}^{1/2}\mathbf{1} = \mathbf{D}^{-1/2}\mathbf{L}\mathbf{1} = 0. \tag{3.28}$$

Another consequence of Theorem 3.2 is that the multiplicity of the first eigenvalue corresponds to the number of connected components. Thus, since we are interested in connected graphs (Box 2.2), we will assume that λ_1 has multiplicity 1.

Relationship between $\mathbf{L}$ and $\mathcal{L}$ Note that other eigenvectors of $\mathcal{L}$ cannot be obtained as in (3.28). Instead, if $\mathbf{u}_i$ is an eigenvector of $\mathbf{L}$ we have that

$$\mathcal{L}\mathbf{D}^{1/2}\mathbf{u}_i = \lambda_i\mathbf{D}^{-1/2}\mathbf{u}_i.$$

The relation between the eigenvectors corresponding to the lowest frequencies of $\mathbf{L}$ and $\mathcal{L}$ shows how the interpretation of frequency changes after symmetric normalization. While $\mathbf{1}$ corresponds to a constant vector, and thus intuitively to a low frequency, the corresponding eigenvector of $\mathcal{L}$ has all positive entries but these are no longer all equal and instead are weighted by $\mathbf{D}^{1/2}$. It is particularly interesting to see how this impacts spectral representations. While all eigenvector entries have equal values for the lowest frequency of $\mathbf{L}$ the ratio between largest and smallest entries of $\mathbf{D}^{1/2}\mathbf{1}$ is $\sqrt{d_{\max}/d_{\min}}$, which can be very large for highly irregular graphs. This idea will be explored further in Section 3.4.5.

3.4.3 Normalized Adjacency Matrices and Random Walk Laplacian

A simple normalization for $\mathbf{A}$ can be obtained by dividing it by the magnitude of its maximum eigenvalue (i.e., λ_1 from Theorem 3.2):

$$\tilde{\mathbf{A}} = \frac{1}{\lambda_1}\mathbf{A},$$

leading to the definition of total variation proposed in [32]:

$$\Delta_{\tilde{A}}(\mathbf{x}) = \|\mathbf{x} - \tilde{\mathbf{A}}\mathbf{x}\|.$$

Since $\tilde{\mathbf{A}}$ and $\mathbf{A}$ have the same eigenvectors, all results derived up to now for adjacency matrices readily apply. In what follows we focus on degree-based normalization, first introduced in Section 2.3.3, which leads to changes in both eigenvalues and eigenvectors.

Row normalization Consider first the random walk normalization $Q = \mathbf{D}^{-1}\mathbf{A}$. According to Theorem 3.2, its largest eigenvalue has all positive entries and, noting that $\mathbf{D}^{-1}\mathbf{A1} = \mathbf{1}$ since the rows of Q sum to 1, we see that $\lambda_1 = 1$ corresponding to the eigenvector $\mathbf{u}_1 = \mathbf{1}$. For a non-bipartite graph, Q has eigenvalues in the interval $(-1, 1]$, so that $|\lambda_i| < 1$ for all $i > 1$, assuming the graph is connected. Thus, successive powers Q^k attenuate signals in all eigenspaces, except the signal corresponding to $\mathbf{u}_1 = \mathbf{1}$. Therefore, for any signal $\mathbf{x}$, if we define the corresponding mean-removed signal

$$\mathbf{x}' = \mathbf{x} - \left(\frac{1}{N}\mathbf{1}^\mathsf{T}\mathbf{x}\right)\mathbf{1},$$

we can see that $\mathbf{x}'$ has no energy along the direction of $\mathbf{1}$, so that $\|Q^k\mathbf{x}'\|$ becomes vanishingly small as k grows. Thus, the higher-degree terms of a graph filter, Q^k, for sufficiently large k, have a negligible effect on mean-removed signals such as $\mathbf{x}'$. This suggests that, even though in theory high-degree polynomials are valid graph filters (see Box 2.6), in practice lower-degree approximations are preferable, since higher-degree polynomials are approximately localized. Also, numerical computation problems may be encountered since Q^k has very small eigenvalues for large k.

Random walk Laplacian $\mathcal{T}$ For a given graph the random walk Laplacian (Definition 2.19, Section 2.4.1) can be defined as:

$$\mathcal{T} = \mathbf{I} - \mathbf{D}^{-1}\mathbf{A} = \mathbf{D}^{-1}\mathbf{L}.$$

Note that although $\mathcal{T}$ is not symmetric it is guaranteed to have real eigenvalues. This can be seen by observing that

$$\mathcal{T} = \mathbf{D}^{-1}\mathbf{L} = \mathbf{D}^{-1/2}\mathcal{L}\mathbf{D}^{1/2},$$

so that for any eigenvector $\mathbf{u}$ of $\mathcal{L}$ with eigenvalue λ we will have that

$$\mathcal{T}\mathbf{D}^{-1/2}\mathbf{u} = \mathbf{D}^{-1/2}\mathcal{L}\mathbf{D}^{1/2}\mathbf{D}^{-1/2}\mathbf{u} = \lambda\mathbf{D}^{-1/2}\mathbf{u};$$

thus $\mathbf{D}^{-1/2}\mathbf{u}$ is an eigenvector of $\mathcal{T}$ with eigenvalue λ. In particular, given that $\mathbf{u}_1 = \mathbf{D}^{1/2}\mathbf{1}$ is an eigenvector of $\mathcal{L}$, $\mathbf{u}_1 = \mathbf{1}$ will be an eigenvector of $\mathcal{T}$. This can also be seen because all the rows of $\mathbf{D}^{-1}\mathbf{A}$ add to 1, so that $\mathbf{D}^{-1}\mathbf{A}\mathbf{1} = \mathbf{1}$ and therefore $\mathcal{T}\mathbf{1} = \mathbf{1} - \mathbf{1} = \mathbf{0}$.

Eigenvalues of $\mathcal{T}$ Since $\mathcal{L}$ has eigenvalues in the interval $[0, 2]$, as will be discussed in Section 3.5, $\mathcal{T}$ will have eigenvalues in the same interval and therefore $\mathbf{D}^{-1}\mathbf{A}$ will have eigenvalues in the interval $[-1, 1]$. In particular, $\lambda = 1$ will always be an eigenvalue of $\mathbf{D}^{-1}\mathbf{A}$, since $\lambda = 0$ is always an eigenvalue of $\mathcal{T}$, while $\lambda = -1$ will be an eigenvalue of $\mathbf{D}^{-1}\mathbf{A}$ if and only if the graph is bipartite (see Section 3.5.3).

Column normalization When $\mathbf{A}$ is symmetric an alternative normalization leads to the graph operator $\mathcal{P} = \mathcal{Q}^{\mathsf{T}}$, which has normalized columns. Note that $\mathcal{P}$ and $\mathcal{Q}$ have the same set of eigenvalues[1] and therefore the eigenvalues of $\mathcal{P}$ are such that $|\lambda| \leq 1$. With a similar reasoning to that followed for $\mathcal{Q}$, $\mathcal{P}^k\mathbf{x}$ will have vanishingly small energy, except along the direction of the eigenvector of eigenvalue $\lambda = 1$, $\mathbf{v}_N$. If the graph is not bipartite, for any signal $\mathbf{x} = \sum_i \alpha_i \mathbf{v}_i$, as k increases $\mathcal{P}^k\mathbf{x} \approx \alpha_N \mathbf{v}_N$.

PROPOSITION 3.1 ($\lambda = 1$ EIGENVECTOR OF $\mathcal{P}$) Define $\mathbf{d} = \mathbf{A}\mathbf{1} = \mathbf{D}\mathbf{1}$, the **degree vector**, and $D_T = \mathbf{1}^{\mathsf{T}}\mathbf{D}\mathbf{1} = \sum_{i=1}^{N} d(i)$. Then the eigenvector $\mathbf{v}_N$ corresponding to $\lambda = 1$ for $\mathcal{P}$ is given by

$$\mathbf{v}_N = \frac{1}{D_T}\mathbf{d}.$$

Proof To find this eigenvector first write

$$Q = \mathbf{D}^{-1}\mathbf{A} = \mathbf{U}\Lambda\mathbf{U}^{-1} = \mathbf{U}\Lambda\mathbf{V};$$

thus

$$\mathcal{P} = \mathcal{Q}^{\mathsf{T}} = \mathbf{V}^{\mathsf{T}}\Lambda\mathbf{U}^{\mathsf{T}}.$$

Therefore $\mathbf{u}_N = \mathbf{1}$ is the last column of $\mathbf{U}$, while $\mathbf{v}_N$ is the last column of $\mathbf{V}^{\mathsf{T}}$, i.e., $\mathbf{v}_N^T$ is the Nth row of $\mathbf{V}$. But given that $\mathbf{U}\mathbf{V} = \mathbf{V}\mathbf{U} = \mathbf{I}$ we must have that $\mathbf{u}_N^{\mathsf{T}}\mathbf{v}_N = 1$, with $\mathbf{u}_N = \mathbf{1}$. Then, from the definition of $\mathbf{d}$,

$$\mathbf{d}^{\mathsf{T}}Q = \mathbf{1}^{\mathsf{T}}\mathbf{D}\mathbf{D}^{-1}\mathbf{A} = \mathbf{1}^{\mathsf{T}}\mathbf{A} = \mathbf{d}^{\mathsf{T}},$$

which shows that $\mathbf{d}^{\mathsf{T}}$ is a right eigenvector of Q and thus a left eigenvector of $\mathcal{Q}^{\mathsf{T}} = \mathcal{P}$. Now divide by D_T, so that $\mathbf{u}_N^{\mathsf{T}}\mathbf{v}_N = \mathbf{1}^{\mathsf{T}}\mathbf{v}_N = 1$ and the entries of $\mathbf{v}_N$ sum to 1. $\square$

3.4.4 Doubly Stochastic Normalization

While $\mathcal{Q}$ and $\mathcal{P}$ are row and column stochastic, respectively, it is possible to obtain a doubly stochastic matrix (both rows and columns sum to 1) from the adjacency matrix $\mathbf{A}$. This idea will be discussed in Section 7.4.3 in the context of image filtering applications, with more details available in [33, 34, 35]. Denoting by $\mathcal{D}$ the doubly stochastic

---

[1] They have the same characteristic polynomial: $\det(\mathbf{Q}^{\mathsf{T}} - \lambda\mathbf{I}) = \det((\mathbf{Q} - \lambda\mathbf{I})^{\mathsf{T}})$.

matrix obtained in this way, note that its largest eigenvalue will still be 1, since $\mathcal{D}$ is row stochastic as well, and the remaining eigenvalues will approximate those of Q while its eigenvectors will be orthogonal thanks to the symmetry property. It can be shown that the eigenvalues of $\mathcal{D}$ approximate those of Q with an error that is bounded by the Frobenius norm of the difference $\mathcal{D} - Q$. Thus, if Q is nearly symmetric then the eigenvalues of $\mathcal{D}$ will be very close to those of Q.

Note that for an undirected graph, for which $\mathbf{A}$ is symmetric, the amount by which $Q = \mathbf{D}^{-1}\mathbf{A}$ deviates from symmetry depends entirely on $\mathbf{D}$. Thus, undirected graphs that are nearly regular (the nodes have similar degrees) will be good candidates for the use of $\mathcal{D}$ as a graph operator. As an example this would be the case for image data, see Section 7.4.3. On the other hand, for highly irregular graphs symmetrization is still possible but can lead to significant differences in the eigenvalues of the resulting $\mathcal{D}$ matrix.

3.4.5 Irregularity, Normalization and the Choice of Inner Product

In this subsection we present a generalization of graph signal processing methods [36] based on the idea that the choice of inner product can be decoupled from the choice of graph signal variation. Up to this point, the standard inner product $\langle \mathbf{x}, \mathbf{y} \rangle = \mathbf{y}^\mathsf{T}\mathbf{x}$ has been used (Section A.3).

Comparing L, $\mathcal{L}$ and $\mathcal{T}$ Consider $\mathcal{L}$ and $\mathcal{T}$ first: $\mathcal{L}$ is symmetric and thus has orthogonal eigenvectors, while this is not true for $\mathcal{T}$. But note that, letting $\mathbf{u}$ be an eigenvector of $\mathcal{L}$, we have that $\mathbf{u}' = \mathbf{D}^{-1/2}\mathbf{u}$ is a corresponding eigenvector of $\mathcal{T}$. Eigenvectors of $\mathcal{T}$ are not orthogonal in the usual sense, since

$$(\mathbf{u}'_i)^\mathsf{T}\mathbf{u}'_j = \mathbf{u}_i^\mathsf{T}\mathbf{D}^{-1}\mathbf{u}_j \neq 0,$$

but they are in fact orthogonal for an inner product defined as $\langle \mathbf{x}, \mathbf{y} \rangle_\mathbf{D} = \mathbf{y}^\mathsf{T}\mathbf{D}\mathbf{x}$. Second, consider $\mathbf{L}$ and $\mathcal{T}$. An eigenpair $(\lambda, \mathbf{u})$ of $\mathcal{T}$ by definition is such that

$$\mathbf{D}^{-1}\mathbf{L}\mathbf{u} = \lambda\mathbf{u} \quad \text{or} \quad \mathbf{L}\mathbf{u} = \lambda\mathbf{D}\mathbf{u},$$

where the second equality shows that $(\lambda, \mathbf{u})$ is a generalized eigenpair of $\mathbf{L}$. From these two observations, the eigenvectors of $\mathcal{T}$ are also generalized eigenvectors of $\mathbf{L}$ under a Q-inner product (see below), for $Q = \mathbf{D}$.

Q-inner product As discussed in detail in Section 1.1.2 we can use one-hop graph operators to quantify graph signal variation. For example, $\mathbf{L}$ leads to the variation $\Delta_\mathbf{L}(\mathbf{x})$ for a signal $\mathbf{x}$. In what follows we denote by $\mathbf{M}$ a generic variation operator which is obviously graph dependent. In contrast, a Q-inner product can be defined that is in general *independent of the graph*.

DEFINITION 3.5   (**Q**-INNER PRODUCT AND **Q**-ORTHOGONALITY)   Let **Q** be a Hermitian positive definite matrix and let **x**, **y** be two vectors in $\mathbb{R}^N$. The **Q**-inner product and the **Q**-norm are defined as

$$\langle \mathbf{x}, \mathbf{y} \rangle_{\mathbf{Q}} = \mathbf{y}^{\mathsf{T}} \mathbf{Q} \mathbf{x} \quad \text{and} \quad \|\mathbf{x}\|_{\mathbf{Q}}^2 = \mathbf{x}^{\mathsf{T}} \mathbf{Q} \mathbf{x}$$

and **x** and **y** are orthogonal if

$$\langle \mathbf{x}, \mathbf{y} \rangle_{\mathbf{Q}} = 0.$$

Note that the standard inner product corresponds to $\mathbf{Q} = \mathbf{I}$ and that in many applications the choice of **Q** will be a diagonal matrix (with positive entries).

With a **Q**-inner product and a variation operator **M** (such as **L**), we can formulate a generalized eigendecomposition problem, where an eigenpair $(\lambda_i, \mathbf{u}_i)$ is such that

$$\mathbf{M}\mathbf{u}_i = \lambda_i \mathbf{Q}\mathbf{u}_i, \tag{3.29}$$

and where successive generalized eigenvectors have to be orthonormal in the **Q**-norm sense, that is, $\langle \mathbf{u}_i, \mathbf{u}_j \rangle_{\mathbf{Q}} = \mathbf{u}_j^{\mathsf{T}} \mathbf{Q} \mathbf{u}_i = \delta(i - j)$. For **M** and **Q** symmetric (an undirected graph and real **Q**), this problem can be solved efficiently [18, Section 8.7].

Note that for any eigenpair $(\lambda_i, \mathbf{u}_i)$ solving (3.29), we have $\mathbf{Q}^{-1}\mathbf{M}\mathbf{u}_i = \lambda_i \mathbf{u}_i$, which shows that the $\mathbf{u}_i$ are invariant under multiplication by $\mathbf{Q}^{-1}\mathbf{M}$. Thus, in analogy with the invariance ideas of Section 2.5, we can define the fundamental operator $\mathbf{Z} = \mathbf{Q}^{-1}\mathbf{M}$, and then (assuming λ_i is simple) $E_i = \text{span}(\mathbf{u}_i)$ is a subspace of signals that are invariant under multiplication by $\mathbf{Z} = \mathbf{Q}^{-1}\mathbf{M}$.

(M, Q)-graph Fourier transform   Let **U** be the matrix of **Q**-orthogonal eigenvectors obtained as a solution of (3.29). Its inverse can be found by using the definition of **Q**-orthogonality, $\mathbf{U}^{\mathsf{T}}\mathbf{Q}\mathbf{U} = \mathbf{I}$, so that $\mathbf{U}^{-1} = \mathbf{U}^{\mathsf{T}}\mathbf{Q}$. This leads to the definition of the GFT.

DEFINITION 3.6   ((**M, Q**)-GRAPH FOURIER TRANSFORM)   The matrix **U** of **Q**-orthogonal eigenvectors, obtained as a solution of (3.29), is the (**M, Q**)-GFT. For any $\mathbf{x} \in \mathbb{R}^N$ we can find its GFT transform $\tilde{\mathbf{x}}$:

$$\mathbf{x} = \mathbf{U}\tilde{\mathbf{x}} \quad \text{where} \quad \tilde{\mathbf{x}} = \mathbf{U}^{\mathsf{T}}\mathbf{Q}\mathbf{x} \quad \text{and} \quad \mathbf{U}^{-1} = \mathbf{U}^{\mathsf{T}}\mathbf{Q}. \tag{3.30}$$

We can define invariant subspaces E_i from the columns of **U**, leading to an $N \times N$ fundamental operator **Z** that can be written as

$$\mathbf{Z} = \mathbf{Q}^{-1}\mathbf{M} = \mathbf{U}\Lambda\mathbf{U}^{\mathsf{T}}\mathbf{Q}. \tag{3.31}$$

Note that this choice leads to a definition of graph filters as polynomials of **Z** that is consistent with Box 3.3. On the basis of (3.31), any polynomial of $\mathbf{Z} = \mathbf{Q}^{-1}\mathbf{M}$ can be expressed as a diagonal operator in the (**M, Q**)-GFT domain. In particular it is easy to show that $\mathbf{Z}^k = \mathbf{U}\Lambda^k\mathbf{U}^{\mathsf{T}}\mathbf{Q}$, so that

$$h(\mathbf{Z}) = \sum_i h_i \mathbf{Z}^i = \mathbf{U}\left(\sum_i h_i \Lambda^i\right)\mathbf{U}^{\mathsf{T}}\mathbf{Q}.$$

From Definition 3.5 and Definition 3.6 we can derive a Parseval relation between $\mathbf{x}$ and $\tilde{\mathbf{x}}$, analogous to that of (A.15)

$$\|\tilde{\mathbf{x}}\|_I^2 = \left(\mathbf{U}^\mathsf{T}\mathbf{Q}\mathbf{x}\right)^\mathsf{T}\mathbf{U}^\mathsf{T}\mathbf{Q}\mathbf{x} = \mathbf{x}^\mathsf{T}\mathbf{Q}\mathbf{U}\mathbf{U}^\mathsf{T}\mathbf{Q}\mathbf{x} = \mathbf{x}^\mathsf{T}\mathbf{Q}\mathbf{x} = \|\mathbf{x}\|_Q^2,$$

where we note that the inner product in the transform domain is the standard one: $\mathbf{Q} = \mathbf{I}$.

The random walk Laplacian operator $\mathcal{T}$ revisited Returning to the discussion at the beginning of this section, for any undirected graph we can see that $\mathbf{Q} = \mathbf{D}$ and $\mathbf{M} = \mathbf{L}$ leads to $\mathbf{Z} = \mathbf{D}^{-1}\mathbf{L} = \mathcal{T}$. Using the relationship between the eigenvectors of $\mathcal{T}$ and $\mathcal{L}$ then shows that $\mathbf{u}_i'$ and $\mathbf{u}_j'$ are $\mathbf{D}$-orthogonal:

$$\langle\mathbf{u}_i', \mathbf{u}_j'\rangle_D = (\mathbf{D}^{-1/2}\mathbf{u}_i)^\mathsf{T}\mathbf{D}(\mathbf{D}^{-1/2}\mathbf{u}_j) = \mathbf{u}_i^\mathsf{T}\mathbf{D}^{-1/2}\mathbf{D}\mathbf{D}^{-1/2}\mathbf{u}_j = \delta(i - j).$$

Comparing the $(\mathbf{L}, \mathbf{I})$-GFT with the $(\mathbf{L}, \mathbf{D})$-GFT we can see that the lowest-frequency vector is $\mathbf{1}$, which is a useful choice corresponding to lowest-frequency mode in conventional signals. This is in contrast with $\mathcal{L}$, for which the lowest-frequency eigenvector is $\mathbf{D}^{1/2}\mathbf{1}$. On the other hand, the fundamental operator $\mathbf{Z} = \mathbf{D}^{-1}\mathbf{L} = \mathcal{T}$ leads to the normalized one-hop filtering operation of (2.21), while for $\mathbf{Z} = \mathbf{L}$ there is no normalization as in (2.22). By using the definitions of the variation operator and the inner product, we can extend the definition of (3.3) to define a normalized variation:

$$\bar{\Delta}_{M,Q}(\mathbf{x}) = \frac{\mathbf{x}^\mathsf{T}\mathbf{M}\mathbf{x}}{\mathbf{x}^\mathsf{T}\mathbf{Q}\mathbf{x}}$$

and note that, from (3.29), for an eigenpair $(\lambda_i, \mathbf{u}_i)$ we have $\bar{\Delta}_{M,Q}(\mathbf{u}_i) = \lambda_i$. Then, defining δ_i, an impulse (delta function signal) centered at node i, the corresponding normalized variations are

$$\bar{\Delta}_{L,I}(\delta_i) = d_i \quad \text{and} \quad \bar{\Delta}_{L,D}(\delta_i) = 1. \tag{3.32}$$

Having the same variation for all impulses, regardless of the local graph structure, is beneficial in settings where the measurements are independent of the graph structure. For example, if a single sensor in a sensor network malfunctions this may be modeled as impulse noise added to the signal. In this context, the fact that $\bar{\Delta}_{L,D}(\delta_i)$ is constant makes $(\mathbf{L}, \mathbf{D})$ a better choice. The choice of $\mathbf{M}$ and $\mathbf{Q}$ is application dependent. As an example $(\mathbf{M} = \mathbf{L}, \mathbf{Q} = \mathbf{D})$ is useful in the design of critically sampled graph filterbanks (Section 5.6.4), as the DC signal $\mathbf{1}$ is useful for signals of interest, while a normalized range of frequencies is needed for filter design. The differences between the various operators are summarized in Table 3.1.

Choice of Q and graph irregularity Note that the difference between the $(\mathbf{L}, \mathbf{I})$-GFT and the $(\mathbf{L}, \mathbf{D})$-GFT depends on the degree distribution. If the graph is exactly regular and all nodes have the same degree, the two definitions are the same and there are no differences in the frequency content of the impulse functions. Conversely, if there are extreme differences in node degree then this would lead to correspondingly large disparities in behavior across nodes: if the highest degree is several orders of magnitude

Table 3.1 Comparison of Laplacian operators

$(\mathbf{M}, \mathbf{Q})$	$\mathbf{Z}$	$\mathbf{u}_1$	Orthogonality	$\bar{\Delta}_{\mathbf{M},\mathbf{Q}}(\delta_i)$
$(\mathbf{L}, \mathbf{I})$	$\mathbf{L}$	$\mathbf{1}$	Standard	d_i
$(\mathcal{L}, \mathbf{I})$	$\mathcal{L}$	$\mathbf{D}^{1/2}\mathbf{1}$	Standard	1
$(\mathbf{L}, \mathbf{D})$	$\mathcal{T}$	$\mathbf{1}$	$\mathbf{D}$	1

larger than the lowest degree, the impulse signal variation will also be several orders of magnitude larger, while the lowest-frequency entries will also vary significantly from node to node for the combinatorial Laplacian.

Other examples Alternative inner product definitions can be useful to reflect specific properties of signals of interest. For example, assume that $\mathbf{x}$ represents a set of sensor network measurements in space. Then, we can choose $\mathbf{Q} = \mathbf{V}$, where $\mathbf{V}$ is a diagonal matrix whose ith entry corresponds to the area of the Voronoi cell[2] around node (sensor) i. In this case $\mathbf{x}^\mathsf{T}\mathbf{V}\mathbf{x}$ is an estimate of the energy of the measured continuous space signal over the area where sensors are placed [36]. For image and video compression a $\mathbf{Q}$-norm related to perceptual quality metrics has been proposed [37].

3.5   Graph Structure and Graph Spectra

In this section we provide insights about how the structure of a graph affects the properties of the corresponding eigenvectors and eigenvalues. Except for very specific cases, the exact spectrum of a graph cannot be written in closed form but there exist numerous results that link node and frequency domain graph properties. An exhaustive review of these results goes beyond the scope of this book (see for example [7] or [8]). Instead we focus on developing intuitions that can be useful to solve practical problems from a GSP perspective.

3.5.1   Spanning Trees and the Matrix Tree Theorem

A tree is a bipartite graph where nodes can be divided into levels, and such that nodes at the ith level of the tree can only connect to nodes at the $(i - 1)$th and $(i + 1)$th levels. Trees are minimally connected, since removing any edge makes the graph disconnected, and contain no loops. Then, for a given arbitrary graph with node set $\mathcal{V}$, a **spanning tree** is any tree that has the same node set and contains only edges in the original edge set $\mathcal{E}$. Enumerating the number of spanning trees $\tau(G)$ is useful for understanding how strongly connected the graph is.

---

[2] Given a set of points $\mathcal{S}$ the Voronoi cell around node $i \in \mathcal{S}$ is a convex region in space where every point in the cell is closer to i than to any of the other points in $\mathcal{S}$.

> THEOREM 3.4   (MATRIX TREE THEOREM)   Let $\mathbf{L}$ be the combinatorial Laplacian of an unweighted graph. Denote by $\mathbf{L}_0$ the matrix obtained by removing one row i and column i from $\mathbf{L}$ (any i can be chosen). Then
>
> $$\tau(G) = \det(\mathbf{L}_0),$$
>
> where the determinant is positive because $\mathbf{L}_0$ is strictly diagonally dominant (the magnitude of the diagonal entry in each row is strictly greater than the sums of the magnitudes of the other entries in that row) and symmetric.

Note that Theorem 3.4 can be applied to analyze the connectivity of a weighted graph by analyzing a corresponding unweighted graph with same edges. Clearly, for two graphs G_1 and G_2 with $|\mathcal{V}_1| = |\mathcal{V}_2|$ and $|\mathcal{E}_1| < |\mathcal{E}_2|$ we expect that $\tau(G_1) < \tau(G_2)$. In Section 3.5.4 we will be able to link the number of edges (in an unweighted graph) to the variation of the signals on that graph. Denser graphs (with more spanning trees) will lead to greater graph signal variation (larger eigenvalues for the corresponding graph Laplacian).

3.5.2   Topology and Eigenvalue Multiplicity

We next consider a few specific examples of graphs with their corresponding frequency representations for given choice of graph operator. Our main goal is to illustrate how different topologies lead to different graph frequencies and also to show that analogies with regular domain signal frequencies are sometimes difficult. We start by considering a complete unweighted graph (Example 3.3) and a star graph (Example 3.4). Both lead to high-multiplicity eigenvalues.

Example 3.3   Complete unweighted graph   For a complete unweighted graph with adjacency matrix

$$\mathbf{A} = \begin{bmatrix} 0 & 1 & 1 & \cdots & 1 \\ 1 & 0 & 1 & \cdots & 1 \\ \vdots & \vdots & \vdots & \ddots & \vdots \\ 1 & 1 & 1 & \cdots & 0 \end{bmatrix}$$

find the eigenvectors of $\mathbf{A}$ and those of its corresponding combinatorial Laplacian, $\mathbf{L}$.

Solution
Recall from Section 3.3.3 that $(\mathbf{A}^2)_{i,j}$ $(i \neq j)$ represents the number of paths of length 2 from i to j, and thus $(\mathbf{A}^2)_{i,i}$ is the degree of i. Then, for $i \neq j$, $(\mathbf{A}^2)_{i,j} = N - 2$ since by going from i to j in two steps one can connect to any $k \neq i, j$ and there are $N - 2$ choices since every node connects to $N - 1$ other nodes. Similarly, $(\mathbf{A}^2)_{i,i} = N - 1$ since every

node has $N - 1$ connections. Thus, we can write

$$\mathbf{A}^2 = \begin{bmatrix} N-1 & N-2 & N-2 & \cdots & N-2 \\ N-2 & N-1 & N-2 & \cdots & N-2 \\ \vdots & \vdots & \vdots & \ddots & \vdots \\ N-2 & N-2 & N-2 & \cdots & N-1 \end{bmatrix}$$

and we have that

$$\mathbf{A}^2 - (N-2)\mathbf{A} = (N-1)\mathbf{I},$$

from which we can obtain the minimal polynomial of $\mathbf{A}$ (the characteristic polynomial of $\mathbf{A}$ will have degree N):

$$x^2 + (N-2)x - (N-1) = 0$$

with roots $\lambda_1 = N - 1, \lambda_2 = -1$. Because the graph is connected we have that λ_1 must be simple by the Perron–Frobenius theorem (Theorem 3.2), and thus the multiplicity of λ_2 must be $N - 1$. Note that $\mathbf{A1} = (N-1)\mathbf{1}$ so that $\mathbf{1}$ is the eigenvector associated with λ_1. As for $\lambda_2 = -1$, we will need to find $N - 1$ eigenvectors orthogonal to $\mathbf{u}_1 = \mathbf{1}$. Thus we will have that $\mathbf{u}_i$, for $i \neq 1$, is such that $\mathbf{u}_i^\mathsf{T}\mathbf{1} = 0$ and, choosing the entries of $\mathbf{u}_i$ to be $0, 1, -1$, we will need to have the same number of 1 and -1 values. For example, for N even, choosing $\mathbf{u}_2 = [1, -1, 1, -1, \ldots]^\mathsf{T}$ we can see that $\mathbf{Au}_2 = -\mathbf{u}_2$. As for any subspace corresponding to an eigenvalue of high multiplicity, there will be an infinite number of possible bases.

The corresponding graph Laplacian

$$\mathbf{L} = (N-1)\mathbf{I} - \mathbf{A},$$

has the same eigenvectors, an eigenvalue of multiplicity 1 for $\lambda' = 0$ and an eigenvalue $\lambda' = N$ of multiplicity $N - 1$. Therefore $\lambda' = (N-1) - \lambda$, where λ is an eigenvalue of the adjacency matrix.

From Example 3.3 we can see that high-multiplicity eigenvalues can make a frequency interpretation challenging. For example, assume we define filters in the spectral domain (see Section 3.3) by assigning a scalar gain to all signals associated with an elementary frequency. Then, for the case of Example 3.3, there are $N - 1$ eigenvectors with eigenvalue $\lambda = -1$, so that essentially most signals in the space (a subspace of dimension $N - 1$ in a space of dimension N) would get the same gain. A filter defined in this way would not be very meaningful.

Example 3.4   Star graph   A star graph has a central node 1 connected to all other $N - 1$

nodes, which are not connected to each other, leading to

$$
A = \begin{bmatrix} 0 & 1 & 1 & \cdots & 1 \\ 1 & 0 & 0 & \cdots & 0 \\ \vdots & \vdots & \vdots & \ddots & \vdots \\ 1 & 0 & 0 & \cdots & 0 \end{bmatrix}
$$

where only the first row and column contain non-zero values. Find the eigenvalues and eigenvectors of A.

Solution

Given A we can see that

$$
A^2 = \begin{bmatrix} N-1 & 0 & 0 & \cdots & 0 \\ 0 & 1 & 1 & \cdots & 1 \\ \vdots & \vdots & \vdots & \ddots & \vdots \\ 0 & 1 & 1 & \cdots & 1 \end{bmatrix}
$$

where the entries of the $(N-1) \times (N-1)$ submatrix excluding the first row and column are all 1, and

$$
A^3 = \begin{bmatrix} 0 & N-1 & N-1 & \cdots & N-1 \\ N-1 & 0 & 0 & \cdots & 0 \\ \vdots & \vdots & \vdots & \ddots & \vdots \\ N-1 & 0 & 0 & \cdots & 0 \end{bmatrix} = (N-1)A.
$$

Thus the minimal polynomial (Sections 2.5.1 and 2.5.3) can be obtained from

$$
A^3 = (N-1)A
$$

with roots $\lambda_1 = \sqrt{N-1}, \lambda_2 = 0, \lambda_3 = -\sqrt{N-1}$. Because the graph is connected, by the Perron–Frobenius theorem (Theorem 3.2) λ_1 has multiplicity 1. As will be shown in Section 3.5.3, because this graph is bipartite, λ_3, which mirrors λ_1, is also simple, while λ_2 has multiplicity $N-2$. Note that $u_1 = [\sqrt{N-1}, 1, 1, \ldots, 1]^T$ is the eigenvector corresponding to λ_1, while $u_3 = [-\sqrt{N-1}, 1, 1, \ldots, 1]^T$ is simply obtained by flipping the sign of one of the two classes of nodes in the bipartite graph (in this case changing the sign of the value assigned to node 1).

Examples 3.3 and 3.4 illustrate how certain graphs have few distinct eigenvalues and some have high multiplicity. In these cases, a frequency analysis of the graph signals may not be very meaningful.

Example 3.5   Regular connected graphs   A regular k-connected graph is such that every node has exactly k neighbors. Regular signals such as 1D sequences and 2D images can be represented as regular graphs (with appropriate circular extensions at the boundaries, as shown in Section 1.3). Find the eigenvalues of a general k-regular graph.

Solution

For a k-regular graph $\mathbf{A}$ will have exactly k entries equal to 1 per row, and thus

$$\mathbf{A}\mathbf{1} = k\mathbf{1},$$

so that $\mathbf{1}$ is an eigenvector corresponding to the eigenvalue $\lambda_1 = k$. Again using Theorem 3.2, this eigenvalue is simple and is the largest eigenvalue (with all entries of the eigenvector positive).

Consider now the particular case $k = 2$, where we extend a line graph so that node 1 and node N are connected. Then $\mathbf{A}$ is a circulant matrix which has eigenvectors corresponding to the discrete Fourier transform (DFT). Denoting by

$$w_k = e^{j2\pi k/N}$$

one of the Nth roots of unity, and defining $\mathbf{x}_k = [w_k, w_k^2, \ldots, w_k^N]^\mathsf{T}$, we have

$$(\mathbf{A}\mathbf{x}_k)_i = w_k^{((i-1)\bmod N)} + w_k^{((i+1)\bmod N)} = (w_k^{+1} + w_k^{-1})w_k^i$$

and thus

$$\mathbf{A}\mathbf{x}_k = (w_k + w_k^{-1})\mathbf{x}_k,$$

so that the eigenvectors are $\mathbf{u}_k = \mathbf{x}_{k-1}$, with eigenvalues $\lambda_{k+1} = w_k + w_k^{-1}$. The first eigenvector corresponds to $k = 0$ and, as expected, $w_0 = 1$ so that $\lambda_1 = 2$. Note also that with this definition $\lambda_{k+1} = 2\cos(2\pi k/N)$ and the eigenvalues are real.

3.5.3 Bipartite Graphs

Bipartite graphs play an important role in graph signal processing owing to their special properties. In a bipartite graph we have two sets of nodes $\mathcal{V}_1$, $\mathcal{V}_2$, such that $\mathcal{V} = \mathcal{V}_1 \cup \mathcal{V}_2$ and $\mathcal{V}_1 \cap \mathcal{V}_2 = \emptyset$, thus all edges in the graph go from $\mathcal{V}_1$ to $\mathcal{V}_2$. Denoting $N_i = |\mathcal{V}_i|$, we can write the adjacency matrix of an undirected bipartite graph (with a suitable permutation of labels) as

$$\mathbf{A} = \begin{bmatrix} \mathbf{0}_{N_1 \times N_2} & \mathbf{A}_0^\mathsf{T} \\ \mathbf{A}_0 & \mathbf{0}_{N_2 \times N_1} \end{bmatrix} \tag{3.33}$$

where $\mathbf{A}_0$ is $N_1 \times N_2$ and represents the connections from $\mathcal{V}_1$ to $\mathcal{V}_2$. Next we can state the following important result.

PROPOSITION 3.2   (BIPARTITE GRAPH SPECTRAL SYMMETRY)   Let $\mathbf{A}$ be the adjacency matrix of a bipartite graph written as in (3.33) and define vectors $\mathbf{u}$ and $\mathbf{u}'$ as:

$$\mathbf{u} = \begin{bmatrix} \mathbf{u}_1 \\ \mathbf{u}_2 \end{bmatrix} \quad \text{and} \quad \mathbf{u}' = \begin{bmatrix} \mathbf{u}_1 \\ -\mathbf{u}_2 \end{bmatrix}, \tag{3.34}$$

where $\mathbf{u}_1$ and $\mathbf{u}_2$ correspond to entries in $\mathcal{V}_1$ and $\mathcal{V}_2$, respectively. Then, if $\mathbf{u}$ is an eigenvector of $\mathbf{A}$ with eigenvalue λ, $\mathbf{u}'$ is also an eigenvector, with eigenvalue $-\lambda$.

Proof Since **u** is an eigenvector, from (3.34) and $\mathbf{Au} = \lambda\mathbf{u}$, we have

$$\begin{bmatrix} \mathbf{A}_0^T \mathbf{u}_2 \\ \mathbf{A}_0 \mathbf{u}_1 \end{bmatrix} = \lambda \begin{bmatrix} \mathbf{u}_1 \\ \mathbf{u}_2 \end{bmatrix};$$

then **u**′ is also an eigenvector, since

$$\begin{bmatrix} \mathbf{0}_{N_1 \times N_2} & \mathbf{A}_0^T \\ \mathbf{A}_0 & \mathbf{0}_{N_2 \times N_1} \end{bmatrix} \begin{bmatrix} \mathbf{u}_1 \\ -\mathbf{u}_2 \end{bmatrix} = \begin{bmatrix} -\mathbf{A}_0^T \mathbf{u}_2 \\ \mathbf{A}_0 \mathbf{u}_1 \end{bmatrix} = (-\lambda) \begin{bmatrix} \mathbf{u}_1 \\ -\mathbf{u}_2 \end{bmatrix}.$$

□

Thus, in a bipartite graph the eigenvalues of **A** come in symmetric pairs, $\lambda, -\lambda$, with corresponding eigenvectors **u** and **u**′, whose entries are related to each other as in (3.34).

The same property can be derived for other elementary operators. For $Q = \mathbf{D}^{-1}\mathbf{A}$, $\lambda = -1$ will be an eigenvalue. For the symmetric normalized Laplacian $\mathcal{L} = \mathbf{I} - \mathbf{D}^{-1/2}\mathbf{AD}^{-1/2} = \mathbf{I} - \mathcal{A}$, if **u** is an eigenvector of $\mathcal{A}$ with eigenvalue λ then $\mathbf{D}^{+1/2}\mathbf{u}$ is an eigenvector of $\mathcal{L}$ with eigenvalue $1 - \lambda$ and therefore $1 + \lambda$ will also be an eigenvalue of $\mathcal{L}$. Note that the smallest eigenvalue of $\mathcal{L}$ is $\lambda = 0$, associated with $\mathbf{D}^{+1/2}\mathbf{1}$, and thus because the graph is bipartite, $\lambda = 2$ will be the largest eigenvalue of $\mathcal{L}$. Therefore, for any eigenvalue λ of $\mathcal{L}$ we have that $\lambda \in [0, 2]$ and there is a symmetric eigenvalue $2 - \lambda$.

3.5.4   Bounds on Eigenvalues

Motivation   From the preceding discussion it should be clear that the range of eigenvalues depends on the structure of the graph. For relatively small graphs it is possible to compute the eigendecomposition, and the exact eigenvalues can be used to design graph filters with desired frequency responses, as described in Section 3.3. For larger graphs, on the other hand, this may not be practical, and alternative designs need to be used. Specifically, the response of a polynomial graph filter $p(\mathbf{Z})$ in the frequency domain is $p(\lambda)$ (see Remark 3.4). Note that $p(\lambda)$ is independent of the frequencies of the graph. If the λ_i are known then we will have the exact frequency response $p(\lambda_i)$ at each frequency. Instead, suppose that $\mathbf{Z} = \mathbf{L}$, so that $\lambda \in [0, \lambda_N]$ depends on the exact values of the frequencies. If estimating all N eigenvalues is impractical then an alternative solution might be to find λ_N (either its exact value or a bound on its value). With knowledge of the interval $[0, \lambda_N]$ we can then design a polynomial with a specific criterion, e.g., such that its passband includes the lower half of the eigenvalue range, without requiring exact knowledge of all the eigenvalues.

In what follows we present a few examples with the goal of illustrating how eigenvalues depend on graph topology. For the graph Laplacians we are primarily interested in the second eigenvalue, λ_2, the so-called Fiedler eigenvalue, and the maximum eigenvalue $\lambda_N = \lambda_{max}$, which is useful to estimate the range of frequencies.

Gerschgorin circle theorem   We start with a general result (valid for any **Z**) and then provide bounds for the eigenvalues of graph Laplacians.

THEOREM 3.5   (**GERSCHGORIN CIRCLE THEOREM**)   Every eigenvalue of a matrix $\mathbf{A}$ lies in at least one of the circles (discs) of center a_{ii} and radius either:

$$R_i = \sum_{j=1,j\neq i}^{N} |a_{ij}|$$

or

$$r_i = \sum_{j=1,j\neq i}^{N} |a_{ji}|.$$

Eigenvalues of graph Laplacians   Theorem 3.5 leads to the following result.

PROPOSITION 3.3   For a graph Laplacian $\mathbf{L}$ we have

$$\lambda_N \leq 2d_{\max}, \tag{3.35}$$

where $d_{\max}$ is the maximum degree of any node in the graph.

Proof   Since $\mathbf{L}$ is symmetric, $R_i = r_i$. Furthermore, by construction $a_{ii} = d(i)$ and $r_i = d(i)$. Therefore a bound of the largest eigenvalue can be obtained by choosing a circle centered at the largest diagonal value of $\mathbf{L}$, i.e., $d_{\max}$, and with radius $d_{\max}$. Since all eigenvalues of $\mathbf{L}$ are real and non-negative the largest value along the real line of this circle is $2d_{\max}$, leading to the bound (3.35). $\square$

Further bounds can be obtained by recalling that, as discussed in Section 3.2.3, successive sets of eigenvalues and eigenvectors can be obtained by minimization of Rayleigh quotients under constraints. Recalling that $\lambda_1 = 0$ and $\mathbf{u}_1 = \mathbf{1}$ for the combinatorial Laplacian $\mathbf{L}$, the second eigenvalue is the solution to

$$\lambda_2 = \min_{\mathbf{x}} \left(\frac{\mathbf{x}^\mathsf{T}\mathbf{L}\mathbf{x}}{\mathbf{x}^\mathsf{T}\mathbf{x}} \right) \quad \text{where } \mathbf{x} \in \mathbb{R}^N, \mathbf{x} \neq \mathbf{0} \text{ and } \mathbf{x}^\mathsf{T}\mathbf{u}_1 = 0, \tag{3.36}$$

and, in general,

$$\lambda_k = \min_{\mathbf{x}} \left(\frac{\mathbf{x}^\mathsf{T}\mathbf{L}\mathbf{x}}{\mathbf{x}^\mathsf{T}\mathbf{x}} \right), \quad \text{where } \mathbf{x} \in \mathbb{R}^N, \mathbf{x} \neq \mathbf{0} \text{ and } \mathbf{x}^\mathsf{T}\mathbf{u}_i = 0 \ \forall i = 1,\dots,k-1. \tag{3.37}$$

We can now state a first bound for λ_2.

PROPOSITION 3.4   Let i and j be two nodes that are not connected, with respective degrees $d(i)$ and $d(j)$. Then

$$\lambda_2 \leq \frac{d(i) + d(j)}{2}.$$

Proof   Choose a signal $x(k) = 0, k \neq i, j$ and $x(i) = 1,\ x(j) = -1$, where i and j are not connected. Clearly this signal is orthogonal to $\mathbf{u}_1 = \mathbf{1}$ and, noting that

$$\mathbf{L}\mathbf{x} = [0,\dots,0,d(i),0,\dots,0,-d(j),0,\dots,0]^\mathsf{T},$$

its variation can be computed as

$$\frac{\mathbf{x}^T \mathbf{L} \mathbf{x}}{\mathbf{x}^T \mathbf{x}} = \frac{d(i) + d(j)}{2};$$

since λ_2 is the minimal variation of any vector orthogonal to $\mathbf{u}_1$ the bound follows. □

Except for complete graphs, it is always possible to find two nodes i, j as in Proposition 3.4. Then λ_2 will be low if the node degrees are low, so that the graph is not strongly connected. As an example, since in a bipartite graph the nodes in $\mathcal{V}_1$, corresponding to one color, connect only to the nodes in $\mathcal{V}_2$, corresponding to the second color, a tight upper bound can be obtained from the pair of lowest-degree nodes in $\mathcal{V}_1$ or $\mathcal{V}_2$. The following bounds require knowledge of the node degrees.

PROPOSITION 3.5 Let $d_{\min}$ and $d_{\max}$ be the minimum and maximum degrees of the nodes in a graph. Then, the eigenvalues λ_2 and λ_N of $\mathbf{L}$ can be bounded as follows:

$$\lambda_2 \le \frac{N}{N-1} d_{\min}$$

and similarly

$$\lambda_N \ge \frac{N}{N-1} d_{\max}.$$

Proof First observe that

$$\sum_{i=1}^{N} \sum_{j=1}^{N} (x(i) - x(j))^2 = \sum_{i=1}^{N} \sum_{j=1}^{N} x(i)^2 + \sum_{i=1}^{N} \sum_{j=1}^{N} x(j)^2 - 2 \sum_{i=1}^{N} x(i) \sum_{j=1}^{N} x(j), \quad (3.38)$$

where the first two terms are equal to $N\mathbf{x}^T\mathbf{x}$. Any eigenvector (other than $\mathbf{u}_1 = \mathbf{1}$) is orthogonal to $\mathbf{1}$, so that $\mathbf{x}^T\mathbf{1} = 0$ or equivalently $\sum_{j=1}^{N} x(j) = 0$. Then, for any vector orthogonal to $\mathbf{1}$, we can simplify (3.38) as

$$\sum_{i=1}^{N} \sum_{j=1}^{N} (x(i) - x(j))^2 = 2N\mathbf{x}^T\mathbf{x},$$

and thus for such a vector the Rayleigh quotient can be rewritten as

$$\frac{\mathbf{x}^T \mathbf{L} \mathbf{x}}{\mathbf{x}^T \mathbf{x}} = 2N \frac{\sum_{i,j,i<j} a_{ij}(x(i) - x(j))^2}{\sum_{i=1}^{N} \sum_{j=1}^{N} (x(i) - x(j))^2}. \quad (3.39)$$

Note that the Rayleigh quotient (3.39) already incorporates the fact that $\sum_{j=1}^{N} x(j) = 0$. In other words, if a term $\alpha\mathbf{1}$ is added to $\mathbf{x}$ it does not change the resulting objective function. Choose $\mathbf{x}$ such that $x(k) = 1$ for some k and $x(i) = 0$ for $i \ne k$. Then we have that

$$\sum_{i,j,i<j} a_{ij}(x(i) - x(j))^2 = \sum_{i} a_{ki} = d(k)$$

and

$$\sum_{i=1}^{N} \sum_{j=1}^{N} (x(i) - x(j))^2 = 2(N-1),$$

because the summation has $2N$ terms of which two are zero (when we have $i = j$) and all the others are equal to 1. Then:

$$\lambda_2 \le \frac{2Nd_{\min}}{2(N-1)} = \frac{N}{N-1}d_{\min}$$

and similarly

$$\lambda_N \ge \frac{N}{N-1}d_{\max}.$$

$\square$

Complement graph   The complement (Definition 2.2) of an unweighted undirected graph has adjacency matrix

$$\mathbf{A}^c = \mathbf{J} - \mathbf{I} - \mathbf{A}$$

and combinatorial Laplacian

$$\mathbf{L}^c = \mathbf{D}^c - \mathbf{A}^c = (N-1)\mathbf{I} - \mathbf{D} - \mathbf{J} + \mathbf{I} + \mathbf{A} = N\mathbf{I} - \mathbf{J} - \mathbf{L}.$$

By construction, $\mathbf{L}$ and $\mathbf{L}^c$ have the same lowest-frequency eigenvector $\mathbf{1}$. Then, denoting by $\mathbf{u}_i$ the ith eigenvector of $\mathbf{L}$, for $i > 1$ we have that $\mathbf{J}\mathbf{u}_i = 0$ since $\mathbf{u}_i^T\mathbf{1} = 0$ and all the rows of $\mathbf{J}$ are equal to $\mathbf{1}^T$. From this it follows that

$$\mathbf{L}^c\mathbf{u}_i = (N - \lambda_i)\mathbf{u}_i,$$

which shows that $\mathbf{L}$ and $\mathbf{L}^c$ share the same eigenvectors, but their order (except for $\mathbf{u}_1 = \mathbf{1}$) is reversed. Denoting by λ_i^c the ith eigenvalue of $\mathbf{L}^c$ we have that $\lambda_{N-i+2}^c = N - \lambda_i$ for any $i > 1$, and λ_{N-i+2}^c and λ_i have eigenvector $\mathbf{u}_i$. The matrix $\mathbf{L}^c$ is positive semidefinite and all its eigenvalues must be greater than or equal to zero, so that

$$\lambda_N \le N.$$

This result can be interpreted as follows. Since $\lambda_N^c = \lambda_2$, we see that the lowest-frequency (non-constant) eigenvector of a graph is the same as the highest-frequency eigenvector of its complement.

Example 3.6   Compute the bounds of Proposition 3.3 and Proposition 3.5 for the graph Laplacians of the following undirected graphs with N nodes: (i) a complete unweighted graph, (ii) an unweighted star graph, (iii) a normalized Laplacian, (iv) a K-regular graph.

Solution
(i) **Complete graphs**   In this case we have $d_{\min} = d_{\max} = N - 1$, leading to

$$N \le \lambda_N \le 2(N-1), \quad \lambda_2 \le N.$$

(ii) **Star graph**   Here $d_{\min} = 1$ and $d_{\max} = N - 1$, so that

$$N \le \lambda_N \le 2(N-1), \quad \lambda_2 \le \frac{N}{N-1}.$$

(iii) **Symmetric normalized Laplacian** With this normalization all nodes have degree 1, so that $d_{\min} = d_{\max} = 1$ and we have

$$\frac{N}{N-1} \le \lambda_N \le 2, \quad \lambda_2 \le \frac{N}{N-1},$$

where $\lambda_N = 2$ if and only if the graph is bipartite.

(iv) **K-regular graph** Here again $d_{\min} = d_{\max} = K$, so that

$$\frac{N.K}{N-1} \le \lambda_N \le 2K, \quad \lambda_2 \le \frac{NK}{N-1},$$

and in the unweighted case we have that $\lambda_N \le \min(2K, N)$.

3.5.5 Nodal Domain Theorems

In Section 3.5.4 we presented properties and bounds linking the eigenvalues of graph operators to characteristics of the corresponding graphs (e.g., the maximum degree as in Proposition 3.3). Since eigenvalues are related to variation properties of the corresponding eigenvectors, this provides some insights about how these eigenvectors capture different graph signal frequency information. In this section we introduce nodal domains, which help to reinforce the frequency interpretation of the eigenvectors of the graph Laplacian. Strong nodal domains are defined as follows.

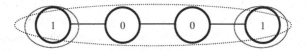

Figure 3.7 Comparison of weak and strong nodal domains. For this graph and signal there is a single weak nodal domain (the ellipse reaching across the diagram), and there are two strong nodal domains (corresponding to the two nodes at the ends of the diagram).

DEFINITION 3.7 (STRONG NODAL DOMAIN) A positive (negative) strong nodal domain of $\mathbf{x} \in \mathbb{R}^N$ in a graph G is a maximally connected induced subgraph $G_1 \subset G$ such that $x(i) > 0$ (respectively $x(i) < 0$), $\forall i \in G_1$. The quantity $ND_s(\mathbf{x})$ is the total number of strong nodal domains of $\mathbf{x}$ (for both the negative and positive cases).

Nodal domains are constructed from *maximal* subgraphs. That is, for any $i \in G_1$, where G_1 is a strong positive nodal domain associated with $\mathbf{x}$, we have that $x(i) > 0$ *and* any node $j \notin G_1$ connected to i must be such that $x(j) \le 0$: all neighbors of i that are not in G_1 must have negative values.

Nodal domains are properties of *signals* on a graph. Thus, different signals will have different nodal domains on a given graph. For example, $\mathbf{x} = \mathbf{1}$ has all positive entries and therefore exactly one strong nodal domain and no negative nodal domains, for any graph. But, since $\mathbf{u}_1 = \mathbf{1}$ and all remaining eigenvectors $\mathbf{u}_i$ of $\mathbf{L}$ are orthogonal to $\mathbf{1}$, any

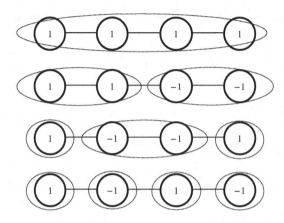

Figure 3.8 Nodal domains for the Hadamard basis vectors $\mathbf{u}_1, \mathbf{u}_2, \mathbf{u}_3, \mathbf{u}_4$, from top to bottom.

$\mathbf{u}_i$, $i > 1$, will have both positive and negative values (since $\mathbf{1}^T\mathbf{u}_i = 0$) and there will be at least two nodal domains. We next introduce weak nodal domains.

DEFINITION 3.8 (WEAK NODAL DOMAIN) A positive (negative) weak nodal domain of $\mathbf{x} \in \mathbb{R}^N$ in a graph G is a maximally connected induced subgraph $G_1 \subset G$ such that $x(i) \geq 0$ (respectively $x(i) \leq 0$), $\forall i \in G_1$. The quantity $ND_w(\mathbf{x})$ is the total number of weak nodal domains of $\mathbf{x}$ (for both the negative and positive cases).

For any $\mathbf{x}$, we have that $ND_w(\mathbf{x}) \leq ND_s(\mathbf{x})$. This is illustrated in the example in Figure 3.7, where we can see that there is only one weak nodal domain, since $x(i) \geq 0$ for all i, but two strong nodal domains, as zero-valued nodes cannot be included in a strong nodal domain. From this it is also easy to see that if $x(i) \neq 0$ for all nodes i then $ND_w(\mathbf{x}) = ND_s(\mathbf{x})$. The quantities $ND_w(\mathbf{x})$ and $ND_s(\mathbf{x})$ are useful for characterizing graph signal frequency. This is illustrated by the following example.

Example 3.7   Figure 3.8 depicts a four-node line graph and four signals $\mathbf{x} = \mathbf{u}_1, \mathbf{u}_2, \mathbf{u}_3, \mathbf{u}_4$ each corresponding to a Hadamard basis vector. Find $ND_w(\mathbf{x})$ and $ND_s(\mathbf{x})$ for $\mathbf{x} = \mathbf{u}_1, \mathbf{u}_2, \mathbf{u}_3, \mathbf{u}_4$ and discuss how they are related to the frequencies of these four signals.

Solution
Since there are no zero values, for any $\mathbf{x}$ we have $ND_w(\mathbf{x}) = ND_s(\mathbf{x})$, and, by inspection,

$$ND_s(\mathbf{u}_1) = 1, \quad ND_s(\mathbf{u}_2) = 2, \quad ND_s(\mathbf{u}_3) = 3, \quad ND_s(\mathbf{u}_4) = 4.$$

We can see that the total number of strong nodal domains increases monotonically with the basis index, from one for $\mathbf{u}_1 = [1, 1, 1, 1]^T$ to four for $\mathbf{u}_4 = [1, -1, 1, -1]^T$. In this case the total number of nodal domains is equal to the number of zero crossings for each of these 1D signals, and therefore increased frequency is linked with increased $ND_s(\mathbf{u}_1)$.

The following theorem formalizes the intuition of the previous example by establishing a link between the number of nodal domains for an eigenvector $\mathbf{u}_k$ and its corresponding frequency λ_k.

> THEOREM 3.6   (NODAL DOMAINS) For a connected graph with eigenvector $\mathbf{u}_k$ corresponding to λ_k with multiplicity r, we have
>
> $$ND_{\mathrm{w}}(\mathbf{u}_k) \leq k,$$
>
> $$ND_{\mathrm{s}}(\mathbf{u}_k) \leq k + r - 1,$$
>
> where $ND_{\mathrm{w}}(\mathbf{u}_k)$ and $ND_{\mathrm{s}}(\mathbf{u}_k)$ denote the total (positive or negative) number of nodal domains associated with $\mathbf{u}_k$.

Note that, from Theorem 3.6, as k increases the number of nodal domains increases, but there is no guarantee that two different eigenvectors will have a different number of nodal domains. Also, for any connected graph and for the combinatorial graph Laplacian $\mathbf{L}$ we have that $\mathbf{u}_1 = \mathbf{1}$ with multiplicity 1, and $ND_{\mathrm{w}}(\mathbf{u}_1) = ND_{\mathrm{s}}(\mathbf{u}_1) = 1$, as noted earlier. Furthermore, $\mathbf{u}_2$ is such that $\mathbf{u}_2^\top \mathbf{u}_1 = 0$, so that $\mathbf{u}_2$ has both positive and negative entries and from Theorem 3.6 it follows that $ND_{\mathrm{w}}(\mathbf{u}_2) = 2$.

In the following examples we consider only connected graphs.

Example 3.8   Line graph For a line graph all the eigenvalues are simple and from Theorem 3.6 we have $ND_{\mathrm{w}}(\mathbf{u}_k) = ND_{\mathrm{s}}(\mathbf{u}_k) = k$. This indicates that the sign pattern of the corresponding eigenvectors will be the same as that of the Hadamard basis functions of Figure 3.8.

Example 3.9   Bipartite graph In a connected bipartite graph, recall that $\mathbf{u}_N$ can be obtained by assigning a positive $\mathbf{u}_1$ by simply flipping the sign of all entries corresponding to $\mathcal{V}_1$ while leaving unchanged those corresponding to $\mathcal{V}_2$. Thus every node in the graph will be a nodal domain for $\mathbf{u}_N$ so that $ND_{\mathrm{w}}(\mathbf{u}_N) = ND_{\mathrm{s}}(\mathbf{u}_N) = N$. This is because we assign the same sign to all nodes in $\mathcal{V}_1$, but nodes in $\mathcal{V}_1$ are by definition not connected to nodes in $\mathcal{V}_2$.

Chapter at a Glance

In this chapter we have shown that graph signals can be represented in terms of elementary graph signals, the eigenvectors of a graph operator $\mathbf{Z}$ which together form the GFT (Section 3.1). For a given graph all graph operators $\mathbf{Z}$ are related but have different properties in terms of graph signal variation (Section 3.2). A GFT can be used to analyze existing graph signals and to design filters with desirable frequency responses. We showed that there is a direct correspondence between node domain and frequency domain filtering. The choice of $\mathbf{Z}$ is usually application dependent. A key factor in this choice is whether a normalization is needed. These differences were discussed in Section 3.4 and a summary of the most common operators is given in Table 3.2. Finally, exact knowledge of the graph frequencies is only possible for small graphs, and we explored various results that link graph properties to graph frequencies (Section 3.5).

Table 3.2 Summary of matrix representations

$\mathbf{L} = \mathbf{D} - \mathbf{A}$	$\lambda_1 = 0$	$\mathbf{u}_1 = \mathbf{1}$	$\lambda_N \le d_{\max}$
$\mathcal{L} = \mathbf{D}^{-1/2}\mathbf{L}\mathbf{D}^{-1/2}$	$\lambda'_1 = 0$	$\mathbf{u}'_1 = \mathbf{D}^{1/2}\mathbf{1}$	$\lambda'_N \le 2$
$\mathcal{T} = \mathbf{I} - \mathbf{D}^{-1}\mathbf{A}$	$\lambda''_1 = 0$	$\mathbf{u}''_1 = \mathbf{1}$	$\mathbf{u}''_i = \mathbf{D}^{-1/2}\mathbf{u}'_i$

The operators $\mathcal{T}$ and $\mathcal{L}$ have the same eigenvalues and their eigenvectors are related by $\mathbf{u}''_i = \mathbf{D}^{1/2}\mathbf{u}''$. After normalization the eigenvalues of $\mathcal{L}$ will be in the interval $[0, 2]$, and $\lambda_N = 2$ if and only if the graph is bipartite. With the frequency representation of Remark 3.4 we can provide a complete picture of the analogy between operations in graph signal processing and in conventional digital signal processing (see Table 3.3).

Further Reading

This chapter has made use of basic results in spectral graph theory. The interested reader can explore further in references such as [31, 8, 7, 38]. Our discussion of bounds follows [8], where many additional results can be found. Spectral graph theory is primarily concerned with understanding the link between graph frequencies and the structural properties of graphs. Instead, we focus on using graph frequency spectra to process graph signals.

Table 3.3 Comparison between DSP and GSP. $p(\lambda)$ is a polynomial in the variable λ with the same coefficients as the polynomial $p(\mathbf{Z})$. The commutativity property $H(z)z = zH(z)$ is equivalent to shift invariance. Note that in this case we assume that $\mathbf{Z}$ is diagonalizable. If $\mathbf{Z}$ is symmetric and there exists a set of orthogonal eigenvectors for $\mathbf{Z}$ then the GFT will be such that $\mathbf{U}\mathbf{U}^\mathsf{T} = \mathbf{I}$ and frequency domain filtering would be written as $\mathbf{y} = \mathbf{U}p(\mathbf{\Lambda})\mathbf{U}^\mathsf{T}\mathbf{x}$.

	DSP	GSP
Fundamental operator	z	$\mathbf{Z}$
Filters	$H(z)$	$p(\mathbf{Z})$
Commutativity	$H(z)z = zH(z)$	$p(\mathbf{Z})\mathbf{Z} = \mathbf{Z}p(\mathbf{Z})$
Transform	DFT $\mathbf{U}^H\mathbf{U} = \mathbf{I}$	GFT $\mathbf{U}^{-1}\mathbf{U} = \mathbf{I}$
Frequency spectrum	$\tilde{\mathbf{x}} = \mathbf{U}^H\mathbf{x}$	$\tilde{\mathbf{x}} = \mathbf{U}^{-1}\mathbf{x}$
Frequency domain filter	$\mathbf{H} = \mathrm{diag}(H(e^{j2\pi k/N}))$	$p(\mathbf{\Lambda}) = \mathrm{diag}(p(\lambda_k))$
Frequency domain filtering	$\mathbf{y} = \mathbf{U}\mathbf{H}\mathbf{U}^H\mathbf{x}$	$\mathbf{y} = \mathbf{U}p(\mathbf{\Lambda})\mathbf{U}^{-1}\mathbf{x}$

GraSP Examples

Several code examples using GraSP are given in Section B.4. First, methods for GFT computation and visualization are considered in Section B.4.1. Visualization techniques, based on the methods of [28], are very important for understanding the properties of graph frequencies, and how they differ from the frequencies of conventional signals. In Section B.4.2 the problem of graph filter design and implementation is revisited from a spectral perspective.

Problems

3.1 *Graph edge weight perturbations*
Consider two weighted graphs G and G' defined over the same set of nodes. Their adjacency matrices are $\mathbf{A}$ and $\mathbf{A}'$, respectively, and we have that

$$\mathbf{A}' = \alpha\mathbf{A},$$

where $\alpha > 0$ is a scalar. We also denote by $\mathbf{D}$ and $\mathbf{D}'$ the respective degree matrices, and by $\mathbf{L}$ and $\mathbf{L}'$ the corresponding graph Laplacians.
a. Express $\mathbf{D}'$ and $\mathbf{L}'$ as functions of $\mathbf{D}$ and $\mathbf{L}$, respectively.
b. Let $\mathbf{u}_k$ and λ_k be the kth eigenvalue and eigenvector of G. Find the eigenvector and eigenvalue $\mathbf{u}'_k$ and λ'_k corresponding to G'.
c. Prove that $\mathcal{L} = \mathcal{L}'$, where $\mathcal{L}$ and $\mathcal{L}'$ are the respective symmetric normalized Laplacians.
d. Assume we have used a spectral method to find the minimum cuts (see the text around Definition 2.11) of G and G'. Would the solutions be the same or different for the two graphs? Would the costs of the solutions be the same or different? Discuss whether your answer depends on the value of α, and if so how.

3.2 *Single edge removal*
Consider a weighted undirected graph G, where $w_{ij} \geq 0$ is the edge weight between nodes i and j. Assume that for a given pair of nodes, k and l, we have $w_{kl} > 0$.
a. Construct a new graph G' with weights w'_{ij} that differs from G in only one edge $w'_{kl} = 0$ and where all other edge weights are unchanged. Denoting by $\mathbf{L}'$ the Laplacian of G' prove that, for any graph signal $\mathbf{x}$,

$$\mathbf{x}^\mathsf{T}\mathbf{L}'\mathbf{x} \leq \mathbf{x}^\mathsf{T}\mathbf{L}\mathbf{x}.$$

b. In each of the following two separate cases, describe how you would select the best edge to remove, w_{mn}, so that the resulting graph G'' provides:
  • for a given signal, $\mathbf{y}$, a maximum decrease in $\mathbf{y}^\mathsf{T}\mathbf{L}''\mathbf{y}$ with respect to $\mathbf{y}^\mathsf{T}\mathbf{L}\mathbf{y}$;
  • for a set of signals, $\mathbf{y}_1, \mathbf{y}_2, \ldots, \mathbf{y}_K$, a maximum decrease in $\sum_{i=1}^{K} \mathbf{y}_i^\mathsf{T}\mathbf{L}''\mathbf{y}_i$ with respect to $\sum_{i=1}^{K} \mathbf{y}_i^\mathsf{T}\mathbf{L}\mathbf{y}_i$.

3.3 *Ideal filterbanks*
For an undirected graph with fundamental graph operator $\mathbf{Z}$, let $\mathbf{H}_0$ and $\mathbf{H}_1$ be ideal low pass and high pass filters, respectively, as defined in Box 3.1. Let $\mathbf{x}$ be a graph signal and define $\mathbf{x} = \mathbf{H}_0\mathbf{x}$ and $\mathbf{x}_1 = \mathbf{H}_1\mathbf{x}$.

a. Prove that the mapping from $\mathbf{x}$ to $\mathbf{x}_1$ is an orthogonal projection. Prove that it is possible to recover $\mathbf{x}$ from $\mathbf{x}_0$ and $\mathbf{x}_1$.

b. Find the minimum degree of an exact representation of $\mathbf{H}_0$ as a polynomial of $\mathbf{Z}$.

3.4 *Frequency domain filtering*

Let G be a weighted undirected graph with Laplacian matrix $\mathbf{L}$, GFT $\mathbf{U}$ and diagonal eigenvalue matrix $\boldsymbol{\Lambda}$, with eigenvalues $\lambda_1 \leq \lambda_2 \leq \cdots \leq \lambda_N$. Write the following spectral domain filters, $h_0(\lambda)$ and $h_1(\lambda)$, in terms of $\mathbf{L}$, i.e., $h_0(\mathbf{L})$ and $h_1(\mathbf{L})$, and provide an interpretation describing the type of filters (low pass, high pass, etc.):

a. $h_0(\lambda) = 1$;

b. $h_1(\lambda) = (\lambda - \lambda_N)^2$.

3.5 *Bipartite graphs*

Let

$$\mathbf{L} = \begin{bmatrix} \mathbf{D}_1 & -\mathbf{A}_0 \\ -\mathbf{A}_0^\mathsf{T} & \mathbf{D}_2 \end{bmatrix}$$

be the Laplacian matrix of a bipartite graph, with two sets of nodes $\mathcal{V}_1$ and $\mathcal{V}_2$ with sizes N_1 and N_2.

a. Find the number of strong and weak nodal domains for the following two signals:

$$\mathbf{x} = \begin{bmatrix} \mathbf{1}_{N_1} \\ \mathbf{0}_{N_2} \end{bmatrix} \quad \text{and} \quad \mathbf{y} = \begin{bmatrix} \mathbf{1}_{N_1} \\ -\mathbf{1}_{N_2} \end{bmatrix}$$

where $\mathbf{1}$ and $\mathbf{0}$ are vectors whose entries are respectively all 1 and all 0, and the subscripts correspond to the lengths of the vectors.

b. Write $\mathcal{L}$, the symmetric normalized version of $\mathbf{L}$, in matrix form. Compute the number of strong and weak nodal domains of $\mathcal{L}\mathbf{x}$ and use the result to justify the kind of filtering achieved by $\mathcal{L}\mathbf{x}$.

3.6 *Gerschgorin circle theorem*

In this problem the goal is to make use of the Gerschgorin circle theorem (Theorem 3.5).

a. Prove that the following graph operators are positive semidefinite (PSD): (i) the combinatorial graph Laplacian ($\mathbf{L} = \mathbf{D} - \mathbf{A}$); (ii) the generalized graph Laplacian $\mathbf{L}' = \mathbf{D} - \mathbf{A} + \mathbf{V}$, where $\mathbf{V}$ is diagonal and contains the positive node weights.

b. Now assume that one of the edge weights is negative. Prove that $\mathbf{L}$ is no longer PSD and propose a choice of node weights $\mathbf{V}$ that can guarantee that $\mathbf{L}'$ will be PSD.

3.7 *Matrix tree theorem*

Let G_n be a complete bipartite unweighted graph, with two sets of nodes $\mathcal{V}_1$ and $\mathcal{V}_2$, where $|\mathcal{V}_1| = 2$ and $|\mathcal{V}_2| = n$, where $n \geq 2$. In such a graph, both nodes in $\mathcal{V}_1$ connect to all nodes in $\mathcal{V}_2$.

a. Find the number of spanning trees, $\tau(G_1)$ and $\tau(G_2)$, and draw the spanning trees in both cases.

b. Prove that in any spanning tree of G_n there is exactly one node in $\mathcal{V}_2$ connected to both nodes in $\mathcal{V}_1$ and all other nodes in $\mathcal{V}_2$ connect to only one node in $\mathcal{V}_1$. Use this to derive $\tau(G_n)$.

3.8 *Nodes with identical neighborhoods*

Let G be an undirected graph with N nodes. Assume that two nodes have exactly the same neighbors and are connected to each other. Without loss of generality let those two nodes be 1 and 2. Denote G' the graph obtained by removing one of those two nodes.

a. Prove that rank($\mathbf{A}$) = rank($\mathbf{A}'$).

b. Let $\mathbf{L}$ be the combinatorial graph Laplacian of G. Prove that $\mathbf{u}_{1,2} = [1, -1, 0, \ldots, 0]^\mathsf{T}$ is an eigenvector of $\mathbf{L}$.

c. Use the previous result to show that, for any other eigenvector $\mathbf{u}_k \neq \mathbf{1}$, we have that $u_k(1) = u_k(2) = 0$.

d. Use these results to suggest how graph signal frequency analysis can be performed after replacing G by G'.

4 Sampling

In some situations all entries of a graph signal may not be available. This may happen because observation of all graph nodes is not possible or is too costly. For example, in a sensor network, the overall energy consumption depends on how many sensors collect and transmit information. Thus, activating only a subset of sensors at a given time can lower energy consumption and increase system lifetime.

In *graph signal sampling*, values (samples) obtained from a subset of nodes are used to infer the values at nodes that were not observed. To do so, we need to consider several questions, including: (i) which nodes are the most informative?; (ii) how to estimate unseen values at nodes that were not measured?; (iii) what are the conditions to recover exactly the information at the unobserved nodes?

If a graph has N nodes, it is clear that we cannot expect to recover an *arbitrary* signal from only $K < N$ observations. Instead, we need to define a class of signals for which recovery is possible, given a sufficient number of samples. We introduce such signal models in Section 4.1. We discuss what *constraints* may apply to various sampling scenarios (Section 4.2). Next we formalize the sampling problem (Section 4.3) and provide an overview of sampling set selection algorithms proposed in the literature (Section 4.4). We conclude the chapter with a discussion of the main insights (Section 4.5).

4.1 Graph Signal Models

In what follows a *model* is a description of a class of graph signals. As we shall see next, this description could be first expressed qualitatively (e.g., smooth graph signals), before being written mathematically (e.g., bandlimited signals for a given graph operator and corresponding GFT). Models allow us to formalize our assumptions about the graph signals we expect to observe. The design of a sampling system (sampling set selection and signal interpolation) and the analysis of its performance (the error estimates as a function of the number of nodes in the sampling set) will depend on the choice of model.

4.1.1 Challenges in Graph Signal Model Selection

Model justification   Several factors make model selection for graph signals particularly challenging. In the case of conventional signals, the most widely used model, that of bandlimited signals, is often well motivated by perceptual criteria. For sampling audio signals, given that humans cannot hear audio frequencies above a certain threshold, it makes sense to prefilter analog signals to limit their range to frequencies within the audible range. For graph signals of interest such assumptions may not apply. Instead a more careful analysis of the available data is needed to determine which model is a reasonable approximation to the observed data. Furthermore, the graph choice itself is part of the modeling, as will be discussed in Chapter 6, and so is the choice of graph operator. Thus, in some scenarios it may be necessary to consider jointly the choice of graph *and* graph signal model, as they are both part of the modeling effort for a specific application. Once a graph has been chosen, the choice of a frequency representation (among the various possible alternatives in Section 3.4) also affects the meaning of the elementary frequencies and thus of the corresponding signal model.

Model mismatch   As for conventional signal sampling, we would like to select signal models that provide a good representation of the observed data, while leading to efficient algorithms with performance guarantees. However, in practice we cannot expect an exact match between the observed signals and the mathematical models we use to represent them. In conventional signal processing, where we have theoretical guarantees if analog signals are exactly bandlimited, practical systems prefilter input signals before sampling and take into account the effect of noise on overall performance. Similarly, there may not be an exact match between the signal modeling assumptions and the observed graph signals and in this case there is a need to develop techniques that minimize some error metric, with methods similar to those used for conventional signals [39].

Models of interest   While there is a growing number of models that can be considered, in what follows we describe those for which our understanding is most advanced and/or which have a clearer connection with graph signals of practical interest.

4.1.2   Bandlimited and Approximately Bandlimited Signals

Bandlimited signals   As a direct extension of the most widely used model for sampling conventional signals, for a graph G and operator $\mathbf{Z}$ with corresponding GFT $\mathbf{U}$, a signal, $\mathbf{x}$, is bandlimited if it can be written as

$$\mathbf{x} = \mathbf{U}_R\,\mathbf{a}, \tag{4.1}$$

where $\mathbf{U}_R$ contains the first R columns of $\mathbf{U}$, and $\mathbf{a} \in \mathbb{R}^R$. Signals consistent with this model are *low pass*, or low frequency, since only their low frequency components can be non-zero. There are similar models where only R out of N frequencies can be non-zero, and the indices of these frequencies are known. If the highest-R or a range of R intermediate frequencies are non-zero, these will give rise to *high pass* and *band pass* signal models, respectively. *R-sparse* signals also have R non-zero frequencies, with their indices known[1] but not necessarily consecutive. As long as the subset of

---

[1] In a more general model of R-sparse signals, the support of the non-zero frequencies is not known.

R non-zero frequencies is known *a priori* all these cases are similar and lead to similar sampling algorithms. Thus, in what follows we can focus on low-frequency bandlimited signals without loss of generality.

The properties of bandlimited signals depend on the choice of graph and operator. As an example, since substantially different definitions of frequency arise from the combinatorial and normalized Laplacians (Section 3.4), it follows that for a given graph different subsets of nodes will be sampled depending on the operator chosen. Moreover, it may be reasonable to expect graph signals to be smooth, but the exact level of smoothness may not be known *a priori*, so that in practice R is not known.

Bandlimited signals in noise   Even if the actual signals are not exactly bandlimited, a bandlimited model can still be useful. For example, the observed signals could be noisy versions of bandlimited signals:

$$\mathbf{x} = \mathbf{U}_R \mathbf{a} + \mathbf{n}, \tag{4.2}$$

where $\mathbf{n}$ represents a perturbation or noise vector, so that the error in approximating a signal by using only R frequencies will depend on the noise.

Frequency priors   So far we have assumed exact knowledge of frequency characteristics (e.g., only R frequencies are non-zero). In practice, we may consider signals having most of their energy in the lower frequencies, without being strictly bandlimited [40]. For example, recall from Remark 3.2 that $\mathbf{x}^\mathsf{T}\mathbf{L}\mathbf{x} = \sum \lambda_i |\tilde{x}_i|^2$, a weighted sum of the frequency components of $\mathbf{x}$. The signals $\mathbf{x}$ belonging to a class defined by having small variation $\mathbf{x}^\mathsf{T}\mathbf{L}\mathbf{x}$ will tend to be smooth even if their high frequencies are not necessarily identically zero. In contrast with the bandlimited signal model of (4.1), we may not be able to describe the signal model as a linear subspace with a known set of basis vectors (R columns of $\mathbf{U}$ in (4.1)). Instead, reconstruction of a signal from its samples may require using an $\mathbf{x}^\mathsf{T}\mathbf{L}\mathbf{x}$ term, which penalizes signals having a higher frequency content without forcing the higher frequencies to be zero. More generally, we may select some filter $\mathbf{W}$ to represent the **prior** (i.e., the assumptions made *a priori*), with the assumption that signals consistent with the model produce smaller outputs when filtered by $\mathbf{W}$, leading to a penalty term $\|\mathbf{W}\mathbf{x}\|$ in the reconstruction algorithm. As an example, the term $\mathbf{x}^\mathsf{T}\mathbf{L}\mathbf{x}$, which penalizes high frequencies, could be replaced by $\|\mathbf{F}\mathbf{x}\|$, where $\mathbf{F}$ is a high pass filter. Both choices will favor the reconstruction of lower-frequency signals but will give different estimated signals. Refer to Section 4.3.5 for a more detailed discussion of reconstruction algorithms.

Integer-valued signals   In some applications the signal of interest might be integer-valued, $\mathbf{x} \in \mathbb{Z}^N$. Examples include: (i) a graph where the nodes represent text documents and vectors of integer-valued word-count data are the corresponding graph signals (one vector of word counts per document), (ii) an online social network where user information is often integer-valued, e.g., age. and (iii) a machine learning application (Section 7.5.3) where for each class there is a *membership function* and the corresponding graph signal is 1 for data points (nodes) belonging to the class and 0 otherwise. Note that smooth integer-valued signals may not be strictly bandlimited, but the integer property can be present along with a bandlimited reconstruction, e.g., by quantizing to the nearest integer each entry of a candidate bandlimited reconstruction signal.

Irregularly sampled continuous space signals   In cases where the graph signal is derived from a continuous-time signal or a space signal (e.g., data measured by a sensor network), it will be important to link the graph-based model to the physical properties in Euclidean space of the signal being observed. In particular, irregular spatial sampling of a smooth, or even bandlimited, signal in continuous space does not lead in general to a bandlimited graph signal. Thus, if a model of the continuous domain signal is available, the selection of graph, graph operator and graph signal model should incorporate knowledge from the continuous domain (see also Section 4.1.4 and Section 7.3).

4.1.3 Piecewise Smooth Signals

Motivation   Piecewise smooth models are popular for conventional signals, such as images, where we encounter smooth regions (e.g., objects) separated by sharp boundaries (image contours). Similarly, a piecewise constant graph signal can be characterized by sets of connected nodes with the same signal value [40]. This can be a good model for the data encountered in machine learning applications (see Section 7.5), where each data point (graph node) is assigned a label, one out of a small set of class labels, and we expect data points that are connected (and thus "similar") to be assigned the same label. Clearly, for this model to be meaningful, these sets of connected nodes, or "regions," have to be sufficiently large, otherwise any discrete graph signal (with one value per node) would be trivially piecewise constant. The main challenge in developing piecewise smooth models, then, is to quantify the size of the connected patches over which signals are expected to be smooth. An intuition of how this can be done is developed in the following box.

Box 4.1   Piecewise constant image models

Consider a conventional digital image, which can be viewed (see also Section 7.4) as a graph where all nodes (pixels) are connected to at least four neighbors. A constant-valued region in an image can be viewed from the graph perspective as a set of pixels (nodes) divided into two classes: (i) *interior* nodes, such that all four neighbors have the same signal values and (ii) *boundary* nodes, for which at least one of the neighbors has a different signal value. Thus, for a piecewise constant model to be suitable for a set of images, constant-valued regions should contain a large number of nodes. In turn, the ratio between the interior and boundary points can provide insights about a region's shape.

Piecewise constant graph signals   Images can viewed as regular unweighted graphs, which makes it easy to introduce the concept of piecewise constant signals as described

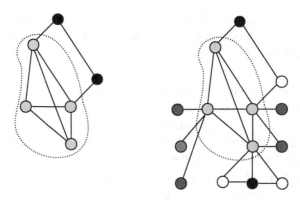

Figure 4.1 Example of piecewise constant regions. In both left and right examples we have the same four nodes with identical values. However, on the left side there are only two connections to other nodes outside of that group. In contrast, on the right side there are 11 connections to other nodes outside of the group.

in Box 4.1. Extending this idea to irregular graphs is not straightforward. Nodes that are "strictly" interior (i.e., with the node and all its one-hop neighbors having the same signal value) may be rare, especially if the nodes have high degree.

To illustrate this point, consider the example of Figure 4.1, where there are four nodes with the same signal value. These four nodes form a **clique**, so that each node connects with three neighbors with the same signal value. On the left in Figure 4.1 there are only a few connections outside the clique, so that declaring that subset of nodes as a piecewise constant region seems reasonable. But on the right of Figure 4.1, the number of connections within the clique is six, while there are nearly twice as many connections to other nodes, so that a piecewise constant model may not be a good choice.

To address this challenge, the authors of [41] propose a definition based on a difference operator over all edges. Denoting by $\mathbf{B}$ the signed incidence matrix (see Definition 2.16), we have that each entry of the vector

$$\mathbf{y} = \mathbf{B}^\mathsf{T}\mathbf{x} \tag{4.3}$$

is the difference in graph signal values across one of the edges in the graph. Then, a signal $\mathbf{x}$ is said to be piecewise constant if $\mathbf{y}$ has small ℓ_0 norm, $\|\mathbf{y}\|_0$. Since the ℓ_0 norm of a vector is equal to the total number of non-zero entries in the vector, if $\|\mathbf{y}\|_0$ is small then only a small subset of edges connect nodes having different values. This is illustrated in the following example.

Example 4.1   Let G be a graph with two subgraphs G_1 and G_2 with N_1 and N_2 nodes, respectively. Consider a signal $\mathbf{x}$ such that the nodes in G_1 and G_2 have signal values x_1 and x_2, respectively, with $x_1 \neq x_2$. Define $\mathbf{y}$ as in (4.3) and let the number of edges between G_1 and G_2 be $|\mathcal{E}_{12}|$, while the numbers of edges within G_1 and G_2 are $|\mathcal{E}_{11}|$ and

$|\mathcal{E}_{22}|$, respectively. Find $\|\mathbf{y}\|_0$ and discuss whether a piecewise constant model is suitable for $\mathbf{x}$, for different values of $|\mathcal{E}_{11}|$, $|\mathcal{E}_{22}|$ and $|\mathcal{E}_{12}|$.

Solution
We can easily see that $\mathbf{y}$ has non-zero values only for connections between G_1 and G_2. Since there are $|\mathcal{E}_{12}|$ such connections we have $\|\mathbf{y}\|_0 = |\mathcal{E}_{12}|$. Since $\mathbf{y}$ has dimension $E = |\mathcal{E}_{11}| + |\mathcal{E}_{22}| + |\mathcal{E}_{12}|$, the total number of edges, we can consider a piecewise constant model to be a reasonable choice if $\|\mathbf{y}\|_0$ is small relative to the dimension of $\mathbf{y}$. Using the criterion in [41], we would have a piecewise constant signal if $|\mathcal{E}_{11}| + |\mathcal{E}_{22}| \gg |\mathcal{E}_{12}|$, which indeed would correspond to small $\|\mathbf{y}\|_0$.

Criteria to decide whether a signal is piecewise constant can be used to classify observed signals and also to generate synthetic piecewise constant signals. This can be done by first identifying subgraphs with desirable properties (e.g., such that cuts between subgraphs are small) and then associating constant (or smooth) signals with each subgraph (see [41]).

Piecewise smooth signals   Note that these ideas can be extended to define piecewise smooth signals. As a first step multiple "pieces" corresponding to smooth regions can be identified, a vector $\mathbf{x}$ can be used as an indicator for each of these regions and $\mathbf{y}$ can be computed to determine whether the indicator functions are truly piecewise constant, using the ℓ_0 norm. Each piece can be declared to be smooth if it is well modeled by an exactly localized smooth basis (Section 5.1.2) or approximated by a basis with strong energy concentration in each of the smooth regions (Section 5.1.3).

4.1.4 Statistical Models

We start with a motivating example, where we compare two models introduced in Section 4.1.2, with expressions given by (4.1) and (4.2).

Example 4.2   Find the vector $\mathbf{a}$ in the signal model (4.1) under the following assumptions: (i) there is no sampling (i.e., the whole signal $\mathbf{x}$ can be observed) and (ii) $\mathbf{U}$ is orthogonal.

Solution
Under the model of (4.1), signals that are exactly bandlimited form a vector space, i.e., a subspace of $\mathbb{R}^N$ of dimension R (see Section A.1). The solution is straightforward:

$$\mathbf{a} = \mathbf{U}_R^\mathsf{T}\mathbf{x},$$

which follows directly from the fact that the columns of $\mathbf{U}_R$ are orthogonal, so that $\mathbf{U}_R^\mathsf{T}\mathbf{U}_R = \mathbf{I}_R$ and $\mathbf{U}_R\,\mathbf{a} = \mathbf{x}$. Moreover under conditions to be described in Section 4.3, it is possible to recover $\mathbf{a}$, assuming we have (at least) R samples of $\mathbf{x}$.

Example 4.3   Repeat for the signal model (4.2).

Solution
In the model (4.2) a "noise" term has been added, but without any additional information this model is essentially useless. If the goal is to recover the noise-free signal $\mathbf{U}_R\mathbf{a}$ we have no way of doing it without making some other assumptions. The least squares solution (Proposition A.4) for this problem is

$$\mathbf{a}_{LS} = (\mathbf{U}_R^\top\mathbf{U}_R)^{-1}\mathbf{U}_R^\top\mathbf{x} = \mathbf{U}_R^\top\mathbf{x},$$

the orthogonal projection of $\mathbf{x}$ onto the columns of $\mathbf{U}_R$. This implies that we are finding the vector in the column space of $\mathbf{U}_R$ that is closest to $\mathbf{x}$, so that $\|\mathbf{n}\|$ is minimized. The error signal $\mathbf{n}$ is in the orthogonal complement of the space spanned by the columns of $\mathbf{U}_R$, that is the span of the last $N - R$ columns of $\mathbf{U}$. Thus, while there is no unique solution, choosing $\mathbf{a}_{LS}$ can be justified by making assumptions about $\mathbf{n}$.

Modeling graph signal likelihood for a given graph   In Example 4.3 we obtained a reconstructed signal $\tilde{\mathbf{x}} = \mathbf{U}_R\mathbf{a}_{LS}$ which led to the minimum $\|\mathbf{n}\| = \|\mathbf{x} - \tilde{\mathbf{x}}\|$. In general, when an observation is affected by a perturbation as in (4.2) there is an infinite number of possible values for $\tilde{\mathbf{x}}$, and any reconstruction algorithm needs to be made on the basis of additional assumptions. These assumptions can be framed using deterministic constraints (e.g., favoring the solution corresponding to the smallest $\|\mathbf{n}\|$) or alternatively, and sometimes equivalently, the problem can be framed by defining statistical graph signal models[2].

In a deterministic setting, a function $f(\cdot)$ can be chosen that associates a scalar penalty with each possible approximation error $\mathbf{n} = \mathbf{x} - \tilde{\mathbf{x}}$, so that solutions with smaller $f(\mathbf{n})$ are favored. In a probabilistic setting, a model is associated with each possible error or to each candidate graph signal $\tilde{\mathbf{x}}$ and a penalty is derived from this model, so that high-probability (high-likelihood) signals are given a lower penalty. In some cases the solution for a given deterministic penalty corresponds to a maximum likelihood solution. For example, the solution of Example 4.3 is optimal under the assumption that $\mathbf{n}$ can be modeled by a multivariate Gaussian distribution with identity covariance.

Following the rationale of our discussion in Box 3.3, deterministic penalties or statistical models should take into account the graph structure. A popular choice for undirected graphs is provided by Gauss–Markov random fields (GMRFs), which compute the likelihood of a signal $\mathbf{x}$ as a function of $\exp(-\mathbf{x}^\top\mathbf{L}\mathbf{x})$, where $\mathbf{x}^\top\mathbf{L}\mathbf{x} = \sum \lambda_i|\tilde{x}_i|^2$ (Remark 3.2). Notice that this leads to a different interpretation of the frequency priors from that discussed in the section after (4.2). Thus, a GMRF model assigns higher likelihood to signals with less variation, corresponding to signals where most of the energy is concentrated in the lower frequencies (i.e., smaller λ_i). See the model (6.20) and related discussions for more details.

---

[2] Note that this book assumes no prior knowledge of probability theory and thus the emphasis is on showing how probabilistic models can be used.

Estimating a model from data   In this section we have described several graph signal models that can be used for sampling and in each case the graph was assumed to be known. In Section 6.3 we will also study the problem of finding a graph that matches a set of observed signals. These techniques allow us to identify a graph corresponding to a GMRF with parameters such that the likelihood of the observed signals is maximized. Thus, solving this problem leads to a graph that matches the observed signal properties. Since in this formulation the starting point is the signals (and no graph is defined) it may seem unclear why a graph is useful for sampling. But, using a graph operator (e.g., graph Laplacian) to approximate the inverse of the covariance matrix leads to a computationally efficient penalty term if the graph is sparse (for example using Remark 3.1).

4.1.5   Graph Signal Stationarity

In the previous section, we sketched two scenarios: (i) a graph is given and the graph signal likelihood or penalty is derived from it and (ii) no graph is given but one is learned from example graph signals (see Section 6.3). We now consider a situation where *both* the graph and graph signals are given. In this setting we discuss the concept of graph signal stationarity, initially introduced in [42]. We start by providing some intuition about the stationarity of time domain signals.

Box 4.2   Stationarity for time domain signals

Weak (or wide-sense) stationarity of a discrete-time random process $x(n)$ holds if the following conditions, on the first and second moments, respectively, are both true: (i) the expected value of the signal at time n, $\mathbb{E}[x(n)]$, is the same as its expected value at time $n + m$, i.e., $\mathbb{E}[x(n)] = \mathbb{E}[x(n + m)] = \mu$ for any $m \in \mathbb{Z}$ and (ii) the covariance of $x(n)$ and $x(n + m)$ depends only on the "distance" between the samples, m. See [17, Section 3.8].

Notice that (i) can be interpreted as stating that the means are not affected by shifting, i.e., replacing the original signal $x(n)$, which has an infinite number of samples, by another signal, $x(n+m)$, with the same values but shifted in time, does not change the mean and covariance. Thus, using the z-transform notation of Box 2.4, the samples of $X(z)$ and $Y(z) = z^{-m}X(z)$ have the same expected values.

Invariance under a shift has important implications: given a filter $H(z) = \sum_i h_i z^{-i}$ and denoting $Y(z) = H(z)X(z)$, if $X(z)$ is wide-sense stationary then $\mathbb{E}[y(n)] = \sum_i h_i \mu$, where $\mathbb{E}[x(n)] = \mu$.

An observation from Box 4.2 is that, for time signals, m represents *both* a "distance" between points in time *and* a signal shift. A major challenge in extending the concept of stationarity to graph signals is that the fundamental operator $\mathbf{Z}$ is different from a time shift. Thus, in the graph case, in general there is no correspondence between observing signal $\mathbf{x}$ at a node j and observing $\mathbf{Zx}$ at some other node i. While in the time domain the shift operation is a simple translation, in the graph domain multiplying by $\mathbf{Z}$ combines data from multiple nodes (see Box 2.4).

Empirical averages for graph signals   From Box 4.2, time signals that are stationary in time can be informally described as having some properties that are the same at any time, i.e., are invariant under shifting. We start by discussing what it would mean for a graph signal (or a series of graph signals) to be "the same" in some sense at every node.

Loosely speaking we would like to determine whether the observed graph signals have the same local properties everywhere on the graph. Assume we have multiple graph signal observations for a given graph. Since condition (i) in Box 4.2 involves the expected value, we start by computing the empirical average at each node. Denoting by x_k the kth observed signal (out of K), we compute the average observation at node i, $\mu_{x,i}$:

$$\mu_{x,i} = \frac{1}{K} \sum_{k=1}^{K} x_k(i). \tag{4.4}$$

If $\mu_{x,i}$ is the same for all i, this will correspond to an intuitive notion of stationarity[3] (i.e., all nodes have similar behavior). But, even in this case it is not easy to link this result to the effect of a shift (such as m in Box 4.2). First consider the following example.

Example 4.4   Let $\mu_{x,i}$ and $\mu_{x,j}$, computed as (4.4), be the empirical means of the values at nodes i and j, respectively and let $\mathbf{Z}$ be the fundamental graph operator. Denote by $y_k = \mathbf{Z}x_k$ a signal multiplied by the elementary operator. We will use this to show that the interpretation of $\mathbf{Z}$ as a "shift" is not straightforward (see also Box 2.4). From these y_k we can define $\mu_{y,i}$ and $\mu_{y,j}$ using (4.4). Assuming that $\mu_{x,i} = \mu_{x,j} = \mu_x, \forall i, j$, under what conditions is it possible for each of the following conditions to hold: (i) $\mu_{y,i} = \mu_{y,j} = \mu_y, \forall i, j$, (ii) $\mu_{y,i} = \mu_{y,j} = \mu_x, \forall i, j$?

Solution
We can write μ_x and μ_y in vector form by defining the $N \times K$ matrix of inputs $\mathbf{X}$, where each column is one of the input graph signals x_k, and then defining $\mathbf{Y} = \mathbf{Z}\mathbf{X}$:

$$\mu_x = \frac{1}{K}\mathbf{X}\mathbf{1}_K, \quad \mu_y = \frac{1}{K}\mathbf{Y}\mathbf{1}_K,$$

where $\mathbf{1}_K$ is the vector of dimension K whose entries are all one. Condition (i) corresponds to $\mu_y = \mu_y \mathbf{1}_N$, while condition (ii) is $\mu_x = \mu_y = \mu_x \mathbf{1}_N$. Using the definition of $\mathbf{Y}$,

$$\mu_y = \frac{1}{K}\mathbf{Y}\mathbf{1}_K = \frac{1}{K}\mathbf{Z}\mathbf{X}\mathbf{1}_K = \mathbf{Z}\mu_x.$$

From this we can see that as long as $\mathbf{Z}\mathbf{1}_N = \mathbf{1}_N$ the two conditions can be met. This could be achieved by selecting $\mathbf{Z} = Q = \mathbf{D}^{-1}\mathbf{A}$.

In the previous example, we can see that averaging and then multiplying the average

---

[3] Note that we are being informal in this discussion. In practice we cannot expect the $\mu_{x,i}$ to be exactly equal. Mathematically sound methods to test for stationarity are beyond the scope of our discussion.

by the graph operator is the same, in some cases, as multiplying by the operator and then averaging, that is:

$$\mu_y = \frac{1}{K}(QX)1_K = Q\mu_x.$$

Note that in this example we consider only the mean values at each node, i.e., μ_x and μ_y, and the result is valid because $\mu_x = \mu_x 1$ and 1 is an eigenvector of the selected $Z = Q$. This may not be true for other graph operators.

A more challenging case arises when we consider the effect of multiplication by Z on the second-moment properties of Box 4.2. Assume that $\mu_x = \frac{1}{K}X1 = \mu_x 1$ and define the empirical covariance as

$$S_x = \frac{1}{K}\sum_k (x_k - \mu_x 1)(x_k - \mu_x 1)^\mathsf{T} = \frac{1}{K}XX^\mathsf{T} - \mu_x^2 11^\mathsf{T} = R_x - \mu_x^2 11^\mathsf{T}, \qquad (4.5)$$

so that we can focus on the correlation matrix R_x and how it is related to R_y:

$$R_y = \frac{1}{K}YY^\mathsf{T} = \frac{1}{K}ZXX^\mathsf{T}Z^\mathsf{T} = ZR_xZ^\mathsf{T}.$$

Invariance of the second moments under multiplication by Z would require that

$$R_y = ZR_xZ^\mathsf{T} = R_x. \qquad (4.6)$$

For arbitrary Z and R_x this property is unlikely to hold in general. In particular, even if R_x is diagonalized by the eigenvectors of Z, the eigenvalues of Z are not identity in general, and $R_y \neq R_x$ (see the discussion of the work in [43] below).

Isometric graph operator   Choosing the complex isometric operator, $\mathcal{I}$, from [42] as the fundamental graph operator provides a solution. Defining the covariance for complex signals as

$$S_x = \frac{1}{K}XX^\mathsf{H} - \mu_x^2 11^\mathsf{H}$$

and assuming that S_x can be diagonalized by the eigenvectors of $\mathcal{I}$, $U_\mathcal{I}$, i.e., $S_x = U_\mathcal{I}\Lambda_x U_\mathcal{I}^\mathsf{H}$, then we have

$$\mathcal{I}S_x\mathcal{I}^\mathsf{H} = U_\mathcal{I}\mathbf{I}U_\mathcal{I}^\mathsf{H}U_\mathcal{I}\Lambda_x U_\mathcal{I}^\mathsf{H}U_\mathcal{I}\mathbf{I}U_\mathcal{I}^\mathsf{H} = S_x,$$

which fulfills the invariance condition of (4.6) for $Z = \mathcal{I}$.

Alternative definitions of stationarity   From the previous discussion, it is not easy to define a class of signals with properties that are invariant with respect to multiplication by Z (which would correspond to the time domain concept of stationarity in Box 4.2). Therefore, alternative definitions of stationarity have been proposed that focus on the structure of the covariance matrix [43] or on local measurements on the graph [44].

In [43] a *localization* operator is first defined. For any polynomial $p(Z)$, the ith row of $p(Z)$, p_i^T, can be interpreted as a kernel that generates the (local) output at node i, that is, if $y = p(Z)x$ then $y(i) = \langle x, p_i \rangle$. Notice that the local kernels, p_i, are different from node to node and adapt to the local graph structure, but they correspond to the same polynomial $p(\mathbf{Z})$. The second-moment stationarity condition proposed by [43] requires $\mathbf{S}_x$ to be diagonalizable by the eigenvectors of $\mathbf{Z}$. This, as noted earlier, does not guarantee that the condition (4.6) will be met, but, since $\mathbf{S}_x$ can be written as a graph filter, there is invariance under the localization operator: the local kernels (corresponding to local correlation) are obtained from a graph filter (a polynomial in $\mathbf{Z}$).

An alternative approach proposed in [44] involves defining a series of normalized local operators at different scales, $y(i, s) = \langle \mathbf{x}, \mathbf{p}_{i,s} \rangle$, where different scales, parameterized by s, can be obtained by using multiple graph filters $p_s(\mathbf{Z})$ with different polynomial degrees. A class of signals can be said to be stationary if these localized measurements are close to constant across all nodes. While both [43] and [44] consider local operations, the conditions for stationarity in [44] are more relaxed. In particular, [43] imposes a condition on the $N \times N$ matrix $\mathbf{S}_x$, while [44] considers only the properties of S N-dimensional vectors, where S is the number of different scales.

Stationarity definition based on node distances   When considering finite-length *time* signals $\mathbf{x}_k \in \mathbb{R}^N$, the empirical covariance, $\mathbf{S}_x$, obtained from $\mathbf{X}$ has the same form as in (4.5). But, because the order of the entries in a finite signal corresponds to the time index, two pairs of entries $(x(i), x(i + m))$ and $(x(j), x(j + m))$ are separated by the same time interval. In contrast, the labeling of graph nodes is arbitrary so that there is no meaning associated with the fact that $i_1 - j_1 = i_2 - j_2$ for two pairs of nodes (i_1, j_1) and (i_2, j_2).

For time domain signals it makes sense to determine whether $\mathbb{E}[x(i)x(i + m)] = \mathbb{E}[x(j)x(j + m)]$ because the distance between the nodes in each pair is the same. For the graph case, an analogous approach would be to define stationarity in terms of the "distance" between nodes. Thus, defining $d(i, j)$ as a suitable distance metric between nodes i and j (see Definition 2.8), the second-moment condition for stationarity would be $\mathbb{E}[x(i)x(j)] = \mathbb{E}[x(i')x(j')]$, where $d(i, j) = d(i', j')$. Note that for a time domain signal with N samples there are $N - 1$ sample pairs that are at a distance 1, but for a weighted graph there may not be two pairs of nodes with the same geodesic distance between them, i.e., such that $d(i, j) = d(i', j')$. Thus in practice an empirical covariance can be estimated by averaging $x(i)x(j)$ terms where the distance between nodes is *approximately* the same.

For the special case of signals obtained from samples in Euclidean space, this would correspond to averaging over multiple pairs of nodes having similar distances. This leads to the notion of empirical "variograms," widely used in the context of statistics for spatial data [45]. This idea can be extended to graphs as shown in [46]: information in different nodes can be aggregated using multiple subgraphs of the original graph (same nodes, non-overlapping edges), where all edge weights in each subgraph are approximately equal and each subgraph connects only those nodes that are within a certain range of distances from each other.

Summary   The operations of (i) shifting a signal and (ii) moving the observation point, are not equivalent for graph signals. This suggests considering two types of definitions of stationarity, based on either (i) the pairwise distance between nodes [45, 46], or (ii) the properties of signals processed by polynomials of the graph operator **Z** [42, 47, 43].

4.2 Sampling Constraints

Graph signal sampling is motivated by situations where there are constraints on signal acquisition. While the most obvious case is where there are constraints on the total number of samples to be observed, we also consider a broader set of cases that may impact sampling set selection.

4.2.1 Total Number of Samples

A natural constraint is to limit only the total number of observed samples. For graph signals obeying a specific model the number of samples to be captured to guarantee reconstruction depends on the model parameters. For example, if the signal is bandlimited to the first R frequencies, at least R samples will be required for exact reconstruction.

In practice, if an accurate model is not available or model parameters are not known, it may be necessary to sample the graph signal incrementally until some desired property is achieved. To illustrate this idea, assume that we consider approximately bandlimited signals, but R is unknown. Then, two possible strategies are the following.

- Observe the change in interpolated values at unobserved nodes as additional nodes are observed. If the interpolated values do not change much, this could be an indication that the sampling process can be stopped without incurring significant error.
- Sample new nodes and compare these observations with previously estimated values at those same nodes. If the newly sampled points agree with the previously estimated values, stopping the sampling process may be acceptable.

4.2.2 Distributed Systems

In a distributed system, where the graph nodes represent data in different locations, the data observed at a given node will have to be transmitted before it can be processed. Thus, there will be a communication cost associated with each sample measured. Moreover this cost may be different from one node location to another. As an example, in a sensor network where information is collected by communicating with a base station, the cost associated with collecting data at a given node will depend on the distance to the base station. Thus, optimizing the sampling set for communication cost may lead to more samples being collected near the base station. In contrast, if communication costs from different nodes are similar, but each node has limited battery power, the sampling algorithm may alternate between subsets of nodes over time, so that energy consumption is balanced across the sensing nodes.

4.2.3   Diffusion and Sampling

In some cases, observed data is known to have propagated through the network. Then, the number of observed nodes can be reduced by increasing the number of observations over time. In this scenario the goal is to recover a signal $\mathbf{x}$ from samples of $\mathbf{x}$, $\mathbf{Zx}$, $\mathbf{Z}^2\mathbf{x}$, etc., as proposed in [48]. The main challenge is to find a graph operator $\mathbf{Z}$ that models the effects of the physical phenomenon being observed, given that a discrete node model is used and time has to be discretized as well.

Developing such a model is particularly challenging for data obtained from continuous physical environments (e.g., data collected by sensor networks) because it entails discretizing, in time and space, what is essentially a continuous phenomenon. If a suitable $\mathbf{Z}$ exists and is known then approaches that exploit diffusion (flows) may be efficient, as they allow more flexibility in deciding which nodes to observe. As an example, in a distributed-sensing setting it may be possible to assign multiple observations to nodes having a favorable property (e.g., lower communication cost).

This kind of sampling may be better suited for settings where the graph captures how the observed signal evolves. For example, if we consider sensors on a road network, the traffic flowing through an intersection will later flow through neighboring intersections. This suggests that a reasonable model can be obtained by considering $\mathbf{x}$ and $\mathbf{Zx}$ to be successive observations, with $\mathbf{Z}$ based on a graph representing the road network.

4.3   Problem Formulation

In Section 4.1 we saw that different models could be used to characterize classes of graph signals to be sampled. Now we formulate the sampling problem for generic signal models and sampling strategies. Define V as a set of graph signals on graph G, which we assume to be a vector subspace of $\mathbb{R}^N$ (Definition A.1). Note that there is no loss of generality in considering only vector subspaces for the majority of models considered in Section 4.1 (one exception would be the set of graph signals with integer entries). Under this assumption, let $\dim(V) = M \leq N$ be the dimension of V and let $\mathbf{U}_S$ be an $N \times M$ matrix whose columns are basis vectors for V.

4.3.1   Node Domain Sampling

Node domain sampling operators measure directly the values at each node. Defining a sampling set of nodes $S \subset \mathcal{V}$, where S is a subset containing R graph nodes, the sampling operation can be represented by an $N \times R$ matrix $\mathbf{S}$ with R linearly independent columns, producing a sampled signal $\mathbf{x}_s \in \mathbb{R}^R$:

$$\mathbf{x}_s = \mathbf{S}^\mathsf{T}\mathbf{x}. \tag{4.7}$$

Direct sampling   In the simplest case the entries of $\mathbf{x}$ can be directly observed.

> DEFINITION 4.1 (DIRECT NODE SAMPLING) Signal entries at the nodes in the set S are observed directly and therefore
>
> $$\mathbf{S} = \mathbf{I}_S,$$
>
> where $\mathbf{S}$ contains the columns of the identity matrix corresponding to the canonical basis vectors associated with the nodes in S.

When sampling as in Definition 4.1 we can write:

$$\hat{\mathbf{x}} = \mathbf{S}\,\mathbf{S}^\mathsf{T}\mathbf{x} = \mathbf{I}_S\,\mathbf{I}_S^\mathsf{T}\mathbf{x}, \tag{4.8}$$

which corresponds to preserving the values at the nodes in S while setting to zero the values in S^c, the complement of S. As an alternative interpretation $\mathbf{S}$ can be seen as a set of R orthogonal vectors that span a subspace of $\mathbb{R}^N$, so that $\hat{\mathbf{x}}$ is the orthogonal projection of $\mathbf{x}$ onto the columns of $\mathbf{S}$.

Filtered sampling In other cases, we may model the observed data as the result of the graph filtering of an unobserved signal that we wish to recover.

> DEFINITION 4.2 (FILTERED NODE DOMAIN SAMPLING) We assume there exists a graph filter, $\mathbf{H}$, see Box 3.3, and we define
>
> $$\mathbf{S}^\mathsf{T} = \mathbf{I}_S^\mathsf{T}\mathbf{H},$$
>
> corresponding to first filtering $\mathbf{x}$ and then applying node domain sampling.

Note that the assumption that $\mathbf{S}$ has linearly independent columns can be made without loss of generality. From Definition 4.1, since $\mathbf{S}$ contains columns from $\mathbf{I}$, it can only have linearly dependent columns if a node is observed more than once, which would be clearly inefficient. As for the case of Definition 4.2, if the columns are linearly dependent then we can define an alternative sampling operator with fewer columns from which the same set of signals can be recovered.

A sampling operator $\mathbf{S}^\mathsf{T}$ maps vectors in $\mathbb{R}^N$ to vectors in $\mathbb{R}^R$, with $R < N$. Thus, sampling cannot be invertible unless we restrict the class of input signals that we wish to recover. In particular, vectors orthogonal to the columns of $\mathbf{S}$ are all mapped to zero. Thus, if there are signals in V that are orthogonal to the columns of $\mathbf{S}$, these cannot be reconstructed. This idea will be developed further in Section 4.3.3.

4.3.2 Frequency Domain Sampling

The idea of frequency domain graph signal sampling was first studied in [49]. It can be viewed as a sampling operation in the frequency representation of $\mathbf{x}$. Letting $\mathbf{U}$ be the matrix of eigenvectors of the chosen graph operator $\mathbf{Z}$, define $\tilde{\mathbf{x}} = \mathbf{U}^\mathsf{T}\mathbf{x}$. Then, frequency domain sampling is based on selecting some of the entries in $\tilde{\mathbf{x}}$.

> DEFINITION 4.3 (FREQUENCY DOMAIN SAMPLING OPERATORS) Frequency domain sampling can be defined as:
>
> $$\tilde{\mathbf{x}}_s = \tilde{\mathbf{S}}^T \tilde{\mathbf{x}},$$
>
> where $\tilde{\mathbf{S}}^T$ is a frequency domain operator. A possible reconstruction after sampling is
>
> $$\hat{\mathbf{x}} = \mathbf{U}\tilde{\mathbf{x}}_S.$$

Note that obtaining $\tilde{\mathbf{x}}$ requires having access to all the entries of $\mathbf{x}$ and thus this approach cannot be viewed as a conventional sampling method, since all data has to be observed first. The approach of Definition 4.3 is primarily useful in the construction of critically sampled graph signal transforms (see Chapter 5).

Relationship between node domain and frequency domain sampling As described in [49], in contrast with conventional signal processing, there is in general no simple relationship between node domain and frequency domain sampling. Thus, a node domain sampling operator (Definition 4.1) does not correspond to a simple operation in the frequency domain. The operator in (4.8) is written $\tilde{\mathbf{S}}^T = \mathbf{U}^T \mathbf{I}_s \mathbf{I}_S^T \mathbf{U}$ in the frequency domain. In general $\tilde{\mathbf{S}}^T$ is a dense matrix and can be interpreted only as a complex mixing of signal frequencies. An exception is the case of bipartite graphs, where the resulting frequency domain sampling can be expressed as the sum of a diagonal and an anti-diagonal matrix (Section 5.6.2). Conversely, a frequency domain sampling (Definition 4.3) cannot be expressed in general by a low-order $p(\mathbf{Z})$ followed by node domain sampling.

4.3.3 General Conditions for Reconstruction

For the sampling methods of Section 4.3.1 and Section 4.3.2, the complete sampling and reconstruction process can be written as

$$\hat{\mathbf{x}} = \mathbf{F}\,\mathbf{S}^T\mathbf{x}, \tag{4.9}$$

where $\mathbf{S}$ is the sampling operator and $\mathbf{F}$ is a reconstruction filter. In order to analyze (4.9), we need to solve two problems: (i) sampling set selection, i.e., choosing $\mathbf{S}^T$ and (ii) interpolation, i.e., defining the linear operation $\mathbf{F}$ that interpolates the missing information.

For node domain sampling (Section 4.3.1), $\mathbf{S}^T$ tells us which nodes should be observed, while $\mathbf{F}$ interpolates information at the unobserved nodes. For frequency domain sampling (Section 4.3.2) the basic operations are frequency selection and interpolation. Notice that, in contrast with conventional signal processing problems, there is no simple systematic approach for sampling set selection. That is, we cannot select "every other node," as can be done for the sampling of a discrete-time signal. The problem of sampling set selection will be studied in Section 4.4.

The general conditions under which a signal in a space V can be reconstructed from its samples can be expressed using the span (Definition A.2), the orthogonal complement (Definition A.7), Proposition A.1 and Proposition A.2.

PROPOSITION 4.1 (CONDITIONS FOR RECONSTRUCTION) Define the following vector spaces: (i) V_F, the span of the columns of $\mathbf{F}$, (ii) V_S, the span of the columns of $\mathbf{S}$ and (iii) $V_S^\perp$, the orthogonal complement of V_S. For any signal $\mathbf{x} \in V$ and $\hat{\mathbf{x}}$ obtained according to (4.9), $\mathbf{x}$ can be recovered from $\hat{\mathbf{x}}$ under the following conditions on $\mathbf{F}$ and $\mathbf{S}$. The sampling operator $\mathbf{S}^\mathsf{T}$ must be such that

$$V \cap V_S^\perp = \{\mathbf{0}\}, \tag{4.10}$$

while for $\mathbf{F}$ we must have

$$V \subset V_F. \tag{4.11}$$

Notice that these are necessary conditions on $\mathbf{F}$ and $\mathbf{S}$ but that these two operators need to be chosen together to guarantee that exact reconstruction is possible. Since $\dim\left(V_S^\mathsf{T}\right) \leq N - M$, from Proposition 4.1, and given that $\dim(V_S) + \dim\left(V_S^\mathsf{T}\right) = N$, it follows that we must have $\dim(V_S) \geq \dim(V) = M$. Thus, in the case of direct node sampling in Definition 4.1, if V has dimension M then we need the sampling set S to contain at least M nodes.

There are cases of practical interest where the conditions of Proposition 4.1 are not met. In particular, if (4.10) is not verified then there will be a non-zero vector $\mathbf{a}$ such that $\mathbf{a} \in V \cap V_S^\perp$. Thus, for any $\mathbf{x} \in V$ we will have $\mathbf{S}^\mathsf{T}(\mathbf{x} + \mathbf{a}) = \mathbf{S}^\mathsf{T}\mathbf{x}$, since $\mathbf{a}$ belongs to the null space of $\mathbf{S}^\mathsf{T}$, and there is no unique solution to the reconstruction problem. In fact the number of solutions will be infinite since any $\mathbf{S}^\mathsf{T}(\alpha\mathbf{a}) = 0$ for any scalar α. Thus we will need to use methods that associate a cost with each of the vectors $\mathbf{x} + \alpha\mathbf{a}$ consistent with the sampled data, and select the vector minimizing the cost. An example of such a metric favors the solution with minimum norm (refer to Section 4.3.3).

4.3.4 Reconstruction of Bandlimited Signals

Let us consider in detail the case where V is the space of bandlimited signals defined by (4.1), and let us use the node domain sampling of Definition 4.1. This was one of the first sampling problems to be studied [50] owing to its analogy with the notion of sampling for conventional signals.

DEFINITION 4.4 (PALEY–WIENER SPACE OF BANDLIMITED SIGNALS) The Paley–Wiener space of bandlimited signals $PW_\omega(G)$ on a graph G for a given frequency ω (and for a given choice of operator $\mathbf{Z}$ and corresponding GFT $\mathbf{U}$) is defined as

$$PW_\omega(G) = \left\{ \mathbf{x} \in \mathbb{R}^N \,\middle|\, \mathbf{x} = \sum_{i=1}^r \tilde{x}_i \mathbf{u}_i = \mathbf{U}_{VR}\tilde{\mathbf{x}}_r \right\}$$

where r is such that $\lambda_r < \omega$ and $\mathbf{U}_{VR}$ is the matrix containing the first r columns of the GFT $\mathbf{U}$.

Conditions for reconstruction Using node domain sampling on a given set S means that V_S is the set of vectors spanned by $\mathbf{I}_S$, and its orthogonal complement can be defined as the set of signals that are identically zero in S:

$$V_S^\perp = L_2(S^c) = \left\{ \mathbf{x} \in \mathbb{R}^N \,\middle|\, \mathbf{x}_S = \mathbf{0} \right\},$$

where $\mathbf{x}_S$ denotes the vector of samples indexed by S. Using (4.10) from Proposition 4.1 we have that a signal can be recovered if and only if

$$V \cap V_S^\perp = PW_\omega(G) \cap L_2(S^c) = \{\mathbf{0}\}, \tag{4.12}$$

which can be interpreted as stating that all signals in $L_2(S^c)$ should have minimum bandwidth (i.e., maximum frequency) greater than ω.

Reconstruction operator   Next consider the reconstruction. Given $\mathbf{I}_S \, \mathbf{I}_S^\mathsf{T} \mathbf{x}$, where all entries corresponding to the nodes in S^c are zero, we would like to see whether it is possible to recover $\mathbf{x}$ exactly. As previously mentioned, this is not possible unless we restrict $\mathbf{x}$ in some way. Assume that $\mathbf{x} = \mathbf{U}_{VR}\mathbf{a}$ (i.e., $\mathbf{x}$ is bandlimited), so that the sampled signal is

$$\mathbf{x}_S = \mathbf{I}_S^\mathsf{T} \mathbf{U}_{VR}\, \mathbf{a} = \mathbf{U}_{SR}\, \mathbf{a}, \tag{4.13}$$

where $\mathbf{U}_{SR}$ is rectangular if $|S| > M$. Using (4.10) for $\mathbf{S} = \mathbf{I}_S$, the condition (4.12) can be expressed as requiring that the relations

$$\mathbf{I}_S^\mathsf{T}\, \mathbf{U}_{VR}\mathbf{a} = \mathbf{U}_{SR}\, \mathbf{a} = 0 \tag{4.14}$$

cannot have a non-zero solution $\mathbf{a}$ (otherwise there would be a bandlimited non-zero signal orthogonal to the columns of $\mathbf{I}_S$). Then, (4.10) guarantees that the columns of $\mathbf{U}_{SR}$ are linearly independent (Definition A.3).

Thus, given (4.13) we can use a least squares (orthogonal) projection (Proposition A.4) to recover $\mathbf{a}$:

$$\mathbf{a} = (\mathbf{U}_{SR}^\mathsf{T}\mathbf{U}_{SR})^{-1}\mathbf{U}_{SR}^\mathsf{T}\mathbf{x}_S, \tag{4.15}$$

and therefore the overall reconstruction from samples is

$$\mathbf{x} = \mathbf{U}_{VR}\mathbf{a} = \mathbf{U}_{VR}(\mathbf{U}_{SR}^\mathsf{T}\mathbf{U}_{SR})^{-1}\mathbf{U}_{SR}^\mathsf{T}\mathbf{x}_S, \tag{4.16}$$

where we note that $\mathbf{U}_{SR}^\mathsf{T}\mathbf{U}_{SR}$ is invertible since $\mathbf{U}_{SR}$ has R independent columns and thus $\mathbf{U}_{SR}^\mathsf{T}\mathbf{U}_{SR}$ is full rank. For this case the operators of Proposition 4.1 are $\mathbf{S} = \mathbf{I}_S$ and $\mathbf{F} = \mathbf{U}_{VR}(\mathbf{U}_{SR}^\mathsf{T}\mathbf{U}_{SR})^{-1}\mathbf{U}_{SR}^\mathsf{T}$.

4.3.5   Alternative Reconstruction Algorithms

Under the condition that (4.14) has no non-zero solutions, $\mathbf{U}_{SR}$ has full column rank and it is possible to recover $\mathbf{x}$ from $\mathbf{x}_S$ using a pseudo-inverse as in (4.16), but this requires exact knowledge of the eigenvectors of the graph operator. Alternative methods can be developed for reconstruction that avoid the complexity associated with computing the pseudo-inverse.

Projection onto convex sets   As an illustrative example we describe a technique based on projections onto convex sets (POCS) [51], where two convex sets of signals are defined and reconstruction alternates projections into each of these sets.

- For a given sampled signal $\mathbf{x}_S$, define P_1, the set of all signals that have the same observed values. Thus, a reconstructed signal $\mathbf{f} \in P_1$ if $\mathbf{f}_S = \mathbf{x}_S$, and any reconstructed signal $\mathbf{f}$ should be in P_1 so that the observed samples are preserved in the reconstructed signal. The set P_1 is clearly convex: if $\mathbf{f}_1$ and $\mathbf{f}_2$ belong to this set then $\alpha \mathbf{f}_1 + (1 - \alpha)\mathbf{f}_2$ is also in the set for any $\alpha \in [0,1]$. Projection onto P_1 can be done efficiently: for any signal $\mathbf{f}$ it is sufficient to change its entries in the sampling set S to match those of $\mathbf{x}$, without changing the remaining entries.
- Define a second set, $P_2 = PW_\omega(G)$, of signals that are bandlimited to the first r frequencies. The set P_2 is a vector space, i.e., closed under addition and scalar multiplication, and thus also convex. Signals that have been projected onto P_1 are not necessarily bandlimited to the first r frequencies. To project a signal onto P_2 its GFT can be computed and then its components corresponding to frequencies greater than r can be set to zero. This can be viewed as defining an ideal filter $h_r(\lambda)$ as $h_r(\lambda) = 1$ if $\lambda \leq \lambda_r$ and $h_r(\lambda) = 0$ otherwise and using this filter for the projection (see Box 3.1).

Note that if the condition of Proposition 4.1 holds then the reconstruction is unique, and thus P_1 and P_2 have a unique signal in their intersection: $P_1 \cap P_2 = \{\mathbf{x}\}$.

Using a polynomial filter   The ideal filter $\mathbf{H}_r = \text{diag}\{h_r(\lambda)\}$ can be replaced by a polynomial approximation, which would only require (approximate) knowledge of the cutoff frequency. A POCS approach to reconstruction would alternate projection onto P_1 and projection onto an approximation to P_2 (owing to the use of a non-ideal filter) until convergence. The convergence performance can be significantly improved by using local set methods [52].

Non-bandlimited signals   If $\mathbf{x}$ is not exactly bandlimited, alternative reconstruction algorithms can be developed under less strict assumptions. For example, if we are given a high pass filter $p(\mathbf{Z})$, the minimization

$$\mathbf{f} = \arg \min_{\mathbf{f} \in \mathbb{R}^N} \left(\|\mathbf{x}_S - \mathbf{S}^\mathsf{T}\mathbf{f}\|_2^2 + \gamma \|p(\mathbf{Z})\mathbf{f}\|_2^2 \right)$$

can find a signal $\mathbf{f}$ whose samples are close to those observed (the first term) but do not have to match exactly (thus accounting for noise), with a second term favoring solutions having most of their energy in the low frequencies [53]. A particular case of interest arises where $\mathbf{Z} = \mathbf{L}$ and the second term is replaced by the Laplacian quadratic form $\mathbf{f}^\mathsf{T}\mathbf{L}\mathbf{f}$, leading to a closed form solution [54]:

$$\mathbf{f} = (\mathbf{S}\mathbf{S}^\mathsf{T} + \gamma\mathbf{L})^{-1}\mathbf{S}\mathbf{x}_S,$$

where the factor $\gamma \geq 0$ controls the relative importance given to the approximation to the observed samples $\mathbf{x}_S$ and to the low-frequency content.

4.4   Sampling-Set-Optimization Algorithms

In what follows we study the sampling set selection problem. Notice that, from (4.15), we can recover an exactly bandlimited signal for *any* sampling set S, as long as the conditions of Proposition 4.1 are met. For a given signal model, if the observed signals are consistent with the model there will be many sampling sets that meet the conditions of Proposition 4.1. We consider this aspect in Section 4.4.1. In practice, however, if the conditions in Proposition 4.1 are not met, and there is a mismatch between the model and the observed signals, sampling set selection becomes an optimization problem where the goal is to minimize the impact of mismatch.

4.4.1   Sampling of Noise-Free Data

Consider again the case of Section 4.3.4 and assume that we are observing noise-free signals. Then, from Proposition 4.1, any sampling set S such that $\mathbf{U}_{SR}$ has full column rank (linearly independent columns) will provide an exact reconstruction, and there will be no advantage in selecting any one among those sets since they all provide the same reconstruction: for any S, as long as the columns of $\mathbf{U}_{SR}$ are linearly independent, (4.15) allows us to recover $\mathbf{a}$. For the noise-free bandlimited scenario, there are generally many possible choices of sampling set S, and since they all lead to the same performance, simple set selection algorithms, including random set selection, are sufficient. As a starting point, note that we can write V and $V_S^\perp$ in (4.12) in terms of their basis vectors as

$$V = \mathrm{span}(\mathbf{u}_1, \mathbf{u}_2, \ldots, \mathbf{u}_r) \quad \text{and} \quad V_S^\perp = \mathrm{span}\left(\{\mathbf{e}_i \mid \forall i \in S^c\}\right),$$

where $V_S^\perp$ is spanned by the canonical basis vectors $\mathbf{e}_i$ of $\mathbb{R}^N$ corresponding to the nodes in S^c. Therefore, for the condition (4.12) on S to hold, we must have that the vectors in $\mathcal{B} = \{\mathbf{u}_1, \mathbf{u}_2, \ldots, \mathbf{u}_r, \mathbf{e}_{i_1}, \mathbf{e}_{i_2}, \ldots\}$ are linearly independent. Thus, given $\mathcal{B}_V = (\mathbf{u}_1, \mathbf{u}_2, \ldots, \mathbf{u}_r)$, the goal is to find $N - r$ vectors $\mathbf{e}_i$ that together with $\mathcal{B}_V$ form an independent set.

Algorithm   The method of [55] constructs $\mathcal{B}$ iteratively, initializing as

$$\mathcal{B}^0 = \{\mathbf{b}_1, \mathbf{b}_2, \ldots, \mathbf{b}_N\} = \{\mathbf{e}_1, \mathbf{e}_2, \ldots, \mathbf{e}_N\},$$

which is equivalent to having $S = \emptyset$. Then, at each iteration one of the $\mathbf{e}_i$ is removed (i.e., i is added to S) and replaced by one of the vectors in $\mathcal{B}_V$, starting with $\mathbf{u}_1$ at the first iteration. At iteration k, we have removed the standard basis vectors in a set S^{k-1} and $\mathcal{B}^{k-1}$ contains vectors $\{\mathbf{u}_1, \mathbf{u}_2, \ldots, \mathbf{u}_{k-1}\}$ as well as the remaining standard basis vectors, i.e., those corresponding to the complement of S^{k-1}. Then, since $\mathcal{B}^{k-1}$ is a basis for $\mathbb{R}^N$, $\mathbf{u}_k$ can be written as

$$\mathbf{u}_k = \sum_{i=1}^{N} \alpha_i \mathbf{b}_i,$$

and we select the index l such that $|\alpha_l|$ is maximum (i.e., we remove the canonical basis vector with the highest correlation to $\mathbf{u}_k$), let $\mathbf{b}_l = \mathbf{u}_k$ and go to the next iteration.

4.4.2 Criteria for Robust Sampling

In practice sampled signals are not exactly bandlimited, e.g., owing to noise or to a mismatch between the signal model assumptions and the actual signals observed. Thus, unlike in Section 4.4.1, the choice of sampling set can have a significant impact on reconstruction. Selecting an optimal sampling set requires defining a metric for the error to be minimized.

Noise model   Consider a noisy signal

$$\mathbf{x}' = \mathbf{x} + \mathbf{n}, \tag{4.17}$$

where $\mathbf{n}$ is a noise term and $\mathbf{x} \in PW_\omega(G)$. Then, denoting by $\mathbf{n}_S$ the noise term restricted to the sampled set S, the reconstruction can be written, using (4.15), as

$$\hat{\mathbf{x}} = \mathbf{U}_{VR}(\mathbf{U}_{SR}^\mathsf{T}\mathbf{U}_{SR})^{-1}\mathbf{U}_{SR}^\mathsf{T}(\mathbf{x}_S + \mathbf{n}_S) = \mathbf{x} + \mathbf{U}_{VR}\mathbf{U}_{SR}^+\mathbf{n}_S, \tag{4.18}$$

where $\mathbf{U}_{SR}^+ = (\mathbf{U}_{SR}^\mathsf{T}\mathbf{U}_{SR})^{-1}\mathbf{U}_{SR}^\mathsf{T}$ is the pseudo-inverse. In order to quantify the effect of noise, define

$$\mathbf{e} = \mathbf{U}_{VR}\mathbf{U}_{SR}^+\mathbf{n}_S$$

and then find the corresponding covariance matrix (assuming for simplicity that the entries of the error vector are independent with zero mean and unit variance).

> PROPOSITION 4.2   (COVARIANCE OF RECONSTRUCTION ERROR)   For the model (4.17) and the interpolation (4.18) the covariance of the reconstruction error is
>
> $$\mathbf{E} = \mathbb{E}\left[\mathbf{e}\,\mathbf{e}^\mathsf{T}\right] = \mathbf{U}_{VR}(\mathbf{U}_{SR}^\mathsf{T}\mathbf{U}_{SR})^{-1}\mathbf{U}_{VR}^\mathsf{T}. \tag{4.19}$$

Proof   First, under the assumption that the noise signal components are independent and identically distributed, we have $\mathbb{E}\left[\mathbf{n}_S\,\mathbf{n}_S^\mathsf{T}\right] = \mathbf{I}$, so that

$$\mathbb{E}\left[\mathbf{e}\,\mathbf{e}^\mathsf{T}\right] = \mathbf{U}_{VR}\mathbf{U}_{SR}^+\mathbb{E}\left[\mathbf{n}_S\,\mathbf{n}_S^\mathsf{T}\right](\mathbf{U}_{SR}^+)^\mathsf{T}\mathbf{U}_{VR}^\mathsf{T} = \mathbf{U}_{VR}\mathbf{U}_{SR}^+(\mathbf{U}_{SR}^+)^\mathsf{T}\mathbf{U}_{VR}^\mathsf{T}.$$

Next, using the definition of the pseudo-inverse, we have

$$\mathbf{U}_{SR}^+(\mathbf{U}_{SR}^+)^\mathsf{T} = (\mathbf{U}_{SR}^\mathsf{T}\mathbf{U}_{SR})^{-1}\mathbf{U}_{SR}^\mathsf{T}\mathbf{U}_{SR}((\mathbf{U}_{SR}^\mathsf{T}\mathbf{U}_{SR})^{-1})^\mathsf{T} = (\mathbf{U}_{SR}^\mathsf{T}\mathbf{U}_{SR})^{-1}$$

since $(\mathbf{U}_{SR}^\mathsf{T}\mathbf{U}_{SR})$ is symmetric and thus its inverse is also symmetric; combining these two equations leads to (4.19). □

Optimality criteria   Note that in (4.19) the term $\mathbf{U}_{VR}$ does not depend on S. Thus, we need to consider only the effect of $\mathbf{U}_{SR}$ on the error metric. We will focus on three criteria related to the properties of $\mathbf{C} = (\mathbf{U}_{SR}^\mathsf{T}\mathbf{U}_{SR})^{-1}$. We start by noting that $\mathbf{C}$ is symmetric and invertible and thus can be diagonalized, has non-zero real eigenvalues $\mathbf{\Lambda}_C$ and orthogonal eigenvectors $\mathbf{U}_C$. Then $\mathbf{E}$ can be written as

$$\mathbf{E} = \mathbf{U}_{VR}\mathbf{U}_C\mathbf{\Lambda}_C\mathbf{U}_C^\mathsf{T}\mathbf{U}_{VR}^\mathsf{T}, \tag{4.20}$$

while the norm of the error $\mathbf{e}$ is

$$\|\mathbf{e}\|^2 = \mathbf{e}^{\mathsf{T}}\mathbf{e} = \mathbf{n}_S^{\mathsf{T}}(\mathbf{U}_{SR}^+)^{\mathsf{T}}\mathbf{U}_{VR}^{\mathsf{T}}\mathbf{U}_{VR}\mathbf{U}_{SR}^+\mathbf{n}_S = \mathbf{n}^{\mathsf{T}}\mathbf{C}\mathbf{n} = \mathbf{n}^{\mathsf{T}}\mathbf{U}_C\mathbf{\Lambda}_C\mathbf{U}_C^{\mathsf{T}}\mathbf{n}, \qquad (4.21)$$

which again shows that the error introduced will depend only on $\mathbf{C}$. Since $\mathbf{U}_C$ is orthogonal it does not affect the norm. Without loss of generality assume the eigenvalues of $\mathbf{\Lambda}_C$ are in increasing order. Then, from (4.20) and (4.21), several optimality criteria can be considered.

A-optimality Minimize the average error incurred over all possible signals, which leads to minimizing the sum of the eigenvalues, i.e., the trace of $\mathbf{\Lambda}_C$, tr($\mathbf{\Lambda}_C$):

$$S_{\text{opt}}^A = \arg\min_{|S|=r}\text{tr}(\mathbf{E}) = \arg\min_{|S|=r}\text{tr}((\mathbf{U}_{SR}^{\mathsf{T}}\mathbf{U}_{SR})^{-1}) = \arg\min_{|S|=r}\text{tr}(\mathbf{\Lambda}_C).$$

E-optimality Minimize the worst-case error, that is, the error amplification due to the largest eigenvalue of $\mathbf{\Lambda}_C$, or maximize the minimum singular value of $\mathbf{U}_{SR}$, i.e.,

$$S_{\text{opt}}^E = \arg\max_{|S|=r}\sigma_{\min}(\mathbf{U}_{SR}) = \arg\min_{|S|=r}\lambda_{\max}(\mathbf{C})$$

D-optimality Minimize the determinant of $\mathbf{C}$, i.e., the product of its eigenvalues:

$$S_{\text{opt}}^D = \arg\min_{|S|=r}\det((\mathbf{U}_{SR}^{\mathsf{T}}\mathbf{U}_{SR})^{-1}) = \arg\min_{|S|=r}\det(\mathbf{\Lambda}_C).$$

See [56] for other design criteria.

4.4.3 Robust Spectral Sampling

Robust sampling approaches [40, 57] use criteria such as those mentioned in the previous section to optimize the sampling set S, with different sets obtained depending on the chosen optimality criterion.

Node selection using the GFT The **greedy** algorithm[4] proposed in [40] for the E-optimality worst-case criterion starts from an empty set S and then selects one row at a time from $\mathbf{U}_{VR}$ to be included in $\mathbf{U}_{SR}$, in such a way that in each iteration the minimum singular value of $\mathbf{U}_{SR}$ is maximized. This approach increases robustness significantly, but still requires computation of the first r basis vectors of the GFT (i.e., $\mathbf{U}_{VR}$).

Alternative procedures were also proposed for A-optimality and D-optimality in [58], where the same greedy approach is followed (adding one node per iteration) but the criterion for selecting a node is different. For A-optimality the choice is based on minimizing the sum of the inverses of the singular values $\mathbf{U}_{SR}$, leading to optimization of the average error. For D-optimality, at each iteration the node leading to the maximum product of the eigenvalues of $\mathbf{U}_{SR}^{\mathsf{T}}\mathbf{U}_{SR}$ is chosen. This product (the determinant) corresponds to the volume of $\mathbf{U}_{SR}^{\mathsf{T}}\mathbf{U}_{SR}$, which is maximized if the columns of $\mathbf{U}_{SR}$ are close to being orthogonal.

---

[4] A node selection algorithm is greedy if it selects nodes iteratively, where at each iteration a node is selected to optimize a metric, and nodes previously selected cannot be replaced. Note that optimizing the metric at each iteration does not guarantee that the solution is globally optimal.

Spectral proxies   To better understand how to optimize the choice of the sampling set S, recall that exact reconstruction is guaranteed under the condition (4.12). Thus, suppose that we have two sets S and S', with $|S| = |S'|$, that we wish to compare. We can then find the bandwidth values R and R' which are the largest values for which (4.12) is met for S and S' respectively. Clearly, we would want to choose S if $R > R'$, and vice versa. Note that if (4.12) is verified for a given S and R, this means that all signals in $L_2(S^c)$ must have bandwidth (maximum non-zero frequency) of at least $R + 1$.

This idea leads to the greedy algorithm given in [57], which does not require explicit computation of $\mathbf{U}_{VR}$. This approach is based on defining a "spectral proxy" that provides an estimate of signal frequency. For a given signal, $\mathbf{x}$, the kth-order spectral proxy is

$$\omega_k(\mathbf{x}) = \left(\frac{\mathbf{x}^\mathsf{T} \mathbf{L}^k \mathbf{x}}{\mathbf{x}^\mathsf{T} \mathbf{x}} \right)^{1/k}. \tag{4.22}$$

If $\omega(\mathbf{x}) = \lambda_j$ is the largest non-zero frequency of $\mathbf{x}$ it can be shown that

$$\forall k, \quad \omega_k(\mathbf{x}) \leq \lim_{i \to \infty} \omega_i(\mathbf{x}) = \omega(\mathbf{x}),$$

and the minimum bandwidth (minimum value of the maximum frequency) of any non-zero signal $\mathbf{x}$ in $L_2(S^c)$ can be defined as

$$\omega_c(S^c) = \min_{\mathbf{x} \in L_2(S^c), \mathbf{x} \neq 0} \omega(\mathbf{x}).$$

Therefore, to compare two sets S and S', as discussed earlier, we can compare $\omega_c(S^c)$ and $\omega_c(S'^c)$ and choose the one leading to the largest minimum bandwidth. Next, notice that any $\mathbf{x}$ in $L_2(S^c)$ has zero values for all entries corresponding to S, so that we can rewrite the spectral proxy definition of (4.22) as

$$\omega_k(\mathbf{x}_{S^c}) = \left(\frac{\mathbf{x}_{S^c}^\mathsf{T} (\mathbf{L}^k)_{S^c} \mathbf{x}_{S^c}}{\mathbf{x}_{S^c}^\mathsf{T} \mathbf{x}_{S^c}} \right)^{1/k}, \tag{4.23}$$

where the subscript S^c indicates that the vector or matrix has been reduced to only the entries in S^c. Recalling the discussion of the Rayleigh quotient in Section 3.2.3 we can see that the solution to (4.23) can be obtained by finding the minimum-variation eigenvalue and corresponding eigenvector, $\mathbf{u}_{S^c}^*$, for $(\mathbf{L}^k)_{S^c}$, which can be done efficiently.

Algorithm based on spectral proxies   In order to find the best sampling set, at the kth iteration, the algorithm in [57] computes $\mathbf{u}_{S^c}^*$ and chooses the coordinate (node) i for which $u_{S^c}^*(i)^2$ is maximal. To understand this approach intuitively, recall from our discussions in Chapter 2 that $\mathbf{L}^k$ represents a k-hop operation. Then $\mathbf{u}_{S^c}^*$ is the graph signal with *minimum variation* on $\mathbf{L}^k$ under the conditions that all values in S must be set to zero. Now, focus on the value of i such that $u_{S^c}^*(i)^2$ is maximal. Since this is the largest value, and $\mathbf{u}_{S^c}^*$ has minimum variation, it follows that the values of $\mathbf{u}_{S^c}^*$ at the neighbors of i in the k-hop graph will be larger than those of the neighbors of other nodes in the graph. On this basis, moving node i from S^c to S maximizes the increase in variation, since all the neighbors of i will then be forced to have a zero-valued neighbor

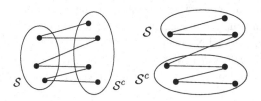

Figure 4.2 Example of sampling set selections: the same graph has been partitioned into two sets S and S^c in two different ways. Intuitively the approach on the left is better, simply because the observed samples (in S) are always "closer," in terms of graph distance, to the unobserved samples (in S_c), as compared with the example on the right.

in their k-hop neighborhood. That is, nodes for which a large value at a neighboring node led to a smooth solution are now forced to have a zero neighbor.

More formally, consider the $k = 1$ case for an unweighted graph. Note that the eigenvector corresponding to the minimum eigenvalue will have all positive entries. If $\mathbf{L}_S$ is irreducible (i.e., the nodes in S^c are still connected) we can use the Perron–Frobenius theorem (Theorem 3.2) as we did for the graph Laplacian. Letting the maximum diagonal entry of $\mathbf{L}_{(S^c)}$ be $d_{\max}$, we can see that $d_{\max}\mathbf{I} - \mathbf{L}_{(S^c)}$ will be a non-negative matrix, with the same eigenvectors as $\mathbf{L}_{(S^c)}$ and, by Theorem 3.2, its first eigenvector will have all positive entries. Thus, for a given S the node i with largest $u^*_{S^c}(i)$ will have fewer neighbors in S because the variation with respect to those nodes would be large (since they all have value zero). Thus the index i corresponds to a node that is mostly connected to other nodes in S^c. Furthermore, of those nodes in S^c, node i will result in the largest increase in variation. This is illustrated by Figure 4.2 and Example 4.5.

Example 4.5   Find the submatrices $\mathbf{L}_{(S)}$ for the two choices of S in Figure 4.2 and for the example on the right of the figure determine what would be the next node to be added to S. Compare the respective minimum frequencies for these two graphs and deduce which is the better sampling choice.

Solution
The corresponding submatrices in this case are

$$
\begin{bmatrix}
1 & 0 & 0 & 0 \\
0 & 2 & 0 & 0 \\
0 & 0 & 2 & 0 \\
0 & 0 & 0 & 1
\end{bmatrix}
\quad \text{and} \quad
\begin{bmatrix}
2 & -1 & 0 & 0 \\
-1 & 2 & -1 & 0 \\
0 & -1 & 2 & -1 \\
0 & 0 & -1 & 1
\end{bmatrix}
$$

for the left and right graphs in Figure 4.2, respectively. The eigenvectors for the first graph are trivial, and those of the second can be found easily with Matlab. For the second graph we can verify numerically that the entries of the lowest eigenvector are monotonically increasing away from the edge connecting S and S^c. The largest entry is the fourth entry, i.e., the node in S^c that is farthest away from S. For the graph on the left the eigenvalues are $\lambda = 1, 2$, by inspection, while for the graph on the right the eigenvalues are in the range $\lambda \in [0.1, 3.6]$, which clearly shows that the sampling approach on the right is worse: if the noise is aligned with the smallest eigenvalue the error in reconstruction can be amplified by close to an order of magnitude. Note that the original graph of Figure 4.2 is a path graph, and thus it should not be surprising that sampling every other node is a better approach (as on the left of Figure 4.2).

Summary In summary, the methods described in this section require the computation of eigenvectors. In the method proposed in [40], the first R columns of $\mathbf{U}$ are computed and stored, and then used for all successive greedy steps. The approach in [57] also requires R eigenvector computations but one is performed per iteration, which requires less storage. An alternative not requiring eigendecomposition can be found in [54], where maximization of a minimum eigenvalue is replaced by maximization of a *bound* on the minimal eigenvalue on the basis of Theorem 3.5.

4.4.4 Random Sampling

In the noise-free case of Section 4.4.1, any sampling set S such that $\mathbf{U}_{SR}$ has linearly independent columns guarantees exact reconstruction. In practice, finding such a set S is not hard. Even random sampling, where each node is chosen with equal probability, can lead to noise-free exact reconstruction, as shown experimentally in [55]. However, under noisy conditions the quality of the reconstruction depends on the properties of $\mathbf{U}_{SR}$, so that significant errors could be incurred if its properties are not favorable, according to the various optimality criteria described in Section 4.4.2.

Node selection probabilities for robust sampling To achieve sampling robustness in the presence of noise, methods have been proposed where the probability of choosing a given graph node is determined by the graph topology [59, 60]. Thus, nodes that would be selected by greedy algorithms such as those in [40] or [57] are given higher probability, so that random sampling is biased towards more informative nodes.

As a representative example of random sampling, the authors of [59] characterize each sample i by the properties of an impulse signal δ_i at i. The quantity

$$p_i = \frac{1}{|R|}\|\mathbf{U}_{VR}\mathbf{U}_{VR}^\mathsf{T}\delta_i\|_2^2 = \frac{1}{|R|}\|\mathbf{U}_{VR}^\mathsf{T}\delta_i\|_2^2 \tag{4.24}$$

determines how much energy from δ_i will be in the span of the first R frequencies in $\mathbf{U}$. Note that, since $\|\delta_i\| = 1$, we have that $0 \le p_i \le 1$. Intuitively, if p_i is small (close to zero) this δ_i will have almost no information in the low frequencies. Conversely if p_i is large (close to 1), δ_i will have most of its energy in the low frequencies. It can also be shown that $\sum_i p_i = 1$ [59]. Thus, the method in [59] associates with node i a sampling probability p_i, so that nodes with larger p_i will be more likely to be selected. Once the p_i have been chosen, this method has very low complexity.

Efficient probability estimation Note, however, that computing the first R eigenvectors of $\mathbf{U}$ is complex, as discussed in Section 4.4.3. Thus, an alternative and more efficient approach is to estimate λ_R, the Rth frequency, and then design a polynomial filter

$P_R(\lambda)$ with a passband that includes λ_R. While computing λ_R exactly might be complex, especially for a large graph where many frequencies are close together, in general an estimate of λ_R is sufficient. Thus, the new probabilities are

$$p_i^a = \|P_R(\mathbf{Z})\delta_i\|_2^2 \qquad (4.25)$$

with $P_R(\mathbf{Z})$ a polynomial of degree K. Note that, once p_i or p_i^a have been computed, random sampling is fast. Its main disadvantage is that performance could vary significantly over different sampling realizations. In particular, since each random node selection is made independently of the previous selections, some chosen nodes could be very close to each other, reducing their effectiveness for sampling. This idea is further developed next.

4.4.5 Sampling Based on Localized Operators

Reconstruction as a function of local operators Note that the vectors $P_R(\mathbf{Z})\delta_i$ and $\mathbf{U}_{VR}\mathbf{U}_{VR}^\mathsf{T}\delta_i$, corresponding to the ith columns of the respective matrices, can be interpreted as interpolators applied to the signal observed at node i. This can be seen by considering the interpolation (4.18) and assuming that $\mathbf{U}_{SR}^\mathsf{T}\mathbf{U}_{SR}$ is close to $\mathbf{I}$ (or close to being diagonal). Then the interpolation can be viewed as an ideal low pass filter with cutoff frequency corresponding to the first R frequencies operating on $\mathbf{x}_S$:

$$\hat{\mathbf{x}} \approx \mathbf{U}_{VR}\mathbf{U}_{VR}^\mathsf{T}\mathbf{x}_S,$$

where $\mathbf{x}_S$ is zero for nodes outside S and the interpolation can be written as

$$\hat{\mathbf{x}} \approx \sum_{i \in S} x(i)\mathbf{U}_{VR}\mathbf{U}_{VR}^\mathsf{T}\delta_i.$$

This illustrates how sampling set quality depends on the properties of the interpolator $\mathbf{d}_i = \mathbf{U}_{VR}\mathbf{U}_{VR}^\mathsf{T}\delta_i$. Notice that δ_i has norm 1 and the columns of $\mathbf{U}_{VR}$ are orthogonal, so that $\|\mathbf{U}_{VR}\mathbf{U}_{VR}^\mathsf{T}\delta_i\|^2 = \|\mathbf{U}_{VR}^\mathsf{T}\delta_i\|^2$ is the norm of the projection of δ_i onto the space of bandlimited signals. This interpretation leads naturally to desirable properties for the interpolator: (i) the chosen i's should be such that their projections are large, i.e., $\|\mathbf{d}_i\|^2$ should be large, because this means a greater percentage of the information provided corresponds to the lower frequencies and (ii) any two chosen interpolators should be as orthogonal as possible, i.e., $\mathbf{d}_j^\mathsf{T}\mathbf{d}_i$ should be close to zero.

Generalization to other localized operators This idea was generalized in [56] by defining a localization operator $\mathbf{T}$, starting with a filter defined as a function of $g(\lambda)$ of the graph frequency λ, typically a low pass filter with some localization properties (e.g., a polynomial of low degree). As in the random sampling case, while a full eigendecomposition is not needed it may be necessary to have a rough estimate of the frequency ranges for the given graph, so that the passband of $g(\lambda)$ matches the estimated frequency range of the signals to be sampled. Once $g(\lambda)$ is chosen, define the operator $\mathbf{T}$ as

$$\mathbf{T} = \mathbf{U}g(\Lambda)\mathbf{U}^\mathsf{T},$$

that is, in the standard form of a graph filter implemented in the frequency domain. The reconstruction of a signal from its samples is

$$\hat{\mathbf{x}} = \left[\mathbf{T}^k\right]_{VS} \left(\left[\mathbf{T}^k\right]_S\right)^+ \mathbf{x}_S,$$

where $\mathbf{T}^k$ corresponds to applying the same operator k times, the square brackets denote a restriction of the $N \times N$ matrix to a subset of rows and columns and $+$ denotes the pseudo-inverse. With this operator one can define A-optimality and E-optimality criteria. In practice, simpler algorithms can be developed by considering the properties of $\mathbf{T}$. Since the operator is localized it can be applied at each node i and a decision on whether to include the node in the sampling set can be based on the local properties. Specifically, denoting by $\mathbf{T}_{g,i}$ the operator at node i, the number of non-zero (or above-threshold) values for $\mathbf{T}_{g,i}$ will give us an idea of the local spread of the interpolating function. Thus, a heuristic can be developed where the i's corresponding to better (more spread out) interpolators are chosen, and where successive choices are made that avoid excessive overlap between interpolators [56].

4.5 Comparisons and Insights

In the previous sections, conditions for reconstruction and approaches for sampling set selection were viewed primarily from a frequency perspective. In this section, our main goal is to provide some insights about how sampling algorithms prioritize nodes to be included in the sampling set. We then use this discussion to compare sampling for normalized and non-normalized graph operators.

Combinatorial Laplacian The definition of sampling probability in (4.24) can be used to provide some insights about node selection based on node properties. Recall that for $\mathbf{Z} = \mathbf{L}$ we have that

$$d_i = \delta_i^\mathsf{T} \mathbf{L} \delta_i = \sum_{k=1}^{N} \lambda_k (\mathbf{u}_k^\mathsf{T} \delta_i)^2 \quad \text{while} \quad p_i = \sum_{k=1}^{R} (\mathbf{u}_k^\mathsf{T} \delta_i)^2 \tag{4.26}$$

and note that since $\mathbf{u}_1 = \mathbf{1}$ we have that $\mathbf{1}^\mathsf{T} \delta_i$ is equal for all nodes. Thus, as a first-order approximation we can use $\mathbf{u}_2^\mathsf{T} \delta_i$ in order to determine how to prioritize graph nodes for sampling. For $\mathbf{u}_2$, assumed to have norm 1 and with corresponding eigenvalue λ_2, we have, using the Rayleigh quotient representation of Section 3.2.3,

$$\lambda_2 = \sum_{i \sim j} w_{ij}(u_2(i) - u_2(j))^2,$$

from which it is clear that a higher-degree node i will have lower $u_2(i)$ than a lower-degree node. Thus nodes that are less connected (lower degree) should be sampled first. This can be seen in the example of Figure 4.3, where nodes in the lower-density cluster (on the right in the figure) have lower degrees and thus are selected earlier in the sampling process (by iteration 20, all have been selected).

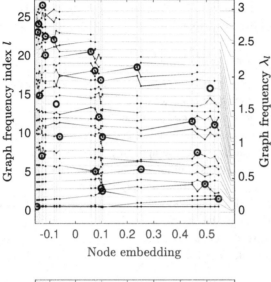

Figure 4.3 Sampling set selected for the combinatorial Laplacian, $\mathbf{Z} = \mathbf{L}$, for the graph of Figure 3.2. This graph contains three clusters. The densest cluster has nodes with higher degree (corresponding to negative values in the embedding), while the sparsest cluster has lower-degree nodes (values close to 0.5 in the embedding). Notice how most of the nodes in the sparsest cluster are chosen first while many nodes in the dense cluster are selected later.

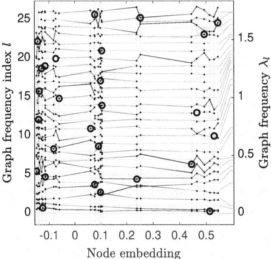

Figure 4.4 For the same graph as in Figure 4.3, the sampling sets selected for the normalized Laplacian $\mathbf{Z} = \mathcal{L}$. Notice that in this case we no longer see a significant difference between clusters regarding how the nodes within each cluster are selected.

This observation can be further justified by noting that, for any signal, the Laplacian quadratic form is a weighted sum of projections along each frequency, with higher frequencies having larger weights. Thus, from (4.26), if $d_i < d_j$ then δ_i has more information in the lower frequencies and should be selected first. Note that in the limit when the degree d_i becomes arbitrarily small, node i becomes disconnected and information at that node must be observed since it cannot be inferred from other nodes. Conversely, when i has high degree and the graph signal is low frequency we would expect node i to be very similar to any of its neighbors, so that it would suffice to observe one of them to achieve a reliable estimate for $x(i)$. As a final note, when applying the method in [54], which maximizes a bound on the minimal eigenvalue on the basis of Theorem 3.5, we can see that this optimization will also start with the lowest-degree node.

Comparing combinatorial and normalized Laplacians From Section 3.4.5 we know that $\delta_i^T \mathcal{L} \delta_i = 1$ for any i, so we can no longer reason about the order in which nodes are selected for sampling on the basis of their degree. Figure 4.3 and Figure 4.4 help us highlight the differences between normalized and non-normalized operators. Recall that the circles in the figure denote, for each frequency in the basis (from low at the bottom to high at the top) the position of the nodes selected by the sampling algorithm. Sampling based on the combinatorial Laplacian (Figure 4.3) selects first those nodes that are more isolated (in sparser clusters), while many nodes in denser clusters are only selected later. In other words, in a densely connected cluster, where the similarity between nodes is high, a few observations are sufficient to estimate a low-frequency signal. Note that, in contrast, for the normalized Laplacian (Figure 4.4), the nodes selected for sampling appear to be distributed uniformly across clusters without nodes in sparser clusters being selected first.

The difference in behavior between $\mathbf{L}$ and $\mathcal{L}$ noted here can have significant implications for specific types of graphs. For example, in a scale-free graph (Section 1.5) the sampling strategies for the two operators will be quite different, with the one for $\mathbf{L}$ approximately based on ordering nodes from low to high degree (with the low-degree nodes being selected first). Thus, as stated at various points in the preceding chapters, the choice of operator is of fundamental importance.

Chapter at a Glance

Graph signal sampling is the process of observing a subset of nodes on a graph and using the observed signals at those nodes to estimate the signal values at other nodes. Since sampling reduces the number of observed nodes, it is only possible to reconstruct sampled signals exactly for subsets of signals characterized by specific models. Thus a sampling strategy is based on choices of graph, graph operator and signal model. Among models that can be considered are bandlimited, approximately bandlimited and piecewise smooth signals. We then formulated the sampling set selection problem as well as the interpolation problem. We reviewed a series of sampling set selection approaches with different properties in terms of performance and complexity, considering in particular whether an eigendecomposition of the graph operator is required. Summarizing the approaches discussed in this chapter, we can classify them as follows (see Table 4.1).

1 Full eigendecomposition required   These methods, such as the one in [40], involve computing the first R columns of $\mathbf{U}$. This can be computationally expensive for large graphs and may require significant storage.
2 Estimated frequencies   These approaches (used, e.g., in [57]) replace exact frequency knowledge by methods analogous to power iterations for eigenvalue computation.
3 Interpolator-based methods   These approaches (used, e.g., in [56]) use a simple interpolator (not requiring eigendecomposition or pseudo-inverse computations) and base sample set selection on the properties of the interpolator.

Table 4.1 Summary of properties of sampling algorithms. Frequency: assuming that a K-bandlimited signal is being sampled what information is required? Interpolator: is the sampling set dependent on a specific interpolator? Search: is the search greedy or random?

	Frequency	Interpolator	Search
Chen et al. [40]	First K eigenvectors	N	Greedy
Anis et al. [57]	One eigenvector per iteration	N	Greedy
Puy et al. [59]	λ_K	N	Random
Sakiyama et al [56]	N	Y	Greedy

4 Random sampling   These methods (used, e.g., in [59]) choose the nodes to be sampled at random, so that an explicit sampling set selection is not required.

Further Reading

As in much of this book, we have selected publications to provide an overview of a few representative methods that illustrate the main ideas, but we refer the readers to a recent overview [53] and references therein for a review of sampling in the context of graph signal processing. For sampling in a broader signal processing context the reader is referred to [61, 39].

GraSP Examples

In Section B.5.1 we provide the code for the sampling example for the graph considered in Figure 4.3 and Figure 4.4. While this example uses the method in [57], it can be modified to make use of alternative sampling techniques.

5   Graph Signal Representations

Up to now we have considered filtering operations from a node domain (Chapter 2) and a frequency domain perspective (Chapter 3). A graph filter (Definition 2.21) can be viewed as a system with an input, the original graph signal, and an output, the filtered graph signal. In some cases, the filtering operation may be invertible while in others the input cannot be recovered from the output.

In this chapter, we study graph signal **representations**, where each signal is seen as a linear combination of elementary signals. Up to this point, we have considered two ways of representing a graph signal: (i) directly as a vector, $\mathbf{x}$, of scalar values, each one associated with a node and (ii) on the basis of its spectrum, $\tilde{\mathbf{x}}$, obtained using one of the definitions of the graph Fourier transform (GFT). In this chapter, we consider representations that can be used as alternatives to the GFT. More precisely, these are functions of $\mathbf{x}$ (or $\tilde{\mathbf{x}}$) from which the original signal can be recovered, exactly (or approximately). Some of these representations can be obtained with **filterbanks**, i.e., a series of filters with different properties. The outputs of all filters or a subset of outputs obtained from sampling (Chapter 4) will provide the "coefficients" of the representation, i.e., the weights associated with each of the elementary functions.

Mathematically, the representations in this chapter are bases, or more generally dictionaries, for signals in vector spaces and we refer the reader to Appendix A for a review of the basic ideas. We start by discussing node and frequency localization in Section 5.1. Then, we introduce desirable properties of graph signal representations (Section 5.2) and use a simple example to illustrate the trade-offs between these properties (Section 5.3). We end by providing an overview of the node domain (Section 5.4) and frequency domain (Section 5.5) representations proposed in the literature.

5.1   Node–Frequency Localization and Uncertainty Principles

Graph signals can be represented directly as vectors of samples (in the node domain) or as linear combinations of elementary frequencies (in the frequency domain). Correspondingly, conventional signals can be represented as a series of samples (in the time domain) or by their Fourier transform coefficients (in the frequency domain). Informally, a signal is localized in the node domain if most non-zero values occur in a small subset of connected nodes (Chapter 2). Similarly, a signal with its energy localized in the frequency domain is expected to have non-zero frequency values in a small range of frequencies. In this section we make these definitions of localization more precise.

Box 5.1 reviews the basic concepts of time–frequency localization, which are the basis for developing similar ideas for graph signals. Because elementary frequency signals have infinite length, perfect localization in frequency (all the energy in one frequency) leads to no localization in time (infinite length). Extending these ideas to graph signals requires defining concepts of spread in the node and frequency domains, equivalent to those proposed for conventional signals (see Box 5.1). But, while good definitions of localization are possible, graph signals, unlike conventional signals, can have compact support in both the node and frequency domains (Section 5.1.2) (i.e., a signal may be identically zero at some nodes and also have zero energy in some frequencies).

Box 5.1   Time–frequency localization

Time localization

Consider an infinite-length real-valued finite-energy signal $\phi(n)$, normalized so that $\|\phi(n)\|^2 = \sum_n \phi(n)^2 = 1$, with "center of mass" in time

$$\mu_{\mathrm{t}} = \sum_{n=-\infty}^{n=+\infty} n\phi(n)^2. \tag{5.1}$$

Since $\phi(n)$ has total energy equal to 1, $p_{\mathrm{t}}(n) = \phi(n)^2$ can be viewed as a probability mass function ($\sum_n p_{\mathrm{t}}(n) = 1$). Then, we can quantify the spread of $\phi(n)$ around its mean by defining

$$\sigma_{\mathrm{t}}^2 = \sum_{n=-\infty}^{n=+\infty} (n - \mu_{\mathrm{t}})^2 \phi(n)^2. \tag{5.2}$$

Note that if we compare two normalized signals $\phi_1(n)$ and $\phi_2(n)$, the signal with larger σ_{t}^2 will have its energy more broadly spread out in the time domain.

Frequency localization

For frequency localization, we follow the approach in [17]. Define the discrete-time Fourier transform (DTFT)

$$\Phi(e^{j\omega}) = \sum_{n=-\infty}^{n=+\infty} \phi(n)e^{-j\omega n} = \langle \phi(n), e^{j\omega n}\rangle,$$

where each $e^{j\omega n}$, for a given parameter ω, is a discrete-time signal and the set $\{e^{j\omega n}\}_{\omega \in [0,2\pi)}$ forms an orthogonal basis set. Given that $\phi(n)$ is real, we have symmetry in the magnitude of the DTFT: $|\Phi(e^{j\omega})| = |\Phi(e^{-j\omega})|$. Then using Parseval's relation (Section A.4) and exploiting this symmetry,

$$\frac{1}{\pi} \int_0^\pi |\Phi(e^{j\omega})|^2 d\omega = 1$$

where we note that each of the elementary signals has infinite length, $p_{\mathrm{f}}(\omega) =$

$(1/\pi)|\Phi(e^{j\omega})|^2$ is a probability density function, the expected value of ω,

$$\mu_f^+ = \frac{1}{\pi} \int_0^\pi \omega |\Phi(e^{j\omega})|^2 d\omega, \tag{5.3}$$

can be seen as the center of mass in the frequency domain and

$$(\sigma_f^+)^2 = \frac{1}{\pi} \int_0^\pi (\omega - \mu_f^+)^2 |\Phi(e^{j\omega})|^2 d\omega \tag{5.4}$$

is a measure of the spread in the frequency domain. Note that $(n - \mu_t)^2$ and $(\omega - \mu_f^+)^2$ in (5.2) and (5.4) measure the distances between locations in time n and frequency ω and the respective centers of mass, μ_t and μ_f.

Uncertainty principle

The uncertainty principle states that for any signal $\phi(n)$ we cannot simultaneously have good localization in time and frequency, or, formally,

$$\sigma_t \sigma_f^+ \geq K, \tag{5.5}$$

where K is a non-zero constant and where "good" localization is achieved when a signal, e.g., a basis function, has low spread, so that σ_t or σ_f are small. From the bound it is not possible for both σ_t and σ_f to be arbitrarily small. The bound given by (5.5) is achieved with equality for Gaussian signals.

Why localization matters As for conventional signals, estimating localization is important to understand graph signals of interest: does an observed signal $\mathbf{x}$ have most of its energy in a specific connected area of the graph? or is it spread out? is the content primarily low frequency and smooth or not? Furthermore, localization is essential to understand elementary basis functions in a signal representation. As described in Section A.3.2, the inner product between a signal and an elementary basis function quantifies the similarity between the two. Thus, if we are observing a signal $\mathbf{x}$ and one of the elementary basis functions is $\mathbf{h}_i$, then if $(1/\|\mathbf{x}\|)|\langle \mathbf{x}, \mathbf{h}_i \rangle|$ is large this is an indication that $\mathbf{x}$ and $\mathbf{h}_i$ have similar localization properties. Since the elementary basis functions or dictionary atoms (see Section A.5) $\{\mathbf{h}_i\}$ are known *a priori*, we can analyze indirectly the localization properties of $\mathbf{x}$ by projecting $\mathbf{x}$ onto each of those $\{\mathbf{h}_i\}$.

5.1.1 Measuring Localization for Graph Signals

We start by describing localization metrics proposed by Agaskar and Lu [62], which extend to graphs the concepts of Box 5.1 while taking into account the unique characteristics of graph signals and graph frequencies.

Node domain localization For node domain localization, the goal is to define quantities similar to (5.1) and (5.2). The first challenge is that there is no way to define a center of mass as in (5.1). Unless the nodes correspond to points in a Euclidean space, there is no such thing as an "average" node. Instead, in [62] it was proposed that local-ization should be defined around each node. Thus, for a specific node a, the localization is measured by computing how the signal decays as our observations move away from a:

$$\Delta_{G,a}^2(\mathbf{x}) = \sum_{i \in V} d(a,i)^2 \frac{x(i)^2}{\|\mathbf{x}\|^2}, \tag{5.6}$$

where $d(a,i)$ is a distance, e.g., the number of hops in the shortest path between a and i (Definition 2.8). Recall from Section 3.3.3 that $(\mathbf{A}^k)_{ai} = 0$ if there are no paths of length k between a and i. Thus, assuming $\mathbf{A}_{ai} = 0$, we have that $d(a,i)$ is the smallest k such that $(\mathbf{A}^k)_{ai} \neq 0$. Similarly to the regular signal case (Box 5.1) we can define $p_i = x(i)^2/\|\mathbf{x}\|^2$ with $\sum_i p_i = \|\mathbf{x}\|^2/\|\mathbf{x}\|^2 = 1$, and thus (5.6) can be viewed as the expected value of the squared distances between a and all other nodes, with weights given by the p_i.

Note that $\Delta_{G,a}^2(\mathbf{x})$ can be defined for any a, whereas the variance in (5.4) is defined with respect to the center of mass. Since $\Delta_{G,a}^2(\mathbf{x})$ can vary significantly for different nodes a, an alternative approach would be to use the node for which this quantity is a minimum as a proxy for the center of mass by defining (see [63])

$$a_{\min}(\mathbf{x}) = \arg\min_{a \in V} \Delta_{G,a}^2(\mathbf{x}).$$

Also note that if d is the diameter of the graph (Definition 2.9) then, for any $\mathbf{x}$ and a, $\Delta_{G,a}^2(\mathbf{x}) \leq d^2$. Computing these quantities for a given reference node a allows us to compare two graph signals $\mathbf{x}$, $\mathbf{y}$ as follows: $\Delta_{G,a}^2(\mathbf{x}) > \Delta_{G,a}^2(\mathbf{y})$ means that $\mathbf{x}$ is more spread out than $\mathbf{y}$, i.e., it has relatively more weight in nodes farther away from a.

Frequency domain localization   The graph frequencies λ_k are discrete and irregularly spread. Thus, the average λ in general does not correspond to an actual graph frequency, unlike in (5.3). Similarly to the node domain example, the spread can be measured with respect to a specific frequency λ_i, so that for a given frequency representation $\tilde{\mathbf{x}}$ of $\mathbf{x}$ the frequency spread can be defined as [62]

$$\Delta_{S,i}^2(\mathbf{x}) = \frac{1}{\|\mathbf{x}\|^2} \sum_{k=1}^{N} |\lambda_i - \lambda_k| \|\tilde{x}(k)\|^2, \tag{5.7}$$

where $\tilde{x}(k)$ is the kth entry of the GFT of $\mathbf{x}$, $\tilde{\mathbf{x}} = \mathbf{U}^T\mathbf{x}$. Defining $p_k = \tilde{x}(k)^2/\|\mathbf{x}\|^2$, from Parseval's relation (Section A.4) we have that $\sum_k p_k = \sum_k \tilde{x}(k)^2/\|\mathbf{x}\|^2 = \|\mathbf{x}\|^2/\|\mathbf{x}\|^2 = 1$, so that p_k can be interpreted as a probability mass function. A simpler expression is possible when using $\lambda_i = \lambda_1$ as the reference for $\mathbf{Z} = \mathcal{L}$:

$$\Delta_{S,1}^2(\mathbf{x}) = \Delta_S^2(\mathbf{x}) = \frac{1}{\|\mathbf{x}\|^2}\mathbf{x}^t\mathcal{L}\mathbf{x} = \frac{1}{\|\mathbf{x}\|^2} \sum_{k=1}^{N} \lambda_k|\tilde{x}(k)|^2, \tag{5.8}$$

where we have used that $\lambda_k > \lambda_1 = 0$ for any $k > 1$. If we use λ_1 as the reference, when comparing two signals $\mathbf{x}$ and $\mathbf{y}$ with similar frequency localization the signal with higher frequencies will lead to a larger $\Delta_{S,1}^2(\cdot)$. For example, for the elementary basis vector $\mathbf{u}_i$ we have $\Delta_S^2(\mathbf{u}_i) = \lambda_i$, so that a larger $\Delta_S^2(\mathbf{u}_i)$ is associated with $\mathbf{u}_i$, corresponding to higher frequencies.

Trade-offs between node and frequency localization To understand the trade-offs in localization it is useful to consider signals that are known to be exactly localized. This is illustrated by the following example (see [62] for more details).

Example 5.1 For $\mathbf{Z} = \mathcal{L}$ and for a given a, find the signal with maximum node domain localization (minimum $\Delta_{G,a}^2(\mathbf{x})$) and also find its frequency localization. Find upper and lower bounds for the frequency localization.

Solution

Perfect node domain localization can be achieved by an impulse centered at a, δ_a, for which (5.6) leads to $\Delta_{G,a}^2(\delta_a) = 0$ since $d(a, a) = 0$ and δ_a has energy only at node a. For this impulse signal, $\delta_a(a) = 1$, from (3.32) we have $\Delta_S^2(\delta_a) = \delta_a^\mathsf{T} \mathcal{L} \delta_a = 1$. The frequency localization in (5.8) uses the lowest frequency $\lambda_1 = 0$ as a reference. Then, the minimum frequency spread is achieved for the eigenvector $\mathbf{u}_1$ corresponding to $\lambda_1 = 0$, while the worst frequency localization is achieved by $\mathbf{u}_N$:

$$\Delta_S^2(\mathbf{u}_1) = 0 = \lambda_1 \leq \frac{\mathbf{x}^t \mathcal{L} \mathbf{x}}{\mathbf{x}^t \mathbf{x}} \leq \lambda_N = \Delta_S^2(\mathbf{u}_N).$$

The general problem of finding feasible regions was studied in [62]. A **feasible region** contains pairs (s, g) for which it is possible to find a signal $\mathbf{x}$ with $\Delta_S^2(\mathbf{x}) = s$ and $\Delta_{G,a}^2(\mathbf{x}) = g$. From Example 5.1.1, we can see that the (s, g) pairs $(0, \Delta_{G,a}^2(\mathbf{u}_1))$ and $(1, 0)$ are both feasible, and correspond to $\mathbf{u}_1$ and δ_a, respectively. From (5.5) (Box 5.1) we have that time and frequency localization cannot both be made arbitrarily small. This brings up the question of whether arbitrarily small $\Delta_S^2(\mathbf{x})$ and $\Delta_{G,a}^2(\mathbf{x})$ can be achieved for some $\mathbf{x}$.

Since $\Delta_S^2(\mathbf{x})$ takes $\mathbf{u}_1$ as a reference, this question is equivalent to asking whether we can have a highly localized signal in the node domain with arbitrarily low frequency. From Theorem 3.2 and the subsequent discussion, we know that the lowest frequency has all positive (non-zero) entries and so will have $\Delta_{G,a}^2(\mathbf{u}_1)$ greater than zero for any a. Lower bounds (away from the origin) for the feasibility region have been developed [62]. Defining a diagonal matrix $\mathbf{P}_a$ with entries $d(a, i)$ ($d(a, a) = 0$), so that (5.6) can be rewritten as

$$\Delta_{G,a}^2(\mathbf{x}) = \frac{1}{\|\mathbf{x}\|^2} \mathbf{x}^t \mathbf{P}_a \mathbf{x},$$

it can be shown that the lower bound of this feasibility region can be described by a series of lines with slope α such that

$$\Delta_{G,a}^2(\mathbf{x}) - \alpha \Delta_S^2(\mathbf{x}) = q(\alpha), \tag{5.9}$$

where $\mathbf{M}(\alpha) = (\mathbf{P}_a^2 - \alpha \mathcal{L})$ and $q(\alpha)$ is the minimum eigenvalue $\mathbf{M}(\alpha)$, with eigenvector $\mathbf{s}(\alpha)$. The bound is achieved by the eigenvector $\mathbf{s}(\alpha)$. Notice that $\mathbf{P}_a^2$ is diagonal, with a zero diagonal entry corresponding to a and other entries having increasing values if they are further away from a. Thus, the resulting optimal trade-off point, corresponding to the minimal eigenvector, is likely to have larger values at a and at nodes close to a, in order to achieve better localization relative to a.

Extension to weighted graphs   The preceding discussion was developed for unweighted graphs. As pointed out in [64], extension to weighted graphs can be problematic. Assume two nodes i and j are connected by an edge of weight ϵ, and that the shortest non-direct path between i and j has weight w. Then as ϵ goes to zero the term $\epsilon^2 x_j^2$ in $\Delta_{G,i}$ will go to zero, but once $\epsilon = 0$ the corresponding term will be $w^2 x_j^2$. As an alternative [64] introduced weights to quantify node domain spread on the basis of similarity, the reciprocal of distance, by defining

$$\forall i, j \in \mathcal{V} \quad \begin{cases} s_{ij} = \infty & \text{if } a_{ij} = 0, \\ s_{ij} = 0 & \text{if } a_{ij} = \infty, \\ s_{ij} = 1/a_{ij} & \text{otherwise.} \end{cases}$$

An alternative approach would be to update the geodesic distance: as ϵ goes to zero, the maximum similarity would not be through the direct connection (of weight ϵ) but rather through the indirect connection (of weight w). In summary, for the localization metric to be meaningful, shortest paths and geodesic distances have to be recomputed every time the graph edge weights change.

5.1.2   Compact Support in the Node and Frequency Domains

Another way to quantify localization is to determine whether some graph signals can have compact support [58]. A signal with compact support is *exactly* localized in a subset of nodes and frequencies, that is, it is exactly zero outside those sets. More formally, given $\mathcal{S}$, a set of nodes where $|\mathcal{S}| < N$ and $\mathcal{F}$, a set of frequencies with $|\mathcal{F}| < N$, the goal is to determine whether a signal can be exactly localized in both $\mathcal{S}$ and $\mathcal{F}$ and thus have zero values in both $\mathcal{S}^c$ and $\mathcal{F}^c$. While simultaneous compactness in time and frequency is not possible for conventional signals (see [17] for example), it is in fact possible for graph signals [58].

Conditions for simultaneous compact support   Define $\mathbf{I}_s$, the indicator matrix for set $\mathcal{S}$, a diagonal matrix with entries equal to 1 for nodes in $\mathcal{S}$ and zero otherwise. Thus, a signal $\mathbf{x}$ such that

$$\mathbf{I}_S \mathbf{x} = \mathbf{x}$$

is *exactly localized in* $\mathcal{S}$ since $\mathbf{I}_S$ preserves the entries in $\mathcal{S}$ and zeroes out the rest. Similarly, for frequency domain localization one can define $\mathbf{I}_F$ as a diagonal matrix with entries equal to 1 for frequencies in the set $\mathcal{F}$ and other diagonal entries equal to zero. Then, with $\mathbf{B}_F = \mathbf{U}\mathbf{I}_F\mathbf{U}^\mathsf{T}$, we can say that a *signal is exactly localized in* $\mathcal{F}$ if we have

$$\mathbf{B}_F \mathbf{x} = \mathbf{x},$$

that is, after taking the GFT, restricting to the frequencies in $\mathcal{F}$ and then inverting, the signal has not changed. It can be shown that a signal $\mathbf{x}$ is perfectly localized in both the node domain and frequency domain if and only if

$$\mathbf{C}(S, \mathcal{F})\mathbf{x} = \mathbf{B}_F \mathbf{I}_S \mathbf{B}_F \mathbf{x} = \mathbf{x}, \qquad (5.10)$$

that is, $\mathbf{x}$ is an eigenvector of $\mathbf{C}(S, \mathcal{F})$ corresponding to eigenvalue 1. Signals that meet the condition of (5.10) belong to the intersection of two subspaces, $\mathcal{V}_S = \mathrm{span}(\mathbf{I}_S)$ and $\mathcal{V}_F = \mathrm{span}(\mathbf{U} \mathbf{I}_F)$, of dimensions $|S|$ and $|\mathcal{F}|$, respectively. Thus, it is easy to see that there will be vectors in the intersection if $|\mathcal{F}| + |S| > N$, i.e., if the total number of degrees of freedom of $\mathcal{F}$ and S is greater than N. As noted in [58], solutions may exist even if this condition is not met.

5.1.3 Graph Slepians

The ideas in Section 5.1.2 also lead to the design of graph Slepians, orthogonal bases that can provide alternatives to the GFT with a design that favors localization in both the node and spectral domains [65]. Denoting, as in Section 5.1.2, S and $\mathcal{F}$ as subsets in the node and frequency domains, with $\mathcal{F}$ chosen to contain the first $|\mathcal{F}|$ frequencies, the goal is no longer to find vectors with constant support in S and $\mathcal{F}$, but rather to design basis vectors with most (but not all) of their energy in those subsets. This can be accomplished by revisiting the Rayleigh quotient, (3.27), and its application in the computation of the eigenvectors of the graph Laplacian, (3.36) and (3.37).

Denoting by $\mathbf{U}_F$ the $N \times |\mathcal{F}|$ matrix containing the first $|\mathcal{F}|$ columns of $\mathbf{U}$, the goal is to find orthogonal basis vectors for the span of $\mathbf{U}_F$ such that their energy is concentrated in the nodes selected by S. Defining $\mathbf{v}_i = \mathbf{U}_F \tilde{\mathbf{v}}_i$, then the set of graph Slepians $\mathbf{v}_i$ can be obtained by solving

$$\tilde{\mathbf{v}}_k = \arg\min_{\tilde{\mathbf{v}}} \left(\frac{\tilde{\mathbf{v}}^T \mathbf{U}_F^T \mathbf{I}_S \mathbf{U}_F \tilde{\mathbf{v}}}{\tilde{\mathbf{v}}^T \tilde{\mathbf{v}}} \right) \qquad (5.11)$$

$$\text{for } \tilde{\mathbf{v}} \in \mathbb{R}^{|\mathcal{F}|}, \ \tilde{\mathbf{v}} \neq \mathbf{0}, \ (\mathbf{U}_F \tilde{\mathbf{v}}_k)^T \mathbf{U}_F \tilde{\mathbf{v}}_i = 0, \ \forall i = 1, \dots, k-1,$$

where the term $\mathbf{U}_F^T \mathbf{I}_S \mathbf{U}_F$ favors the concentration of energy in S but leads to solutions $\mathbf{v}_i$ that are not compactly supported on S. An example application is related to piecewise smooth signals, described in Section 4.1.3. If we are given a series of regions on the graph, S_1, S_2, etc., a graph Slepian that combines smoothness and node localization can be designed for each region S_i. Thus, graph Slepians can be used to determine whether a signal is piecewise smooth over a region S_i: if it is, then its projection onto vectors $\tilde{\mathbf{v}}$ corresponding to S will contain most of its energy in the nodes in S.

5.2 Representing Graph Signals: Goals

A graph signal representation is completely described by a *dictionary*, $\mathbf{V}$, an $N \times M$ matrix where each column is an elementary signal in the representation. The number of columns, M, is not necessarily equal to the number of rows. Refer to Section A.2 and Section A.5 for a review of basic concepts related to signal representations.

Any graph signal $\mathbf{x} \in \mathbb{R}^N$ can be written as

$$\mathbf{x} = \mathbf{V}\mathbf{y}, \tag{5.12}$$

where $\mathbf{y} \in \mathbb{R}^M$ is the vector of expansion coefficients, with each entry of $\mathbf{y}$ providing the weights for the corresponding column in $\mathbf{V}$. If $M > N$ there are infinitely many choices of $\mathbf{y}$ for a given $\mathbf{x}$ (see Section A.5). In some cases $\mathbf{y}$ is found by applying a linear transformation $\tilde{\mathbf{V}}$, an $M \times N$ matrix,

$$\mathbf{y} = \tilde{\mathbf{V}}\mathbf{x}, \tag{5.13}$$

while in other cases $\mathbf{y}$ in (5.12) is the solution to an optimization problem (Section A.5). We now discuss desirable properties for these transforms, some general and applicable to regular domain signals, others specific to graph signals.

5.2.1 Structured Dictionaries

Similarly to our discussion about graph filters in Box 3.3, defining a *graph* signal representation requires choosing elementary vectors (the vectors in the dictionary) that have a node domain (Section 5.4) and/or graph spectral domain (Section 5.5) interpretation. A natural way to achieve this goal is by designing graph filterbanks, where $\mathbf{y}$ is obtained from the output of a series of graph filters and $\mathbf{x}$ is recovered by applying another set of filters to $\mathbf{y}$. Thus, both $\mathbf{V}$ and $\tilde{\mathbf{V}}$ can be written in terms of polynomials of $\mathbf{Z}$, and indeed the dictionaries do not need to be expressed in matrix form (Section 5.5.1 and Section 5.6.4 give examples of these types of structured dictionaries). The main advantage of structured bases is that their elementary operations can be easily interpreted. For example, if a representation uses two filters, one low pass and one high pass, we can easily analyze signals based on filter outputs. Section 5.3 provides a simple example of a structured representation based on one-hop operators.

5.2.2 Exact Reconstruction

A transformation given by (5.12) and (5.13) provides an exact reconstruction, i.e., $\mathbf{x}$ can be recovered from $\mathbf{y}$. Just as $\tilde{\mathbf{x}}$, the GFT of $\mathbf{x}$, provides information about the elementary frequency components in $\mathbf{x}$, a signal representation $\mathbf{y}$ makes it possible to analyze $\mathbf{x}$ using the properties of its elementary vectors, the columns of $\mathbf{V}$. In applications such as machine learning even non-invertible representations can be useful: a good representation needs to preserve only the information needed for classification or recognition. In particular, no exact reconstruction is possible if $\mathbf{V}$ is such that $M < N$. In this case the dimension of the column space of $\mathbf{V}$ is less than N, so that we cannot represent all signals in $\mathbb{R}^N$. This would be an example of dimensionality reduction.

5.2.3   Orthogonality

In an orthogonal representation $\mathbf{V}$ is a square matrix such that

$$\mathbf{V}^T\mathbf{V} = \mathbf{I} \quad \text{and} \quad \tilde{\mathbf{V}} = \mathbf{V}^T.$$

Orthogonality is desirable because the transform coefficients can be efficiently computed through an orthogonal projection onto the columns of $\mathbf{V}$, $y(i) = \langle \mathbf{x}, \mathbf{v}_i \rangle$, and from Parseval's relation the norm is preserved, $\|\mathbf{y}\| = \|\mathbf{x}\|$ (see Section A.2 for a review).

In a biorthogonal representation $\mathbf{V}$ is square, $\tilde{\mathbf{V}} = \mathbf{V}^{-1}$ but the norm is not preserved. If $\mathbf{V}$ is biorthogonal then $\mathbf{V}^T\mathbf{V} \neq \mathbf{I}$, but biorthogonal transformations can be designed to be nearly orthogonal, so that $|\langle \mathbf{v}_i, \mathbf{v}_j \rangle|$ is small for $i \neq j$ and can be useful if an orthogonal transform with desirable properties cannot be found (see Section 5.6.4 for an example).

5.2.4   Critical Sampling and Overcomplete Representations

A transform is critically sampled if the dictionary $\mathbf{V}$ is a square matrix, $M = N$, with $\tilde{\mathbf{V}} = \mathbf{V}^{-1}$ if $\mathbf{V}$ is invertible. If $M < N$ we will not be able to recover arbitrary signals, but this kind of "compressive" representation may be useful in applications where dimensionality reduction is needed (e.g., in a machine learning setting, as mentioned earlier).

In an overcomplete representation (Section A.5) $M > N$, so that a larger number of transform coefficients is needed to represent a signal. In a structured filterbank dictionary, the representation will be overcomplete if all filter outputs are preserved. For example, if we use two filters, each will produce N outputs and without any further sampling we will have $M = 2N$. A critically sampled representation can be achieved by selecting a subset of the filter outputs. Section 5.5.1 and Section 5.6.4 give examples of filterbank-based overcomplete and critically sampled representations, respectively.

While an overcomplete representation appears to be less efficient than a critically sampled method, there are two main reasons why overcomplete representations are useful. First, they give us more flexibility in designing the dictionary, i.e., choosing $\mathbf{V}$, in such a way that the elementary vectors have desirable properties. If we are interested in basis vectors with good localization properties, for example, an overcomplete $\mathbf{V}$ can include an elementary basis with different localization characteristics (k-hop-localized elementary vectors for various choices of k and centered at different nodes i). This may not be so easily done if we need to choose vectors for $\mathbf{V}$ having those localization properties while remaining linearly independent. Second, for a given overcomplete dictionary, an infinite number of representations $\mathbf{y}$ are possible for a given signal $\mathbf{x}$ (Section A.5). Thus, algorithms can be developed to choose the more informative representation among those available. For example, choosing a sparse $\mathbf{y}$ (few non-zero entries) allows us to interpret a signal using just a small set of elementary vectors.

5.2.5   Localization

Similarly to transforms for regular domain signals, we can analyze the node domain and graph frequency domain localization of graph signals. As discussed in Section 5.1, such an extension is possible but many results for regular domain signals do not extend directly. Since a transformation as described by (5.13) and (5.12) involves computing inner products with the rows of $\hat{\mathbf{V}}$ in (5.13) and thus measuring similarity, as shown in Section A.3.2, we can also characterize the localization of these transforms (in the node and frequency domains) in terms of the localization of the basis (the columns of $\mathbf{V}$ and rows of $\tilde{\mathbf{V}}$).

As we shall see, node domain designs (Section 2.4) allow us to control localization in the node domain directly, while frequency domain designs allow us to control spectral localization. However, it is important to keep in mind that the differences with respect to conventional signal processing designs are substantial, and thus it may be possible to have several basis vectors representing a narrow range of frequencies while also being localized in the node domain (see Section 5.1).

5.2.6   Graph Types Supported, Graph Approximation and Simplification

As discussed throughout this book, graphs of interest can have very different structures. Some types of transforms can be used for any graph, while others require graphs to have specific characteristics. In those cases, we may choose to approximate the original graph G by a graph G' having the desired properties. If a transformation operates on G then every node domain operation for any given node uses all the neighboring nodes in G with their corresponding edge weights. Equivalently, the filtering operation can be written in matrix form as a function of the fundamental matrix $\mathbf{Z}$ of G.

When a transform cannot be applied directly to a given graph G, the graph may be "simplified" by maintaining the same set of nodes but removing certain edges from the graph, leading to G'. Note that removing edges leads to a different spectral representation, and simplifications need to be performed carefully if one wishes to preserve frequency interpretations provided by the original graph (see Section 6.1).

5.3   Building a Simple Representation Based on One-Hop Operations

To illustrate the choices described in the previous section, we present a simple example in detail. Linear graph signal representations can be written directly in matrix form as in (5.12) and (5.13). Here we consider a simple structured design, where $\mathbf{V}$ and $\tilde{\mathbf{V}}$ are designed using graph filters.

Selecting a filter   Let us start with the random walk Laplacian $\mathcal{T} = \mathbf{D}^{-1}\mathbf{A}$ and denote

$$\mathbf{y}_1 = (\mathbf{I} - \mathcal{T})\mathbf{x}, \tag{5.14}$$

which can be interpreted as a high pass filter $\mathbf{H}_1$ acting on the input and written as a polynomial of $\mathcal{T}$:

$$\mathbf{H}_1 = h_1(\mathcal{T}) = \mathbf{I} - \mathcal{T}. \tag{5.15}$$

In the frequency domain $\mathcal{T} = \mathbf{U}\Lambda\mathbf{U}^{-1}$, $\tilde{\mathbf{x}} = \mathbf{U}^{-1}\mathbf{x}$ and $h_1(\mathcal{T}) = \mathbf{U}(\mathbf{I} - \Lambda)\mathbf{U}^{-1}$. Notice that the filter (5.15) has a frequency response corresponding to a polynomial of λ:

$$h_1(\lambda) = 1 - \lambda \tag{5.16}$$

with root $\lambda = 1$, which indeed corresponds to the eigenvector $\mathbf{1}$. The corresponding one-hop transform at node i can be written as

$$y_1(i) = x(i) - \frac{1}{d_i} \sum_{m=1}^{N} a_{im} x(m), \tag{5.17}$$

which assigns to each node i the difference with respect to the average of its neighbors. Clearly, since $h_1(\mathcal{T})\mathbf{1} = \mathbf{0}$ this transformation is not invertible.

Adding a second filter   As a first step towards creating an invertible transform we can select a second filter $h_0(\mathcal{T})$, and, since $h_1(1) = 0$, we should choose this new filter so that $h_0(1) \neq 0$. For example, let us choose

$$\mathbf{H}_0 = h_0(\mathcal{T}) = \mathbf{I} + \mathcal{T}, \tag{5.18}$$

corresponding to $h_0(\lambda) = 1 + \lambda$ with root $\lambda = -1$. We can then combine the two transformations to obtain a two-channel filterbank:

$$\mathbf{y} = \begin{bmatrix} \mathbf{y}_0 \\ \mathbf{y}_1 \end{bmatrix} = \mathbf{T}_a \mathbf{x} = \begin{bmatrix} \mathbf{I} + \mathcal{T} \\ \mathbf{I} - \mathcal{T} \end{bmatrix} \mathbf{x}, \tag{5.19}$$

where we can see that, since $(\mathbf{I} + \mathcal{T})\mathbf{1} \neq \mathbf{0}$, the columns of $\mathbf{T}_a$ are linearly independent and $\mathbf{T}_a$ has rank N. To invert (5.19) and recover $\mathbf{x}$ from $\mathbf{y}$ observe that

$$\mathbf{T}_a^\mathsf{T}\mathbf{T}_a = 2(\mathbf{I} + \mathcal{T}^\mathsf{T}\mathcal{T}) \tag{5.20}$$

has rank N and thus will be invertible. This shows that $\mathbf{x}$ can be recovered since (see also Section A.5)

$$\mathbf{x} = (\mathbf{T}_a^\mathsf{T}\mathbf{T}_a)^{-1}\mathbf{T}_a^\mathsf{T}\mathbf{y} = (\mathbf{T}_a^\mathsf{T}\mathbf{T}_a)^{-1}\mathbf{T}_a^\mathsf{T}\mathbf{T}_a\mathbf{x}. \tag{5.21}$$

Thus, we can view the system in terms of (5.13) and (5.12) by making the identifications

$$\tilde{\mathbf{V}} = \mathbf{T}_a, \quad \mathbf{V} = (\mathbf{T}_a^\mathsf{T}\mathbf{T}_a)^{-1}\mathbf{T}_a^\mathsf{T}. \tag{5.22}$$

Note that this simple design can be easily generalized by choosing other filters $\mathbf{H}_0$ and $\mathbf{H}_1$. The only requirement is for these filters not to have a common zero, which can be expressed as $\mathcal{N}(\mathbf{H}_0) \cap \mathcal{N}(\mathbf{H}_1) = \{\mathbf{0}\}$. If $\mathbf{H}_0$ and $\mathbf{H}_1$ are polynomials $\mathcal{T}$ (or in general $\mathbf{Z}$), $h_0(\lambda)$ and $h_1(\lambda)$ cannot have a common root. That is, there is no eigenvalue λ_i of $\mathcal{T}$ such that $h_0(\lambda_i) = h_1(\lambda_i) = 0$.

Observations   This example leads to several important observations that will be useful in the rest of the chapter.

1 The transform defined by (5.22) is not critically sampled, since we are representing a signal $\mathbf{x} \in \mathbb{R}^N$ by a vector $\mathbf{y} \in \mathbb{R}^{2N}$. In general, overcomplete representations are easier to construct and, as we just saw, a reconstruction is possible under mild conditions (no common zeros) on $\mathbf{H}_0$ and $\mathbf{H}_1$.

2 Achieving critical sampling for the given filtering operations $\mathbf{H}_0$ and $\mathbf{H}_1$ would require selecting N entries of $\mathbf{y}$ while guaranteeing reconstruction. This is equivalent to selecting N rows of $\mathbf{T}_a$, to obtain $\mathbf{T}'_a$, while ensuring that the rank of $\mathbf{T}'_a$ is N. This can be done with some of the sampling techniques in Chapter 4.

3 However, $\mathbf{T}'_a$ will no longer have a simple frequency domain interpretation. We can write (5.20) as

$$\mathbf{T}_a^\mathsf{T}\mathbf{T}_a = \mathbf{U}(2\mathbf{I} + \mathbf{\Lambda})\mathbf{U}^{-1}$$

in the frequency domain defined by the operator $\mathbf{Z} = \mathcal{T}$ but, in general, $\mathbf{T}'_a$ cannot be diagonalized by $\mathbf{U}$.

4 The reconstruction operator $(\mathbf{T}_a^\mathsf{T}\mathbf{T}_a)^{-1}$ is not a simple polynomial of $\mathcal{T}$. It can be written as follows:

$$(\mathbf{T}_a^\mathsf{T}\mathbf{T}_a)^{-1} = \mathbf{U}((2\mathbf{I} + \mathbf{\Lambda})^{-1})\mathbf{U}^{-1}$$

and thus as a polynomial of $\mathcal{T}$. Generally, though, this polynomial will not have low degree. Consequently, while the input filters are one-hop localized, the reconstruction filters are not (the degree could be as high as that of the minimal polynomial of $\mathcal{T}$).

The designs we describe in Section 5.4 and Section 5.5 illustrate how different operating points can be achieved in terms of the properties of Section 5.2: structure, oversampling, exact reconstruction, orthogonality and applicability to specific graphs.

5.4 Node Domain Graph Signal Representations

Node domain signal representations are computed as linear combinations of information at a node and its immediate neighbors, similarly to the one-hop operation of (5.17) in Section 5.3 but operating locally over larger neighborhoods around each node. If these local operations can be expressed in terms of $\mathbf{Z}$ then there will be a corresponding spectral interpretation.

5.4.1 Node Domain Graph Wavelet Transforms (NGWTs)

Node-domain graph wavelet transforms (NGWTs) [66], one of the first proposed transforms for graph signals, are overcomplete node domain transforms for unweighted graphs built by extending node domain operations such as those of (5.17) beyond one-hop. An important difference with respect to the filterbank methods in Section 5.5 is that NGWTs cannot be expressed in terms of polynomials of $\mathbf{Z}$.

The idea is that NGWTs consider the neighborhood of each node in terms of a series of rings. Denoting by $\mathcal{N}_k(i) \in \mathcal{V}$ the subset of nodes that are up to k-hops away from node i, where $\mathcal{N}_0(i) = \{i\}$ and $\mathcal{N}_1(i) = \mathcal{N}(i)$, a ring $\mathcal{N}'_k(i)$ is defined as

$$\mathcal{N}'_k(i) = \mathcal{N}_k(i) - \mathcal{N}_{k-1}(i)$$

and contains nodes whose shortest path to i is exactly k hops long. Denote by $r_k(i) = |\mathcal{N}'_k(i)|$ the number of nodes in the kth ring around node i and by $k_{\max}(i)$ the number of hops for the maximum non-empty k-hop ring around i. Note that in an unweighted graph the shortest path from a node i to a node j will be k if there is a node j' that is an immediate neighbor of j whose shortest path to i has length $k - 1$. Thus, when we reach an empty ring $\mathcal{N}'_{k_{\max}(i)+1}$ around i there will be no other rings around node i. Note also that $k_{\max}(i)$ is in general different for each node i. As a simple example for a star graph (see Example 3.4), $k_{\max}$ will be 1 for the central node and 2 for any other node.

Filters are then designed with the same weight, α_k, applied to all nodes in each ring:

$$y_1(i) = \sum_{k=0}^{K} \alpha_k \sum_{m \in \mathcal{N}'_k(i)} x(m). \tag{5.23}$$

The filter coefficients α_k can be derived directly from those of a time domain signal (see Box 5.2). They are chosen so that

$$\sum_{k=0}^{K} \alpha_k(i)|\mathcal{N}'_k(i)| = 0, \tag{5.24}$$

where $K \leq k_{\max}(i)$ and K controls the "scale." Note that under this condition the lowest-frequency signal, i.e., $\mathbf{1}$, produces a zero output at every node, since, for any ring $\mathcal{N}'_k(i)$, we have $\sum_{m \in \mathcal{N}'_k(i)} 1 = |\mathcal{N}'_k(i)|$. Thus this transformation can be characterized as a "high pass" filter with response equal to zero at the lowest frequency, which is similar to wavelet filter design for regular domain signals [67]. As in Section 5.3, the filter corresponding to (5.23) can be written as a matrix $\mathbf{H}_1$, where the coefficients in (5.23) correspond to the ith row. The matrix $\mathbf{H}_1$ is such that $\mathbf{1}$ belongs to its null space. Thus, a low pass filter associated with a matrix $\mathbf{H}_0$ can be designed to ensure an overcomplete invertible transform, as in Section 5.3.

Box 5.2 NGWT filter design

As an example, we can take a continuous domain filter $\psi(t)$ with compact support in the interval $[0, T]$ and such that

$$\int_0^T \psi(t)dt = 0.$$

Then we divide the interval into $K + 1$ intervals of equal length $T_K = T/(K + 1)$, denote by ψ_k the average value of the filter within the interval $[kT_K, (k + 1)T_K]$ and then choose $\alpha_k = \psi_k/|\mathcal{N}'_k(i)|$. Examples of these filters can be found in the wavelets literature [67].

Node domain graph wavelet transforms are low complexity and can be used for any unweighted graph, but there is no clear way to extend them to weighted graphs and they do not have a straightforward frequency interpretation. The reason is that our frequency representation is based on an operator, for example $\mathbf{A}$, and, therefore, for filters to have a frequency interpretation they have to be written as polynomials of the operator. However, rings cannot be constructed via linear operations on the $\mathbf{A}^k$ terms. To see this, recall that in an unweighted graph (Section 3.3.3) the i, j entry in $\mathbf{A}^k$ represents the number of paths of length k between i and j. Thus, in a graph filter with coefficient $a_k \mathbf{A}^k$, the contribution of the value at j to the output at node i is non-zero as long as there exists a path of length k. Assume that the shortest path between i and j has length 2; then $\mathbf{A}_{ij} = 0$ and $(\mathbf{A}^2)_{ij} \neq 0$ but in general $(\mathbf{A}^k)_{ij} \neq 0$ for $k > 2$ and we cannot replicate per-ring weights through linear operations on the $\mathbf{A}^k$ terms of a polynomial.

5.4.2   Graph Lifting Transforms

In Section 5.3 we combined two filters that did not have common zeros to obtain an invertible overcomplete representation. To modify such a system and achieve critical sampling would require selecting exactly N rows of $\mathbf{T}_a$ in (5.19) while preserving the full column rank. Instead of designing filters first and then selecting rows, lifting designs [68], which are extensions of similar designs for regular signal domains [69], guarantee critical sampling and invertibility *by construction*.

Prediction   A graph lifting transform is constructed on the basis of two elementary node domain operations, prediction and update, which correspond to high pass and low pass filtering, respectively. This is similar to the design in Section 5.3, but with the key difference that at each node either low pass or high pass filtering is applied (but not both).

An important limitation of graph lifting is that the graph is assumed to be bipartite (Section 3.5.3). If the original graph is not bipartite then it is first simplified (edges are removed) so that it becomes bipartite (see Section 5.6.4 for a discussion of this approximation). In this situation, while the lifting transform is invertible, it cannot be expressed in terms of graph filters on the original graph.

Assuming that we have a bipartite graph, any vector $\mathbf{x}$ and the normalized graph Laplacian can be written, without loss of generality, as

$$\mathbf{x} = \begin{bmatrix} \mathbf{x}_1 \\ \mathbf{x}_2 \end{bmatrix} \quad \text{and} \quad \mathcal{L} = \begin{bmatrix} \mathbf{I}_{N_1 \times N_2} & -\mathcal{A}_0 \\ -\mathcal{A}_0^\mathsf{T} & \mathbf{I}_{N_2 \times N_1} \end{bmatrix},$$

where $\mathbf{x}_1$ and $\mathbf{x}_2$ are the entries corresponding to the two sets of nodes, $\mathcal{V}_1$ and $\mathcal{V}_2$, in the bipartition. Then the filtered signal $\mathcal{L}\mathbf{x}$ can be written as:

$$\mathcal{L}\mathbf{x} = \mathcal{L} \begin{bmatrix} \mathbf{x}_1 \\ \mathbf{x}_2 \end{bmatrix} = \begin{bmatrix} \mathbf{x}_1 - \mathcal{A}_0 \mathbf{x}_2 \\ \mathbf{x}_2 - \mathcal{A}_0^\mathsf{T} \mathbf{x}_1 \end{bmatrix}, \tag{5.25}$$

where, recalling that this graph only has connections between $\mathcal{V}_1$ and $\mathcal{V}_2$, the term $\mathbf{x}_1 - \mathcal{A}_0 \mathbf{x}_2$ can be interpreted as subtracting from $\mathbf{x}_1$ a prediction based only on one-hop neighbors in $\mathcal{V}_2$. The **prediction** operation in lifting is written in terms of an operator $\mathbf{H}_\mathrm{p}$, defined by

$$\mathbf{x}_\mathrm{p} = \mathbf{H}_\mathrm{p} \begin{bmatrix} \mathbf{x}_1 \\ \mathbf{x}_2 \end{bmatrix} = \begin{bmatrix} \mathbf{I} & -\mathcal{A}_0 \\ \mathbf{0} & \mathbf{I} \end{bmatrix} \begin{bmatrix} \mathbf{x}_1 \\ \mathbf{x}_2 \end{bmatrix} = \begin{bmatrix} \mathbf{x}_1 - \mathcal{A}_0 \mathbf{x}_2 \\ \mathbf{x}_2 \end{bmatrix}, \tag{5.26}$$

where the values in $\mathcal{V}_1$ are predicted from those in $\mathcal{V}_2$, and the prediction residual is stored in the $\mathcal{V}_1$ nodes, while the values in $\mathcal{V}_2$ are left unchanged. We can view this as a "high pass" filter, because the output stored in the $\mathcal{V}_1$ nodes will be zero if the input $\mathbf{x}_1$ was exactly predicted by the information available in $\mathbf{x}_2$. Notice that (5.26) can be interpreted as filtering $\mathbf{x}$ as in (5.25) and then "downsampling," so that the output of (5.26) on the nodes in $\mathcal{V}_1$ is preserved, along with the original samples in $\mathcal{V}_2$. This idea will be further explored in the design of general filterbanks (Section 5.6.4).

Invertibility The prediction of (5.26) is easily invertible. Since $\mathbf{x}_2$ has not been modified, $\mathbf{x}_1$ can be recovered from $\mathbf{x}_p$ by computing $\mathbf{x}_1 - \mathcal{A}_0 \mathbf{x}_2 + \mathcal{A}_0 \mathbf{x}_2$:

$$\begin{bmatrix} \mathbf{x}_1 \\ \mathbf{x}_2 \end{bmatrix} = \begin{bmatrix} \mathbf{I} & \mathcal{A}_0 \\ \mathbf{0} & \mathbf{I} \end{bmatrix} \begin{bmatrix} \mathbf{x}_1 - \mathcal{A}_0 \mathbf{x}_2 \\ \mathbf{x}_2 \end{bmatrix} \tag{5.27}$$

so that invertibility can be guaranteed for any bipartite graph:

$$\mathbf{H}_p^{-1} = \begin{bmatrix} \mathbf{I} & \mathcal{A}_0 \\ \mathbf{0} & \mathbf{I} \end{bmatrix}. \tag{5.28}$$

Update Since this prediction operation can be viewed as "high pass," we next define a low pass operator, called **update** in the lifting literature,

$$\mathbf{H}_u = \begin{bmatrix} \mathbf{I} & \mathbf{0} \\ \mathcal{A}_0^\mathsf{T} & \mathbf{I} \end{bmatrix}, \tag{5.29}$$

which can be easily seen to be invertible as well:

$$\mathbf{H}_u^{-1} = \begin{bmatrix} \mathbf{I} & \mathbf{0} \\ -\mathcal{A}_0^\mathsf{T} & \mathbf{I} \end{bmatrix}. \tag{5.30}$$

Since each application of prediction and of update is invertible, more complex transforms can be built by simply cascading elementary prediction and update operations. Also, $\mathbf{H}_p$ and $\mathbf{H}_u$ are defined in (5.26) and (5.29) as simple functions of $\mathcal{A}_0$, but we can replace $\mathcal{A}_0$ and $\mathcal{A}_0^\mathsf{T}$ by other matrix operations and still preserve invertibility. The invertibility of $\mathbf{H}_p$ and $\mathbf{H}_u$ results from their block structure, so that any operations where $\mathbf{x}_1$ is changed on the basis of the values in $\mathbf{x}_2$, or vice versa, will be invertible. Furthermore, filterbanks on bipartite graphs such as those in Section 3.5.3 can be represented as a series of lifting steps [70].

Extension to general graphs If the original graph is bipartite, the node domain operations of (5.26) and (5.29) have a spectral interpretation, and correspond to special cases of the filterbank designs presented in Section 5.6.4. To apply a lifting transform to signals on a non-bipartite graph it is necessary to first approximate the original graph by a bipartite graph. Approaches to performing this approximation will be discussed in Section 5.6.4. A particular case where graph lifting has been used in practice is described in Box 5.3.

> **Box 5.3 Lifting for mesh graphs**
>
> One of the earliest applications of graph-based representations is in computer graphics, where objects are represented as triangular meshes with attributes (e.g., colors) associated with each triangle. It is important to compress these meshes, including both their geometry and attributes, for storage and transmission.
>
> An undirected, unweighted graph can be constructed where each triangle corresponds to one node, and two nodes are connected if their respective triangles share one edge. We can then view attribute information as a graph signal (one attribute associated with each triangle/node) to which a graph-based transform can be applied prior to compression. Note that most triangles share a side with three other triangles, so that the node degrees of the corresponding graph are close to regular.
>
> For lifting-based compression of triangular meshes, two sets of nodes, $\mathcal{V}_1$ and $\mathcal{V}_2$, are chosen and a bipartite graph is created so that only connections across these sets exist (see [71, 72]). Lifting can be combined with a multiresolution representation, where a coarser mesh based on the nodes in $\mathcal{V}_2$ can also inherit the update coefficients as attributes, i.e., a low pass filtered version of the original attribute. Multiresolution approaches are particularly useful for storing large models (containing many triangles) which can be rendered at different resolutions.

5.4.3 Subgraph Filterbanks

For any graph we can always use the corresponding GFT to analyze graph signals. But this representation involves a global operation on the graph. In contrast the simple filterbank of Section 5.3 uses one-hop local operations but is not critically sampled. Subgraph filterbanks combine critical sampling and locality by dividing a graph into subgraphs.

Block transforms and subgraph filterbanks   Subgraph filterbanks [73] have similarities with the block transforms used in conventional image and video processing. If we think of an image as a regular grid graph (Section 1.3.2), a block-based transform divides the grid graph into blocks (e.g., 8×8 pixels) and applies a transform to each block: instead of treating the image as an $N \times N$ graph, it is processed as a collection of disconnected 8×8 subgraphs. In [73], a given graph with N nodes is split into K subgraphs. For subgraph k an invertible transformation is applied. To simplify the discussion, we assume that the GFT corresponding to the subgraph, $\mathbf{U}_k$, is used but other transforms are possible. We note that each subgraph contains a set of N_k connected nodes, and in general N_k is different for each subgraph. Notice also that normally $\mathbf{U}$ (the GFT for the original graph) cannot be expressed easily in terms of the $\mathbf{U}_k$ (for the K subgraphs).

Subgraph decomposition and relationship to lifting   To understand the difference between the subgraph approach and lifting, consider the case where $K = 2$. Notice that both methods partition the original set of nodes $\mathcal{V}$ into two subsets $\mathcal{V}_1$ and $\mathcal{V}_2$, which is equivalent to approximating (after permutation) the adjacency matrix of the original graph $\mathbf{A}$ by a block diagonal matrix. In contrast, the bipartition strategy in lifting uses only connections between $\mathcal{V}_1$ and $\mathcal{V}_2$. Accordingly the partition criteria are different. For partitioning into subgraphs, $\mathcal{V}_1$ and $\mathcal{V}_2$ should be chosen to minimize the number of connections between the two sets (i.e., to minimize the number of edges, or their total weight). In contrast, a lifting approximation maximizes the number of edges between sets. These two criteria are closely linked to the problems of identifying minimum and maximum cuts (see Definition 2.11 and the related discussion).

Multi-scale filterbanks   The subgraph filterbank is critically sampled, since each subgraph with N_k nodes produces exactly N_k values. Moreover, a multi-scale representation can be constructed. If K subgraphs are used, a new graph with K nodes can be constructed where each node corresponds to one of the original subgraphs, and two nodes are connected if the corresponding subgraphs were connected. A multi-scale transform can be obtained by creating a series of graph signals associated with this new graph, the first containing the lowest-frequency outputs of all subgraph transforms $\mathbf{U}_k$. Note that the subgraphs have different sizes and thus they produce a different number of frequencies. Denoting by $N_{\min}$ the minimum subgraph size, there will be $N_{\min}$ signals for the coarse resolution graph, but subsequent graphs will be smaller (fewer than K nodes) and will contain only outputs for subgraphs such that $N_k > N_{\min}$. Refer to [73] for details of this multi-scale construction.

5.5   Frequency (Spectral) Domain Graph Signal Representations

We have just discussed node domain filter design. While such approaches are intuitive and computationally efficient, they often lack a frequency interpretation.

5.5.1   Spectral Graph Wavelet Transforms (SGWTs)

Spectral graph wavelet transforms (SGWTs) [74] are among the most popular transforms for graph signals. These transforms are frames, overcomplete representations which guarantee that signals can be recovered (Section A.5). We focus on the article [74] as it gives a popular design, but there are other examples, including the early work [75], based on similar principles. Some of these alternative methods will be discussed later in this chapter.

Bandpass filter prototype   The core idea in [74] is to consider a basic filter prototype, a bandpass filter $g(\lambda)$, defined in the frequency domain. Focusing on Laplacian operators, for which the lowest frequency corresponds to $\lambda = 0$, a bandpass filter is such that (i) $g(0) = 0$ (i.e., the lowest frequency is removed) and (ii) $g(\lambda)$ goes to zero as λ goes to infinity (in practice this means that $g(\lambda)$ should be close to zero for the maximum frequency λ_N). The corresponding filter can be written as a function of $\mathbf{Z}$:

$$\mathbf{G} = g(\mathbf{Z}) = \mathbf{U}g(\mathbf{\Lambda})\mathbf{U}^\mathsf{T},$$

where $g(\Lambda)$ is defined on the basis of the chosen bandpass filter as follows:

$$g(\Lambda) = \mathrm{diag}\{g(\lambda_1), g(\lambda_2), \ldots, g(\lambda_N)\}.$$

Recall that the invertibility of $\mathbf{T}_a$ defined in (5.19) (Section 5.3) required us to select filters without common zeros. The same principle is applied here: additional filters are added in such a way that there is no single frequency λ_i where *all* the filters are zero.

Filterbank design   Given $g(\lambda)$, a complete filterbank is designed by introducing a series of scaling parameters, $t_1, t_2, \ldots, t_J$ and corresponding filters, so that for scaling t we define $g_t(\lambda) = g(\lambda t)$, $g_t(\Lambda)$ and $\mathbf{G}_t = \mathbf{U}g_t(\Lambda)\mathbf{U}^\mathsf{T}$. Note that, for any t, $g_t(\lambda)$ remains a bandpass filter since $g_t(0) = 0$ for any t. Also, choosing $t > 1$ leads to $g_t(\lambda)$ becoming a bandpass filter with a higher-frequency passband than the original filter $g(\lambda)$.

Since all filters $g_t(\lambda)$ with parameters $t_1, t_2, \ldots, t_J$ are zero at $\lambda = 0$, following the discussion in Section 5.3 the transformation cannot be invertible. Thus, an additional filter, the scaling or low pass filter $h(\lambda)$, is chosen having the property $h(0) > 0$. Using filters $h(\lambda)$, $g_{t_i}(\lambda)$, for $t_i \in \{t_1, t_2, \ldots, t_J\}$ we can obtain $J + 1$ outputs with N values, thus oversampling by a factor $J + 1$. With $\mathbf{H}$, $\mathbf{G}_{t_i}$ defined as above, the overall operation can be written in matrix form as

$$\mathbf{y} = \mathbf{T}_a\mathbf{x} = \begin{bmatrix} \mathbf{H} \\ \mathbf{G}_{t_1} \\ \mathbf{G}_{t_2} \\ \vdots \\ \mathbf{G}_{t_J} \end{bmatrix} \mathbf{x}, \tag{5.31}$$

where $\mathbf{T}_a$ is the $(J + 1)N \times N$ matrix of the transformation.

Invertibility   In order to recover $\mathbf{x}$ from $\mathbf{y}$ in (5.31), this overcomplete representation has to be a frame (see Section A.5) and there exist frame bounds $0 < A \le B$ such that

$$A\|\mathbf{x}\|^2 \le (\mathbf{x}^\mathsf{T}\mathbf{T}_a^\mathsf{T})(\mathbf{T}_a\mathbf{x}) \le B\|\mathbf{x}\|^2 \tag{5.32}$$

for all $\mathbf{x} \ne 0$. Alternatively, defining

$$G(\lambda) = h^2(\lambda) + \sum_{i=1}^{J} g^2(t_j\lambda), \tag{5.33}$$

$$A = \min_{\lambda \in [0,\lambda_N]} G(\lambda) \quad \text{and} \quad B = \max_{\lambda \in [0,\lambda_N]} G(\lambda), \tag{5.34}$$

the frame condition (5.32) will be verified as long as $A > 0$. This condition is met if the columns of $\mathbf{T}_a$ are linearly independent, so that $\mathbf{T}_a^\mathsf{T}\mathbf{T}_a$ has rank N and is invertible as in the example of Section 5.3. Then, we can recover $\mathbf{x}$ as in (5.21):

$$\mathbf{x} = (\mathbf{T}_a^\mathsf{T}\mathbf{T}_a)^{-1}\mathbf{T}_a^\mathsf{T}\mathbf{y}, \tag{5.35}$$

where for large graphs an iterative conjugate gradient method can be used to compute the pseudo-inverse and so avoid having to find the inverse of $\mathbf{T}_a^\mathsf{T}\mathbf{T}_a$ [74].

The SGWT design, and its extensions, to be described next, selects filters in the frequency domain with the goal of achieving good frame bounds (i.e., A and B are almost equal). But a good spectral design leads to filtering operations in the frequency, so that the GFT has to be computed first, and the resulting filters are not likely to be local in the node domain. For this reason, SGWT filterbanks are designed in the frequency domain and then approximated by low-degree polynomials, so that they can be applied directly in the node domain. This can be done using Chebyshev polynomials (Section 5.5.2).

Tight frame design   Generally it is not difficult to find $g(\lambda)$ and scaling parameters $t_1, t_2, \ldots, t_J$ that guarantee that the SGWT is invertible. A more challenging task is to select filters such that the spectral response is as close to flat as possible, such that $A \approx B$. As an alternative a tight frame design was proposed in [76]. By construction, the filters used in [76] are such that the frame bounds in (5.32) are equal, $A = B > 0$, and reconstruction does not require a pseudo-inverse, since $\mathbf{T}_a^{\top}\mathbf{T}_a = \mathbf{I}$; therefore

$$\mathbf{x} = \mathbf{T}_a^{\top}\mathbf{y}.$$

Reconstruction is exact unless the original filters are approximated by polynomials, which leads to a frame that is not exactly tight and to some error in the reconstruction.

Frequency adaptive design   Note that the scaling parameters chosen in SGWT, $g_t(\lambda) = g(\lambda t)$, are chosen by considering λ to be a continuous variable. However, we have N discrete frequencies, in a range upper-bounded by $2d_{max}$ for the combinatorial Laplacian, and in general these frequencies are not evenly distributed. Thus, if we choose $t = 2$, which roughly halves the range of frequencies of $g_2(\lambda)$ with respect to $g(\lambda)$, this does not mean that the number of discrete frequencies in the range of $g_2(\lambda)$ will be divided by two. Indeed, in some extreme cases there could be bandpass filters containing no discrete frequencies. To address this issue, the authors of [77] proposed the use of spectral adapted methods, where the frame design takes into consideration the distribution of discrete frequencies.

5.5.2 Chebyshev Polynomial Implementation

In the previous discussion, $g(\lambda)$ and $h(\lambda)$ were defined in the spectral domain. A direct implementation of these transformations would require one first to compute the GFT, $\mathbf{U}$, and then compute $\mathbf{Ux}$ and perform filtering in the spectral domain. Recall from Remark 3.4 that if $p(\lambda)$ is a polynomial of the variable λ then the spectral domain filter with frequency response $p(\lambda)$ can be implemented in the node domain using a polynomial filter $p(\mathbf{L})$, avoiding GFT computation.

Since $g(\lambda)$ and $h(\lambda)$ are not polynomial in general, in [74] a local implementation was proposed based on polynomial filters $p_g(\lambda)$ and $p_h(\lambda)$ that approximate $g(\lambda)$ and $h(\lambda)$. For sufficiently high degree the approximation can be arbitrarily good. While other approaches can be used to find an approximation (for example **Remez exchange techniques** can be employed to find a minimax solution of a given degree), Chebyshev polynomials have been shown to be particularly efficient [74]. Denoting by $T_k(\lambda)$ the

kth Chebyshev polynomial (of the first kind), $T_{k+1} = 2\lambda T_k(\lambda) - T_{k-1}(\lambda)$, with $T_0(\lambda) = 1$ and $T_1(\lambda) = \lambda$, any function $h(\lambda)$ can be written as

$$h(\lambda) = \frac{1}{2}c_0 + \sum_{k=1}^{+\infty} c_k T_k(\lambda),$$

where the coefficients $c_0, c_1, \ldots, c_k$ can be computed directly from $h(\lambda)$. The main advantage of the Chebyshev polynomial approximation is that it does not require a specific optimization for a given polynomial degree. Instead, the coefficients $c_0, c_1, \ldots, c_k$ can be computed up to a certain degree and then an approximation can be selected by truncating the polynomial at a suitable k. See Box 5.4.

Box 5.4   Computation of Chebyshev coefficients

The coefficients do not depend on the order, K, of the polynomial approximation because the Chebyshev polynomials form an orthogonal basis and each c_k is obtained through an orthogonal projection of $h(\cdot)$:

$$c_k = \langle h, P_k \rangle = \begin{cases} \displaystyle\int_{-1}^{1} h(x) T_k(x) \frac{1}{\sqrt{1-x^2}} dx & \text{first kind,} \\[3ex] \displaystyle\int_{-1}^{1} h(x) U_k(x) \sqrt{1-x^2} dx & \text{second kind,} \end{cases}$$

where P_k is a Chebyshev polynomial of the first or second kind, T_k and U_k respectively. Note that above it is assumed that the interpolation interval is $[-1, 1]$, i.e., all eigenvalues of the graph operator $\mathbf{Z}$ are in that interval. Given the actual range of frequencies $\lambda \in [0, \lambda_{\max}]$, a linear transformation will be needed to compute the coefficients.

In practice, one does not need to compute explicitly each of the polynomials T_k (or $T_k(\lambda)$ or $T_k(\mathbf{Z})$ or $T_k(\mathbf{Z})\mathbf{x}$). For example, once $c_0, c_1, \ldots, c_{k+1}$ have been found through the orthogonal projections described in Box 5.4, the output of the graph filter $h(\mathbf{Z})$ given an input $\mathbf{x}$ can be computed recursively:

$$\mathbf{x}_0 = \mathbf{x}, \qquad\qquad\qquad \mathbf{y}_0 = c_0\mathbf{x}_0,$$
$$\mathbf{x}_1 = \mathbf{Z}\mathbf{x}, \qquad\qquad\qquad \mathbf{y}_1 = \mathbf{y}_0 + c_1\mathbf{x}_1,$$
$$\mathbf{x}_2 = 2\mathbf{Z}\mathbf{x}_1 - \mathbf{x}_0, \qquad\qquad \mathbf{y}_2 = \mathbf{y}_1 + c_2\mathbf{x}_2,$$
$$\vdots \qquad\qquad\qquad\qquad \vdots$$
$$\mathbf{x}_{k+1} = 2\mathbf{Z}\mathbf{x}_k - \mathbf{x}_{k-1}, \qquad \mathbf{y}_{k+1} = \mathbf{y}_k + c_{k+1}\mathbf{x}_{k+1},$$

Example 5.2   Finding a Chebyshev polynomial approximation for a filter   Define a filter $f(\lambda) = \exp\left(-40 \times (\lambda - 0.2)^2\right)$ and find its decomposition into Chebyshev polynomials.

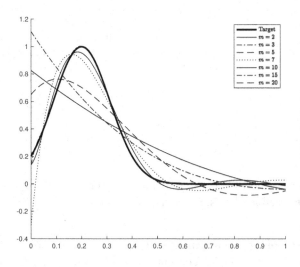

Figure 5.1 Chebyshev approximation for the filter of Example 5.2. Note that as m, the degree of the polynomial, increases, the accuracy of the approximation also improves. However, an increased m also decreases the locality of the filtering operations. Thus in choosing m these two factors should be considered.

Code Example 5.1 `Matlab` code to generate Chebyshev polynomials

```
%% Function to approximate
f = @(x) exp(-40 * (x - 0.2) .^ 2);

%% Chebyshev approximations
for m = [2 3 5 7 10 15 20]
h = grasp_filter_cheb(f, m, 'cheb_kind', 2);
end
```

Solution

This can be done with Code Example 5.1. The frequency responses of the functions obtained using a polynomial approximation are shown in Figure 5.1.

5.6 Critically Sampled Graph Filterbanks

While graph lifting (Section 5.4.2) has built-in critical sampling, this comes at the cost of having to approximate a graph by a bipartite subgraph (which has the same nodes but only a subset of the edges). In contrast, SGWT designs (Section 5.5.1) can be applied to any graph and use filters with well-defined frequency responses, but they are not critically sampled. In this section we study the design of critically sampled transforms and show that designs are possible for bipartite graphs.

Critical sampling is achieved by using standard graph filtering followed by downsampling. If the original graph signal is defined on a set of nodes $\mathcal{V}$, then filtered outputs are preserved on either $\mathcal{V}_0$ or $\mathcal{V}_1$, the two subsets of nodes. Each downsampled signal can be seen as a graph signal on a reduced graph, with nodes on $\mathcal{V}_0$ or $\mathcal{V}_1$. Combining filtering and downsampling leads to a graph filterbank, as shown in Figure 5.2. In

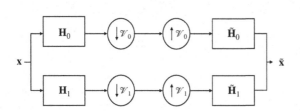

Figure 5.2 Two-channel graph filterbank: two graph filters $\mathbf{H}_0$ and $\mathbf{H}_1$ are applied to the input graph signal $\mathbf{x}$. The outputs in $\mathcal{V}_0$ and $\mathcal{V}_1$, respectively are kept, so as to make this a critically sampled system. The two sets form a partition of the original set of nodes $\mathcal{V}$.

what follows we will study how filters and sampling operators can be designed so as to guarantee exact reconstruction (that is, $\tilde{\mathbf{x}} = \mathbf{x}$ in Figure 5.2).

Graph signal downsampling and multirate systems   A system such as that of Figure 5.2 is analogous to a filterbank in conventional signal processing. Owing to the presence of downsampling and upsampling, these are *multirate* systems. For conventional signals multiple rates means that there are sequences with different number of samples per second, as discussed in Box 5.5. For graph signals, rather than signals with multiple "rates" we have signals defined on multiple graphs.

Box 5.5   Multirate systems in digital signal processing (DSP)

Define a finite-energy discrete-time signal $x(n) \in l_2(\mathbb{Z})$. Downsampling $x(n)$ by a factor of 2 means keeping only samples at even locations. Upsampling by a factor of 2 brings us back to the original rate. Denoting by $x'(n)$ the signal obtained after a downsampling–upsampling (DU) operation, we have that

$$x'(n) = x(n) \quad \text{if} \quad n = 2k, \quad \text{and} \quad x'(n) = 0 \quad \text{if} \quad n = 2k + 1.$$

From this we can see that $x'(n)$ can be written as

$$x'(n) = \frac{1}{2}(x(n) + (-1)^n x(n)), \tag{5.36}$$

so that

$$X'(z) = \frac{1}{2}(X(z) + X(-z))$$

and

$$X'(e^{j\omega}) = \frac{1}{2}\left(X(e^{j\omega}) + X(e^{j(\omega+\pi)})\right), \tag{5.37}$$

where $X(z)$ and $X(e^{j\omega})$ denote the z-transform and the discrete-time Fourier transform, respectively. From (5.37), the DU operation creates *aliasing*, by adding a shifted version of the original signal spectrum $X(e^{j(\omega+\pi)})$. In general $X(e^{j\omega})$ and $X(e^{j(\omega+\pi)})$ overlap, so it is not possible to recover $x(n)$ from $x'(n)$.
If $x(n)$ is bandlimited and such that

$$X(e^{j\omega}) = 0, \quad \forall \omega \in [-\pi, -\pi/2] \cup [\pi/2, \pi],$$

then it is easy to see that $X(e^{j\omega})$ and $X(e^{j(\omega+\pi)})$ do not overlap and thus it is possible to apply an ideal low pass filter that cancels out the energy in the range $[-\pi, -\pi/2] \cup [\pi/2, \pi]$ and thus to recover $x(n)$ exactly.

5.6.1 Downsampling–Upsampling Operations for Graph Signals

We now study downsampling–upsampling (DU) operations similar to those of Box 5.5 but for graph signals, where for simplicity we consider only the case of downsampling and upsampling by a factor of 2. As noted in Chapter 4 there is no obvious way of defining "every other" node on a graph, since the nodes are not ordered in any way. Thus, for DU operations we define two disjoint subsets of nodes $\mathcal{V}_0$, $\mathcal{V}_1$, such that $\mathcal{V}_0 \cup \mathcal{V}_1 = \mathcal{V}$ and $\mathcal{V}_0 \cap \mathcal{V}_1 = \emptyset$. Then, the DU signal $\mathbf{x}'$ maintains the original values in the entries corresponding to set $\mathcal{V}_0$ while setting to zero those corresponding to set $\mathcal{V}_1$:

$$x'(i) = x(i) \quad \text{if} \quad i \in \mathcal{V}_0, \quad \text{and} \quad x'(i) = 0 \quad \text{if} \quad i \in \mathcal{V}_1, \tag{5.38}$$

where the actual sampling rate is $|\mathcal{V}|/|\mathcal{V}_1|$. The relations (5.38) can be written more compactly by defining $\Delta_\beta = \text{diag}(\boldsymbol{\beta})$, a diagonal matrix where $\beta(i)$ is the ith diagonal entry, with $\beta(i) = 1$ if $i \in \mathcal{V}_0$ and $\beta(i) = -1$ if $i \in \mathcal{V}_1$, so that

$$\mathbf{x}' = \frac{1}{2}(\mathbf{I} + \Delta_\beta)\mathbf{x} = \mathbf{I}_{\mathcal{V}_0}\mathbf{I}_{\mathcal{V}_0}^{\mathsf{T}}\mathbf{x}, \tag{5.39}$$

which corresponds to (5.36) in Box 5.5 and where the last term expresses the DU operation directly in terms of node sampling, as in Definition 4.1 and (4.8).

DU in the frequency domain Is it possible to write a simple frequency domain expression for graph DU that is similar to (5.37) in Box 5.5? Denoting $\tilde{\mathbf{x}} = \mathbf{U}^{\mathsf{T}}$ and $\tilde{\mathbf{x}}' = \mathbf{U}^{\mathsf{T}}\mathbf{x}'$, we can rewrite (5.39) as

$$\tilde{\mathbf{x}}' = \mathbf{U}^{\mathsf{T}}\mathbf{x}' = \frac{1}{2}\mathbf{U}^{\mathsf{T}}(\mathbf{I} + \Delta_\beta)\mathbf{U}\tilde{\mathbf{x}} = \frac{1}{2}(\mathbf{I} + \mathbf{U}^{\mathsf{T}}\Delta_\beta\mathbf{U})\tilde{\mathbf{x}} = \frac{1}{2}(\tilde{\mathbf{x}} + \mathbf{U}^{\mathsf{T}}\Delta_\beta\mathbf{U}\tilde{\mathbf{x}}), \tag{5.40}$$

where $\mathbf{U}^{\mathsf{T}}\Delta_\beta\mathbf{U}\tilde{\mathbf{x}}$ introduces an error in the frequency domain. In Box 5.5 we observed aliasing, where the frequency ω after the DU operation contains information from exactly two frequencies (ω and $\omega + \pi$) in the original signal. For something similar to occur in the graph case, the kth frequency of $\tilde{\mathbf{x}}'$ would need to include two frequencies, the original one, $\tilde{x}(k)$, and another term coming from $\mathbf{U}^{\mathsf{T}}\Delta_\beta\mathbf{U}\tilde{\mathbf{x}}$. This can only happen if $\mathbf{U}^{\mathsf{T}}\Delta_\beta\mathbf{U}$ is a permutation matrix. In general this is not the case, with the important exception of bipartite graphs, which we discuss next.

5.6.2 Bipartite Graphs

In Section 3.5.3 we showed that for the symmetric normalized Laplacian $\mathcal{L}$ of a bipartite graph, if $\mathbf{u}_k$ is an eigenvector with eigenvalue λ_k then $\mathbf{u}_{N-k+1}$ has eigenvalue $2 - \lambda_k$, and we can write

$$\mathbf{u}_k = \begin{bmatrix} \mathbf{u}_k^0 \\ \mathbf{u}_k^1 \end{bmatrix} \quad \text{and} \quad \mathbf{u}_{N-k+1} = \begin{bmatrix} \mathbf{u}_k^0 \\ -\mathbf{u}_k^1 \end{bmatrix}, \tag{5.41}$$

where $\mathbf{u}_k^0$ and $\mathbf{u}_k^1$ contain the entries corresponding to $\mathcal{V}_0$ and $\mathcal{V}_1$, respectively. Note that this mirroring operation means that an eigenvector with eigenvalue $\lambda = 1$ matches itself (refer to Box 5.6). For the DU operation where values for nodes in $\mathcal{V}_0$ are preserved while those in $\mathcal{V}_1$ are set to zero (the DU operation in the bottom part of Figure 5.2) the corresponding Δ_β is such that $\beta(i) = -1$ for $i \in \mathcal{V}_1$. Thus

$$\Delta_\beta \mathbf{u}_{N-k+1} = \Delta_\beta \begin{bmatrix} \mathbf{u}_k^0 \\ -\mathbf{u}_k^1 \end{bmatrix} = \begin{bmatrix} \mathbf{u}_k^0 \\ \mathbf{u}_k^1 \end{bmatrix} = \mathbf{u}_k, \tag{5.42}$$

so that we have

$$\mathbf{U}^\mathsf{T} \Delta_\beta \mathbf{U} = \mathbf{J},$$

where $\mathbf{J}$ is the reversal matrix, an anti-diagonal permutation of the identity matrix, which maps an input at position k in the input vector to position $N - k + 1$ in the output vector:

$$\mathbf{J} = \begin{bmatrix} 0 & 0 & \cdots & 0 & 1 \\ 0 & 0 & \cdots & 1 & 0 \\ \vdots & \vdots & \ddots & \vdots & \vdots \\ 1 & 0 & \cdots & 0 & 0 \end{bmatrix}.$$

In summary, in the frequency domain we have

$$\tilde{\mathbf{x}}' = \frac{1}{2}(\mathbf{I} + \mathbf{J})\tilde{\mathbf{x}} = \frac{1}{2}(\tilde{\mathbf{x}} + \tilde{\mathbf{x}}_r), \tag{5.43}$$

where the entries in $\tilde{\mathbf{x}}_r$ are the same as in $\tilde{\mathbf{x}}$ but appear in the reverse order.

Extensions to DU by a factor of M   So far we have focused on DU by a factor of 2, concluding that DU can be interpreted as a simple spectral domain operation only for bipartite graphs. A similar result was derived for M-channel filterbanks in [78, 79], where a complete theory was obtained for downsampling and upsampling by a factor of M. This work showed that the corresponding DU operations can only be interpreted as simple spectral operations for M-block cyclic graphs. These are directed M colorable graphs, such that after dividing nodes into sets $S_1, S_2, \ldots, S_M$, and for a suitable labeling, all connections are from S_i to $S_{(i+1) \bmod M}$.

5.6.3   Ideal Filterbank Design

We can now revisit the design ideas of Section 5.3 for the bipartite graph case. An overcomplete system can be inverted as long as the filters do not share zeros. Assume for simplicity that $\lambda = 1$ is not an eigenvalue (see Box 5.6). Then, all eigenvalues come in pairs $(\lambda_k, 2 - \lambda_k)$, with corresponding eigenvectors from (5.41), and since there are

N eigenvectors, and they come in pairs, N must be even and $|\mathcal{V}_0| = |\mathcal{V}_1| = N/2$. From (5.43), if a signal $\mathbf{x}_0$ has energy only in the low frequencies, then

$$\tilde{\mathbf{x}}_0 = \begin{bmatrix} \tilde{\mathbf{x}}_0^0 \\ \mathbf{0} \end{bmatrix} \quad \text{and} \quad \tilde{\mathbf{x}}_0' = \frac{1}{2}\begin{bmatrix} \tilde{\mathbf{x}}_{0,}^0 \\ \tilde{\mathbf{x}}_{0,r}^0 \end{bmatrix}, \tag{5.44}$$

where $\tilde{\mathbf{x}}_{0,r}^0$ has the entries of $\tilde{\mathbf{x}}_0^0 \in \mathbb{R}^{N/2}$ in reverse order. Thus, $\mathbf{x}_0$ can be recovered exactly from $\tilde{\mathbf{x}}_0'$ by applying an ideal low pass filter to recover $\tilde{\mathbf{x}}_0^0$.

Following Box 3.1, denote by $\mathbf{H}_0$ and $\mathbf{H}_1$ ideal low pass and high pass filters, respectively, and by $\mathbf{x}_0 = \mathbf{H}_0 \mathbf{x}$ and $\mathbf{x}_1 = \mathbf{H}_1 \mathbf{x}$ the respective outputs when $\mathbf{x}$ is the input. Then, for $i = 0, 1$,

$$\mathbf{x}_i = \mathbf{H}_i \mathbf{I}_{\mathcal{V}_i} \mathbf{I}_{\mathcal{V}_i}^\mathsf{T} \mathbf{x}_i = \mathbf{H}_i \mathbf{I}_{\mathcal{V}_i} \mathbf{I}_{\mathcal{V}_i}^\mathsf{T} \mathbf{H}_i \mathbf{x},$$

where $\mathbf{I}_{\mathcal{V}_i} \mathbf{I}_{\mathcal{V}_i}^\mathsf{T}$ represents the DU operation. If the ideal filters $\mathbf{H}_0$ and $\mathbf{H}_1$ exactly complement each other (so that the gain for each frequency is 1 in one filter and 0 in the other) then $\mathbf{x}_0 + \mathbf{x}_1 = \mathbf{x}$ and a critically sampled filterbank can be constructed using these ideal filters:

$$\mathbf{x} = (\underbrace{\mathbf{H}_0 \mathbf{I}_{\mathcal{V}_0} \mathbf{I}_{\mathcal{V}_0}^\mathsf{T} \mathbf{H}_0}_{\text{low pass}} + \underbrace{\mathbf{H}_1 \mathbf{I}_{\mathcal{V}_1} \mathbf{I}_{\mathcal{V}_1}^\mathsf{T} \mathbf{H}_1}_{\text{high pass}})\mathbf{x}$$

where the low pass and high pass terms in the equation correspond to the top and bottom parts of the system in Figure 5.2, respectively.

Box 5.6 DU effect on signals at $\lambda = 1$

Let us examine in more detail the case where a graph has an eigenvalue $\lambda = 1$, in which case the frequency mirroring of (5.41) maps $\lambda = 1$ back onto itself. Denote by $\mathbf{u}$ an eigenvector corresponding to $\lambda = 1$. Then by definition $\mathcal{L}\mathbf{u} = \mathbf{u}$ and

$$\begin{bmatrix} \mathbf{I}_0 & -\mathcal{A}_0 \\ -\mathcal{A}_0^\mathsf{T} & \mathbf{I}_1 \end{bmatrix}\begin{bmatrix} \mathbf{u}^0 \\ \mathbf{u}^1 \end{bmatrix} = \mathbf{u} \quad \text{so that} \quad \mathbf{I}\mathbf{u} + \begin{bmatrix} \mathbf{0} & -\mathcal{A}_0 \\ -\mathcal{A}_0^\mathsf{T} & \mathbf{0} \end{bmatrix}\begin{bmatrix} \mathbf{u}^0 \\ \mathbf{u}^1 \end{bmatrix} = \mathbf{u}$$

and therefore

$$\mathcal{A}_0 \mathbf{u}^1 = 0, \quad \mathcal{A}_0^\mathsf{T} \mathbf{u}^0 = 0.$$

The dimension of the $\lambda = 1$ subspace is a function of the rank of $\mathcal{A}_0$. Let $N_i = |\mathcal{V}_i|$ and without loss of generality assume that $N_0 \geq N_1$. Then $r = \text{rank}(\mathcal{A}_0) = \text{rank}(\mathcal{A}_0^\mathsf{T}) \leq N_1 \leq N_0$, and the multiplicity of λ_1 is $m_1 = N - 2r$. The smallest m_1 is achieved when r is largest, i.e., $r = N_1$. Therefore, $m_1 \geq |N_0 - N_1|$.

Whenever $N_0 > N_1$, we will have $m_1 > 1$. From Box 3.2, when designing ideal filters we should assign the complete $\lambda = 1$ subspace to either the low pass or the high pass channel (the top and bottom in Figure 5.2, respectively). Thus, denoting $\mathbf{x}_0 = \mathbf{H}_0 \mathbf{x}$, we design $\mathbf{H}_0$ as follows: (i) $\tilde{x}_0(k) = 0$ for all k such that $\lambda_k \geq 1$, or (ii) $\tilde{x}_0(k) = 0$ for k such that $\lambda_k > 1$. Choice (i) excludes $\lambda = 1$ from $\mathbf{x}_0$ and thus we should sample $\mathbf{x}_0$ on the smaller set $\mathcal{V}_1$, while for case (ii) we should sample $\mathbf{x}_0$ on $\mathcal{V}_0$.

5.6.4 Graph Filterbanks

Section 5.5.1 described a filterbank for which invertibility can be guaranteed under simple conditions on the filters, but without downsampling. We have just described, for the bipartite graph case, a critically sampled filterbank providing exact reconstruction with ideal low pass and high pass filters. We now formulate the general problem of designing graph filterbanks with critical sampling and exact reconstruction.

Problem formulation   To achieve critical sampling from an overcomplete expansion such as that in (5.31), we need to select $|\mathcal{V}| = N$ filter outputs, while preserving our ability to recover the input signal $\mathbf{x}$. Thus, we need to choose N out of the $N \times (J + 1)$ rows of $\mathbf{T}_a$, while ensuring that the rank of the resulting $N \times N$ matrix, $\mathbf{T}_s$, remains as N. Consider a two-channel filterbank without downsampling, with filters $h_0(\lambda)$ and $h_1(\lambda)$:

$$\mathbf{T}_a = \begin{bmatrix} \mathbf{H}_0 \\ \mathbf{H}_1 \end{bmatrix}. \tag{5.45}$$

From (5.32) and (5.33), $\mathbf{T}_a$ has full column rank as long as $h_0(\lambda)$ and $h_1(\lambda)$ do not share any zeros. As in noise-free sampling (Section 4.4.1), choosing N out of $2N$ rows of $\mathbf{T}_a$ so that $\mathrm{rank}(\mathbf{T}_s) = N$ is not difficult. Selecting one row from $\mathbf{H}_0$ or $\mathbf{H}_1$ corresponds to sampling the output at the corresponding node in the graph. Rows chosen from $\mathbf{H}_0$ give us nodes to assign to the set $\mathcal{V}_0$, while those chosen from $\mathbf{H}_1$ will be in $\mathcal{V}_1$. We next address several important questions.

- What subset of nodes should be chosen?
- Should the subset be the same for the low pass and high pass channels?
- If $\mathbf{H}$ and $\mathbf{G}$ are polynomials of $\mathbf{Z}$ can the reconstruction also be a polynomial?

General case   Sampling can be written in matrix form (Definition 4.1) once $\mathcal{V}_0$ and $\mathcal{V}_1$ are chosen, where $|\mathcal{V}_0| + |\mathcal{V}_1| = N$ for critical sampling. Then the resulting critically sampled filterbank can be written in matrix form as

$$\mathbf{T}_s = \begin{bmatrix} \mathbf{I}_{\mathcal{V}_0}^\mathsf{T} \mathbf{H}_0 \\ \mathbf{I}_{\mathcal{V}_1}^\mathsf{T} \mathbf{H}_1 \end{bmatrix},$$

where as long as $\mathbf{T}_s$ is invertible we can simply multiply the output generated by $\mathbf{T}_s$ by $\mathbf{T}_s^{-1}$ in order to recover the input. However, note that even if $\mathbf{H}_0$ and $\mathbf{H}_1$ are polynomials $\mathbf{T}_s^{-1}$ may not be polynomial in general. Thus, for large graphs (N large) computing $\mathbf{T}_s^{-1}$ and then applying it for reconstruction is likely to be computationally expensive.

Exact reconstruction   Let us formulate the conditions for exact reconstruction using graph filters $\tilde{\mathbf{H}}_0$ and $\tilde{\mathbf{H}}_1$, which are applied after DU operations along each of the channels of the filterbank (see Figure 5.2). The end to end response of the system is

$$\tilde{\mathbf{x}} = \left(\tilde{\mathbf{H}}_0 \mathbf{I}_{\mathcal{V}_0} \mathbf{I}_{\mathcal{V}_0}^\mathsf{T} \mathbf{H}_0 + \tilde{\mathbf{H}}_1 \mathbf{I}_{\mathcal{V}_1} \mathbf{I}_{\mathcal{V}_1}^\mathsf{T} \mathbf{H}_1 \right) \mathbf{x} \tag{5.46}$$

and therefore the goal is to design $\tilde{\mathbf{H}}_0, \mathbf{H}_0, \tilde{\mathbf{H}}_1, \mathbf{H}_1, \mathcal{V}_0, \mathcal{V}_1$ in such a way that

$$\tilde{\mathbf{H}}_0 \mathbf{I}_{\mathcal{V}_0} \mathbf{I}_{\mathcal{V}_0}^\mathsf{T} \mathbf{H}_0 + \tilde{\mathbf{H}}_1 \mathbf{I}_{\mathcal{V}_1} \mathbf{I}_{\mathcal{V}_1}^\mathsf{T} \mathbf{H}_1 = \mathbf{I}. \tag{5.47}$$

Recall from our observations after (5.22) that a filter can be written as a polynomial of $\mathbf{Z}$ as long as it can be diagonalized by the GFT. This turns out not to be possible for general graphs [80] because, in accordance with (5.39), the following operators cannot be diagonalized by the GFT (except for a bipartite graph):

$$\mathbf{I}_{V_0}\mathbf{I}_{V_0}^{\mathsf{T}} = \frac{1}{2}(\mathbf{I} + \Delta_0) \text{ and } \mathbf{I}_{V_1}\mathbf{I}_{V_1}^{\mathsf{T}} = \frac{1}{2}(\mathbf{I} + \Delta_1),$$

where Δ_i corresponds to sampling set V_i; see (5.39).

Conditions in the frequency domain To rewrite the condition (5.47) in terms of the spectral characteristics of the filters, denote by $x(\lambda)$ the frequency domain representation of signal $\mathbf{x}$, and define

$$\mathbf{y}_0 = \mathbf{I}_{V_0}\mathbf{I}_{V_0}^{\mathsf{T}}\mathbf{H}_0\mathbf{x} = \frac{1}{2}(\mathbf{I} + \Delta_0)\mathbf{H}_0\mathbf{x} \text{ and } \mathbf{y}_1 = \mathbf{I}_{V_1}\mathbf{I}_{V_1}^{\mathsf{T}}\mathbf{H}_1\mathbf{x} = \frac{1}{2}(\mathbf{I} + \Delta_1)\mathbf{H}_1\mathbf{x},$$

which can be written in the frequency domain as

$$y_0(\lambda) = \frac{1}{2}\left(h_0(\lambda)x(\lambda) + y_0'(\lambda)\right) \text{ and } y_1(\lambda) = \frac{1}{2}\left(h_1(\lambda)x(\lambda) + y_1'(\lambda)\right),$$

where $y_0'(\lambda)$ and $y_1'(\lambda)$ are the GFTs of $\Delta_0\mathbf{H}_0\mathbf{x}$ and $\Delta_1\mathbf{H}_1\mathbf{x}$, respectively. Thus the condition for exact reconstruction can be written in the frequency domain as

$$\tilde{h}_0(\lambda)y_0(\lambda) + \tilde{h}_1(\lambda)y_1(\lambda) = x(\lambda). \tag{5.48}$$

Solutions for bipartite graphs In this case V_0 and V_1 are chosen to be the two sets of the bipartition, i.e., there are no edges connecting nodes within V_0, and the same is true for nodes within V_1. From (5.42) we have that, multiplying by Δ_0 the eigenvector with eigenvalue λ_k we get the eigenvector corresponding to $2 - \lambda_k$. Then if $\mathbf{x}' = \Delta_0\mathbf{x}$ and $\mathbf{x}'' = \Delta_1\mathbf{x}$ we will have that

$$x'(\lambda) = x(2 - \lambda) \text{ and } x''(\lambda) = -x(2 - \lambda),$$

which can be interpreted as inducing a "frequency folding," mapping the information carried at frequency λ to the frequency $2 - \lambda$. From this we can then see that

$$y_0(\lambda) = \frac{1}{2}\left(h_0(\lambda)x(\lambda) + h_0(2 - \lambda)x(2 - \lambda)\right),$$

$$y_1(\lambda) = \frac{1}{2}\left(h_1(\lambda)x(\lambda) + h_1(2 - \lambda)x(2 - \lambda)\right),$$

which allows us to represent the conditions for exact reconstruction in terms of signals, filters and their frequency folded versions, by grouping terms corresponding to $x(\lambda)$ and $x(2 - \lambda)$. The GFT of $\tilde{\mathbf{x}}$ in (5.46) can then be written as

$$\tilde{x}(\lambda) = \frac{1}{2}\underbrace{\left(\tilde{h}_0(\lambda)h_0(\lambda) + \tilde{h}_1(\lambda)h_1(\lambda)\right)}_{A(\lambda)}x(\lambda)$$

$$+ \frac{1}{2}\underbrace{\left(\tilde{h}_0(\lambda)h_0(2 - \lambda) - \tilde{h}_1(\lambda)h_1(2 - \lambda)\right)}_{B(\lambda)}x(2 - \lambda) \tag{5.49}$$

under the assumption that the samples in $\mathcal{V}_0$ and $\mathcal{V}_1$ are kept in the low pass and high pass channels, respectively. Thus, the goal is to design filters $\tilde{h}_0(\lambda)$, $h_0(\lambda)$, $\tilde{h}_1(\lambda)$ and $h_1(\lambda)$ such that

$$B(\lambda) = \tilde{h}_0(\lambda)h_0(2 - \lambda) - \tilde{h}_1(\lambda)h_1(2 - \lambda) = 0 \qquad (5.50)$$

and

$$A(\lambda) = \tilde{h}_0(\lambda)h_0(\lambda) + \tilde{h}_1(\lambda)h_1(\lambda) = 2. \qquad (5.51)$$

Orthogonal filter design   An orthogonal solution to meet the conditions of (5.50) and (5.51) can be designed [81] by first selecting,

$$\tilde{h}_0(\lambda) = h_0(\lambda) \text{ and } \tilde{h}_1(\lambda) = h_1(\lambda)$$

so that the analysis and synthesis filters are the same, and next defining

$$h_1(\lambda) = h_0(2 - \lambda),$$

which converts a low pass filter $h_0(\lambda)$ into a high pass filter $h_0(2 - \lambda)$, as can be noted by observing that if $h_0(2) = 0$ then $h_1(0) = 0$. With this choice we have $B(\lambda) = 0$ and we can write the condition on $A(\lambda)$ as

$$A(\lambda) = h_0^2(\lambda) + h_0^2(2 - \lambda) = c^2, \quad \forall \lambda, \qquad (5.52)$$

where only a single filter $h_0(\lambda)$ needs to be designed. For example $h_0(\lambda)$ could be an ideal filter (see Box 3.1 and Section 5.6.3) with a flat response within the range $\lambda \in [0, \ 1]$ and zero elsewhere (see also Box 5.6 for a discussion of the case where an eigenvalue equals 1).

Alternative designs that are similar to the design approaches used in Meyer wavelets [67] can soften the response. Refer to [81] for detailed examples. It can be shown that the condition (5.52) cannot be met by polynomial filters. Thus a possible solution is to find a filter that provides an exact solution for (5.52) and then find sufficiently good polynomial approximations to that filter.

Biorthogonal filter design   As an alternative, it is possible to design filters that are compactly supported by giving up the orthogonality condition. These biorthogonal solutions achieve $A(\lambda) = 0$ by choosing

$$\tilde{h}_0(\lambda) = h_1(2 - \lambda), \quad h_0(2 - \lambda) = \tilde{h}_1(\lambda), \qquad (5.53)$$

where again replacing λ by $2 - \lambda$ "modulates" a filter, converting a low pass into a high pass and vice versa. Then we need to design such filters to guarantee that $B(\lambda) = 2$, that is,

$$B(\lambda) = \tilde{h}_0(\lambda)h_0(\lambda) + \tilde{h}_0(2 - \lambda)h_0(2 - \lambda) = 2,$$

and, defining $p(\lambda) = \tilde{h}_0(\lambda)h_0(\lambda)$, all we need to do is to design $p(\lambda)$ such that

$$p(\lambda) + p(2 - \lambda) = 2, \quad \forall \lambda,$$

and then factor $p(\lambda)$ to obtain $\tilde{h}_0(\lambda), h_0(\lambda)$ [82].

Alternative filter designs The filter designs described so far in this section have limitations. In the orthogonal case, while $A(\lambda)$ is designed to be constant, a polynomial approximation of the filters could lead to significant deviation from a flat frequency response. In the biorthogonal case exact reconstruction is guaranteed with polynomial filters, but we have no control over the responses at each frequency, which means that very different gains can be associated with each frequency. These problems can be addressed by better filter designs, such as those of [83] where solutions are derived from existing filters for conventional signals.

Solutions for non-bipartite graphs Note that the critically sampled filterbanks described above were designed for bipartite graphs. If a graph is not bipartite, it can be approximated by a bipartite graph but then, as in Section 5.4.2, signals will be processed on a graph that differs from the original one. As an alternative, the original graph can be represented by a series of bipartite subgraphs. This idea will be discussed in Section 6.1. Note also that while we have considered critically sampled designs in this section, oversampled approaches such as in [84] have been proposed that lead to less constrained filter designs.

Efficient implementation A naive implementation of a filterbank following Figure 5.2 would be inefficient in that half the samples are discarded after being computed. Even if only the outputs in the sampled set are computed, this still requires operating with the larger graph. An alternative computation strategy is possible [70], where the normalized off-diagonal terms $\mathcal{A}_0$ and $\mathcal{A}_0^\mathsf{T}$ of $\mathcal{L}$ are used to create two subgraphs, with node sets $\mathcal{V}_0$ and $\mathcal{V}_1$ and with respective adjacency matrices $\mathbf{A}_0 = \mathcal{A}_0^\mathsf{T}\mathcal{A}_0$ and $\mathbf{A}_1 = \mathcal{A}_0\mathcal{A}_0^\mathsf{T}$ (in the undirected case). Then all filtering operations can be defined using subsampled signals $\mathbf{x}^0 \in \mathcal{V}_0$ and $\mathbf{x}^1 \in \mathcal{V}_1$ and filters based on $\mathbf{A}_0$ and $\mathbf{A}_1$. In practice, this means that it is possible to obtain the same output by using four filters (instead of two) but applied on the downsampled signals $\mathbf{x}^0$ and $\mathbf{x}^1$ on smaller subgraphs, with N_0 and N_1 nodes, respectively. This approach is a graph counterpart of the polyphase domain implementation of conventional filterbanks [67].

5.6.5 Pyramid Transforms

Spectral graph wavelet transforms (Section 5.5.1) can be applied to any graph and have a frequency interpretation, but SGWTs are overcomplete so that the number of outputs is $N_f \times N$ where N_f is the number of filters. Conversely, critically sampled methods (Section 5.6.4) produce N samples in total but can only be applied to bipartite graphs or require a decomposition of the graph into a series of bipartite subgraphs.

The multi-scale pyramid of [85] provides an intermediate design which can be applied to any graph, as can the SGWT but with much lower oversampling. In contrast with the subgraph approach of [73] (Section 5.4.3), where subgraphs were selected prior to filtering, the pyramid approach applies filtering prior to downsampling. Similar to the Laplacian pyramid for image processing of [86], the challenge of designing filters that do not share zeros and that guarantee reconstruction is overcome by designing a single filter and storing the error with respect to an interpolated signal.

Define $\mathbf{H}_0$ as a low pass graph filter and let $\tilde{\mathbf{H}}_0$ be a suitable interpolation filter.[1] Then, with $\mathcal{V}_0$ the set of sampled nodes, any signal $\mathbf{x}$ is approximated by a signal $\mathbf{x}_1$:

$$\mathbf{x}_1 = \mathbf{I}_{S_0}^\mathsf{T} \mathbf{H}_0 \mathbf{x},$$

defined only on the set $\mathcal{V}_0$, and by an approximation error:

$$\mathbf{e}_0 = \mathbf{x} - \tilde{\mathbf{H}}_0 \mathbf{I}_{S_0} \mathbf{x}_1,$$

which takes values on all nodes in $\mathcal{V}$. Note that this representation requires $N + |\mathcal{V}_0|$ values instead of the $2N$ required if we had used two filters without downsampling as in (5.45). This operation can be repeated at successive levels of the pyramid, i.e., $\mathbf{x}_2$ and $\mathbf{e}_1$ can be obtained from $\mathbf{x}_1$ given a new sampling set $\mathcal{V}_1 \subset \mathcal{V}_0$ and filters $\mathbf{H}_1$, $\tilde{\mathbf{H}}_1$. In [85], the sampling sets were chosen using the highest-frequency eigenvectors of the graph Laplacian and, after downsampling, graphs containing a subset of nodes (e.g., S_0) are reconnected using the Kron reduction (Section 6.1.5) followed by a sparsification (Section 6.1.2).

5.7 Diffusion Wavelets

Diffusion wavelets [87] are designed using a different approach that combines both spectral and node domain characteristics, but without providing exact localization in either of these domains.

Recall the definition of the symmetric normalized Laplacian:

$$\mathcal{L} = \mathbf{D}^{-1/2} \mathbf{L} \mathbf{D}^{-1/2} = \mathbf{I} - \mathcal{A},$$

where $\mathcal{A} = \mathbf{D}^{-1/2} \mathbf{L} \mathbf{D}^{-1/2}$ is a symmetric **diffusion operator** such that $\mathcal{A}^\mathsf{T} = \mathcal{A}$ and with eigenvalues $\lambda \in [-1, 1]$, where $\lambda_1 = 1$ corresponds to the lowest frequency of $\mathcal{L}$ ($\lambda = 0$) while the frequency closest to -1 corresponds to the highest frequency of $\mathcal{L}$ ($\lambda \leq 2$). Note that, unless the graph is bipartite, all eigenvalues of $\mathcal{A}$ except λ_1 are such that $|\lambda| < 1$.

ϵ-**Span**   The key observation in the design of diffusion wavelets is that successive powers of $\mathcal{A}$, or some other diffusion operator, such as Q, will have increasingly lower numerical rank. This follows because the eigenvalues of $\mathcal{A}^k$ are λ^k and these can become arbitrarily small as k increases, since $|\lambda| \leq 1$.

This leads to the definition of the ϵ-span of a set of vectors, $\Phi_v = \{\mathbf{v}_1, \mathbf{v}_2, \ldots, \mathbf{v}_N\}$, which could be for example the columns of $\mathcal{A}^k$. Define a new set of vectors $\Phi_u = \{\mathbf{u}_1, \mathbf{u}_2, \ldots, \mathbf{u}_j\}$, with $j \leq N$. Then Φ_u ϵ-spans Φ_v if, for all $i = 1, \ldots, N$,

$$\|\mathbf{P}_{\Phi_u} \mathbf{v}_i - \mathbf{v}_i\|_2 \leq \epsilon,$$

---

[1] Note that exact reconstruction can be achieved for any chosen interpolation filter. Thus, a "suitable" interpolation filter is one that can provide a good approximation. For example, sampling a low pass $\tilde{\mathbf{H}}_0$ could be a good choice for reconstruction if $\tilde{\mathbf{H}}_0$ is a low pass filter.

Table 5.1 Summary of properties of representations. Note that all methods included can achieve some localization in the node domain. Methods with a frequency interpretation are such that the filters can be written in the frequency domain. Representations that are not oversampled are critically sampled: any graph signal with N entries is represented by N transform coefficients. Also note that, while filterbank representations are valid for bipartite graphs only, it is possible to approximate any arbitrary graph by a series of bipartite subgraphs

	Frequency	Orthogonal	Oversampled	Graph type
Lifting [68]	No	No	No	Any
Subgraphs [73]	No	No	No	Any
Frames [74, 77, 76]	Yes	No	Yes	Any
Pyramids [85]	Yes	No	Yes	Any
Filterbanks [81, 82]	Yes	Yes	No	Bipartite
Diffusion [87]	Yes	Yes	No	Any

where $\mathbf{P}_{\Phi_u}$ computes the projection of a vector onto the span of Φ_u. Intuitively if Φ_u ϵ-spans Φ_v with $j < N$ this means that not much of an error is made by approximating the span with a smaller set of vectors.

Diffusion wavelet construction In the design of diffusion wavelets, ϵ-spans are used by selecting sets:

$$\Phi_{A^{2^i}} = \{\lambda_{t,1}^{2^i}\mathbf{v}_1, \ldots, \lambda_{t,b}^{2^i}\mathbf{v}_N\} \tag{5.54}$$

where $\{\mathbf{v}_1, \ldots, \mathbf{v}_N\}$ are the eigenvectors of $\mathcal{A}$. Then, in the design, $V_0 = \mathbb{R}^N$ and $V_i = \text{span}(\Phi_i)$, where Φ_i ϵ-spans $\Phi_{A^{2^i}}$. At any stage we can then find a subspace W_i such that $V_i \oplus W_i = V_{i-1}$. Choosing a specific i we write

$$V_0 = V_i \oplus W_i \oplus W_{i-1} \oplus W_{i-2} \oplus \cdots \oplus W_1$$

so that the basis vectors for the W_i spaces form the orthogonal wavelets and the basis vectors for V_i correspond to the scaling function. The choice of 2^i in the exponents of (5.54) is meant to mimic the behavior of wavelet transforms for conventional signals, where going from higher to lower resolution (from V_{i-1} to V_i and W_i) corresponds to dividing the frequency range of V_{i-1} in two.

Chapter at a Glance

Many representations have been proposed for signals on graphs. In Section 5.1 we formalized the problem of defining localization in the node and frequency domains, a key observation being that we should not take for granted insights developed for conventional signals, i.e., that signals with good localization in both node and frequency domains can be found. As for the various designs introduced in this chapter, the main message is that it is not possible in general to achieve all the desirable properties defined in Section 5.2 simultaneously. In particular, some designs are restricted to only some types of graphs, while others may have inefficient reconstruction methods.

Further Reading

An overview of uncertainty principles can be found in [63]. Descriptions of the various methods presented in this section are found in the references cited in the text (see also Table 5.1). A recent overview paper [88] provides additional references and details.

GraSP Examples

In Section B.6.1 an example is provided of the SGWT applied with the random walk Laplacian. This also includes the application of Chebyshev polynomials to achieve a local implementation with polynomial filters. This section also shows an example of the elementary basis vectors in the SGWT dictionary.

6   How to Choose a Graph

Up to this point, we have developed graph signal processing tools on the assumption that a graph is given. Both node domain operations (Chapter 2) and spectral domain operations (Chapter 3) strongly depend on the choice of graph. In many applications there is an "obvious" graph within the system to be analyzed. For example, when studying a social, transportation or communication network, a graph with known topology may be available and can be used for analysis. Instead, in this chapter we consider scenarios where either a graph is not available or we wish to modify an existing graph. We address three different but related cases.

- **Graph approximation** (Section 6.1)   We already have a graph, but we would like to modify it to have more desirable properties, such as being sparser (fewer edges) or smaller (fewer nodes), or having a specific topology (e.g., bipartite).
- **Constructing graphs from node attributes** (Section 6.2)   We wish to construct a graph using node attributes, such as for example the sensor positions in a sensor network. An underlying assumption is that nodes that are "similar" on the basis of those attributes should be connected in the graph.
- **Learning graphs from signal examples** (Section 6.3)   We have collected signal observations over time and would like to design a graph that in some way "matches" the observed signals.

Graph selection as a modeling problem   Selecting a graph is a fundamental step for graph signal analysis, since the definitions of frequency and node domain operations are dependent on this choice. In deciding which graph is chosen, we need to consider the application, as well as other factors. On the one hand, desired frequency or node domain properties (e.g., frequency spacing or edge sparsity) determine the effectiveness and computational complexity of GSP tools. On the other hand, achieving those properties may require graph approximations that modify the topology significantly or may not be possible if the number of parameters to be learned (estimated) is excessive relative to the number of available training examples. Thus, it is important to view graph selection as a *modeling task*, where the results achieved with the model and the fidelity of its representation of the data are both important. While in Section 4.1 we assumed a graph was given and considered models for signals, here we take a broader view of the modeling problem and allow the graph itself to be selected.

In this chapter we provide an overview of methods, highlighting their corresponding underlying assumptions, under the premise that there is no "right" graph: the choice

Table 6.1 Graph learning problems

Problem	Nodes	Edge weights	Data	Criterion
Graph approximation	Known	Known	Maybe	Preserve spectrum
Node similarity	Distances known	Distance-based	N/A	Distance-based
Data-driven	Known	N/A	Yes	Optimized for data

has to take into account multiple criteria. Note that we consider the three cases above in separate sections, but in some cases it may be useful to combine some of these approaches. In particular, even if signal examples are available and can be used directly to learn a graph, the resulting graph may be combined with a graph construction derived from node similarity (see Box 6.1).

Box 6.1   Combining the node similarity and data-driven approaches

As an example where techniques based on node similarity (Section 6.2) and graph signal observations (Section 6.3) might be combined, consider a sensor network for which we have information about node locations and a series of signal observations. A graph construction based on the methods in Section 6.2 assigns edge weights solely on the basis of distance between nodes, while the methods in Section 6.3 connect nodes using statistical correlation. Since the latter methods do not take into account node location, they may strongly connect nodes that are far away in space. Thus, a combination of both methods has the potential to incorporate both the observed data behavior (using statistical correlation) and physical system characteristics (the distances between nodes). Combining both sources of information can potentially lead to a more interpretable model.

Another scenario where different types of similarity can be combined is described in Section 7.4.2, where graphs used for image processing combine both pixel location and pixel intensity information.

In this chapter, we assume that a set of nodes, $\mathcal{V} = \{1, \ldots, N\}$, is given and we connect them by choosing edge weights, w_{ij}, leading to an adjacency matrix $\mathbf{A}$ and a selected fundamental graph operator $\mathbf{Z}$. Graph selection will be based on various criteria, summarized in Table 6.1 and to be discussed in what follows.

6.1   Graph Approximation

In this scenario, both the graph nodes and edges are already available, so that $\mathbf{A}$ is known, but we wish to approximate the known graph by another one having some desired properties (e.g., sparsity, a specific topology, etc.). If a graph is already known then a reasonable criterion would be to find the most "similar" graph having the desired property. Similarity between graphs can be quantified in many ways: by comparing some of their node and frequency domain characteristics or their performance in a given application.

We organize the discussions in this section by considering in turn four possible tasks, namely, sparsification, computational simplification, topology selection and graph reduction. We start by introducing methods that enable us to understand the effect of edge removal.

6.1.1 Edge Removal and Graph Spectrum

Edge removal and eigenvalues Intuitively, removing edges with small weights should lead to small changes in the frequency spectrum. We now justify this idea more formally. Let us focus on the undirected-graph case with $\mathbf{Z} = \mathbf{L}$. Removing one edge means setting to zero two entries in $\mathbf{L}$. Thus, for any number of edges removed we obtain an undirected graph $\tilde{G}$ with Laplacian operators $\tilde{\mathbf{L}}$:

$$\tilde{\mathbf{L}} = \mathbf{L} + \mathbf{E} \tag{6.1}$$

where $-\mathbf{E}$ is also a graph Laplacian (see Example 6.1).

From Theorem 3.5, we know that eigenvalues of $\mathbf{L}$ and $\tilde{\mathbf{L}}$ are contained in the union of N circles, where the ith circle is centered at d_i and has radius d_i. Since removing an edge of weight w_{ij} changes the degree of node i from d_i to $d_i - w_{ij}$, it follows that removing the smallest-weight edge leads to the smallest change in d_i (and thus the smallest change in the eigenvalue bounds, from Theorem 3.5). A more general result can be stated using the following theorem (see [19, Theorem 3.8, p. 42]).

THEOREM 6.1 (SYMMETRIC PERTURBATION OF SYMMETRIC MATRICES) For a symmetric matrix $\mathbf{L}$ and a symmetric perturbation as in (6.1) the eigenvalues of $\tilde{\mathbf{L}}$ can be bounded by

$$|\tilde{\lambda}_i - \lambda_i| \leq \|\mathbf{E}\|_2,$$

where $\|\mathbf{E}\|_2$ is the largest singular value of $\mathbf{E}$.

Notice that $-\mathbf{E}$ is a graph Laplacian and has real non-negative eigenvalues that can be bounded using Theorem 3.5. This is illustrated by the following example.

Example 6.1 Let $\mathbf{L}$ be the combinatorial Laplacian for an undirected graph G and assume that a new graph is obtained by setting to zero an edge with weight e in G. Find a bound on the error for any eigenvalue λ_i of $\mathbf{L}$.

Solution
Define $\tilde{\mathbf{L}} = \mathbf{L} + \mathbf{E}$. Without loss of generality assume that the removed edge is the edge between nodes 1 and 2 (we can do this by labeling the nodes appropriately, since the

numbering of the nodes is arbitrary as discussed in Box 2.1). Then we can write

$$-\mathbf{E} = \begin{bmatrix} e & -e & \cdots & 0 & 0 \\ -e & e & \cdots & 0 & 0 \\ \vdots & \vdots & \ddots & \vdots & \vdots \\ 0 & 0 & \cdots & 0 & 0 \end{bmatrix},$$

which makes it obvious that $-\mathbf{E}$ is a graph Laplacian. Therefore using Proposition 3.3 we see that the maximum eigenvalue of $-\mathbf{E}$ is bounded by $\lambda_{max} = 2e$. Thus we have that $|\tilde{\lambda}_i - \lambda_i| \le 2e$.

This idea can be generalized as follows.

Remark 6.1   (EIGENVALUE APPROXIMATION)   Let $-\mathbf{E}$ be the difference graph Laplacian of (6.1). Let e_{max} be its maximum degree. Then, for any eigenvalue λ_i of $\mathbf{L}$ and corresponding eigenvalue $\tilde{\lambda}_i$ of $\tilde{\mathbf{L}}$, we have

$$|\tilde{\lambda}_i - \lambda_i| \le 2e_{max}. \tag{6.2}$$

Proof   We are given that $-\mathbf{E}$ is a graph Laplacian, with maximum degree e_{max}, corresponding to the node whose degree has been most changed in the simplification. Then applying the bound Proposition 3.3 and the result Theorem 6.1 we obtain (6.2).   □

Note that e_{max} is determined by the node that has lost the most edges (for an unweighted graph) or has lost a greater cumulative edge weight (for a weighted graph). Thus, this remark suggests that a strategy that spreads out edge removals across many nodes, while keeping e_{max} small, can be more effective to reduce the effect of graph simplification on the frequency spectrum.

Edge removal and eigenvectors   To extend these results to eigenvectors, we can use the following result [19, Theorem 3.16, p. 51].

THEOREM 6.2   (EIGENVECTOR PERTURBATION)   Let $\Delta\lambda$ be the absolute value of the minimum spacing between the eigenvalues of $\mathbf{L}$. Let $\mathbf{u}$ and $\tilde{\mathbf{u}}$ be the eigenvectors of $\mathbf{L}$ and $\tilde{\mathbf{L}}$ defined as in (6.1), where $\mathbf{u}$ and $\tilde{\mathbf{u}}$ correspond to simple eigenvalues with the same index, e.g., the kth eigenvalues of $\mathbf{L}$ and $\tilde{\mathbf{L}}$. Then θ, the angle between $\mathbf{u}$ and $\tilde{\mathbf{u}}$ (see (A.9)), is such that

$$\sin\theta \le \frac{\|\mathbf{E}\|_2}{\Delta\lambda}. \tag{6.3}$$

Notice that this result holds for eigenvectors corresponding to simple eigenvalues with matching indices in $\mathbf{L}$ and $\tilde{\mathbf{L}}$. Thus, if the minimum eigenvalue spacing is very small, a small perturbation can have a significant effect. As an extreme case, consider the case where we have a high-multiplicity eigenvalue. The corresponding eigenspace has dimension greater than 1 and, as discussed in Chapter 2, there exists an infinite number of orthogonal bases of eigenvectors. Essentially this means that we can "perturb" an eigenvector maximally (so that $\sin\theta = 1$), without changing the eigenvalue.

Let us consider the case when two simple eigenvalues, $\lambda_k < \lambda_{k+1}$, are very close, each corresponding to orthogonal subspaces spanned by vectors $\mathbf{u}_k$ and $\mathbf{u}_{k+1}$. A small perturbation can change the order of the eigenvalues without much affecting the corresponding subspaces. Thus, if we match the kth eigenvalue of the original graph to the kth eigenvalue of the perturbed graph, we could be matching $\mathbf{u}_k$ to a vector close to $\mathbf{u}_{k+1}$, namely $\tilde{\mathbf{u}}_k$. Since $\mathbf{u}_k$ and $\mathbf{u}_{k+1}$ are orthogonal, it is possible that $\mathbf{u}_k^\mathsf{T}\tilde{\mathbf{u}}_k$ will be close to zero and thus $\sin\theta$ can be close to 1, as (6.3) suggests.

In summary, while according to (6.3) a small perturbation can lead to significant changes when $\Delta\lambda$ is small, such changes may not be particularly relevant, precisely because the frequencies are so close to each other. This idea is further explored in Section 6.1.3.

6.1.2 Graph Sparsification

The goal of graph sparsification is to reduce the number of edges in a given graph, while preserving some of its properties. For an unweighted graph, a simple approach is to remove edges with small weights.

Sparsification and processing complexity Reducing the number of edges leads to lower processing complexity: since the elementary graph filtering operation is $\mathbf{Z}\mathbf{x}$, whose complexity is a function of the number of one-hop neighbors (i.e., the number of edges), removing some edges leads to less computation. Matrix–vector multiplication for vectors of size N has complexity of the order of N^2, but a sparse graph with $M \ll N^2$ edges can be represented as an adjacency list (see the discussion in the last paragraph before Section 2.3.2), leading to computations that scale with M.

Locality and interpretation Reducing the number of edges increases the graph filtering locality, which depends on the overall graph topology and the polynomial degree k chosen for the filter. Clearly, removing graph edges increases geodesic distances, so that radius and diameter can also increase (see Definition 2.8 and Definition 2.9). Thus, for a given degree k, processing will be more localized after edges have been removed, and both sparsification and reduction of k are alternative approaches to achieving local processing (Section 2.2.2).

Box 6.2 Sparsification of similarity graphs

In some applications, such as the processing of sensor network data, each node has an attribute (e.g., node location), and similarity graphs are constructed on the basis of the distance between nodes (see Section 6.2), with the corresponding edge weights decreasing as the distance increases. Because there always exists a distance between two nodes in Euclidean space, these similarity graphs are by nature complete (i.e., there are edges between all pairs of nodes). If the original weights are used, i.e., we use a complete weighted graph for processing, this would mean that filter operations involving **L** or **A** are not exactly localized (i.e., they do not have compact support), and are computationally expensive. Sparsification techniques make use of several intuitive ideas.

- Edges with smaller weights have less effect on the variation (and thus frequency) so they should be removed first.
- The density of nodes in space is not necessarily constant, so that pairwise distances can be smaller in some areas and larger in others. In order to preserve overall connectivity it may be better to ensure there is at least some minimum number of neighbors per node.

K-Nearest-neighbor (K-NN) sparsification K-NN is one of the most widely used approaches. It follows the principles described in Box 6.2. Instead of deciding on some threshold, T, and then removing all edges with weights w_{ij} such that $w_{ij} \leq T$, K-NN preserves local topology (and avoids a disconnected graph) by keeping for each node i the K connections with highest weight. A major advantage of K-NN is that it maintains a regular topology (albeit with unequal weights). A limitation of K-NN is that its sparsity target (K) is the same around every node, while the local graph properties may be quite different (e.g., higher-degree nodes are more strongly connected to their neighbors). Methods that can adapt local neighborhood characteristics are discussed in Section 6.2.

Spectral sparsification   Note that K-NN does not explicitly aim to preserve the spectral characteristics of the adjacency matrix **A**. Spectral sparsification [89, 90] is an alternative approach in which decisions are made per edge (rather than per node as in K-NN), by using random edge sampling. In [90] the probability of selecting an edge is taken to be proportional to $p_{ij} = w_{ij} r_{ij}$, where w_{ij} is the edge weight and r_{ij} is the **effective resistance** between the corresponding nodes. Recall the definition of geodesic distance (Definition 2.8), which shows how distance is obtained by adding the reciprocal of the weights along a path. Each of these reciprocals can be viewed as a "resistance" and thus distance increases with resistance. For example, assuming that i and j are not connected and that there is a single path (going through k) between two nodes i and j:

$$r_{ij} = r_{ik} + r_{kj} = \frac{1}{w_{ik}} + \frac{1}{w_{ik}}. \tag{6.4}$$

More generally, assuming that there are multiple paths between i and j including the direct path, r_{ij} incorporates the effect of all these paths, following the same rules as are used for electric circuits. In the previous example, if $w_{ij} \neq 0$ and the path through k is the only other connection between i and j then we would have

$$\frac{1}{r_{ij}} = w_{ij} + \frac{1}{r_{ik} + r_{kj}}. \tag{6.5}$$

Notice that when multiple paths exist, the effective resistance decreases. To understand the effect of this sparsification criterion consider two examples. First, assume that the only connection between i and j is the direct one, with weight w_{ij}. Since there are no other paths from i to j, $r_{ij} = 1/w_{ij}$. Therefore $p_{ij} = 1$, the maximal possible value, and the edge ij is chosen with high probability. This is justified because in this case the graph would become disconnected if ij were removed, leading to a significant change in its spectrum.

Second, assume that there are only two ways of going from i to j, as in (6.5), and that w_{ij} is small while r_{ik}, r_{kj} are also small. Then the second term in (6.5) is much larger than the first one, so that $r_{ij} \cong r_{ik} + r_{kj}$ and p_{ij} is small. This allows the edge ij to be removed with high probability (i.e., it has a low probability of being chosen). Intuitively, the direct connection can be removed because indirect connections have higher similarity (lower resistance).

In summary, w_{ij} depends only on the edge between i and j, while r_{ij} depends on all the connections between i and j. The maximum resistance is $r_{ij} = 1/w_{ij}$; this occurs when there is only one connection between i and j and any additional paths between i and j lead to lower r_{ij} and thus to a lower probability of the edge between i and j being kept.

6.1.3   Graph Simplification and Graph Fourier Transform Computation

As we have seen, graph sparsification is useful as a way to reduce the cost of node domain processing, e.g., graph filtering with the polynomials of a graph operator. In this section we focus on reducing the complexity of computing the GFT, $\mathbf{U}^\mathsf{T}\mathbf{x}$, for arbitrary $\mathbf{x}$. In general, N^2 operations are required for this computation for an arbitrary graph with N nodes. Fast transforms are possible for some graphs, e.g., path graphs (Section 1.3.3), for which the GFT is the discrete cosine transform (DCT). Since N^2 computation may not be practical for large N, there is an ongoing interest in identifying properties that can lead to faster GFT computation, along with algorithms to modify graphs (or their GFTs) to achieve computation cost savings. Two such methods are described next.

Graph symmetry and fast GFT   The size-eight DCT, for which a fast algorithm is available, is the GFT of the path graph in Figure 1.8, where all seven edges have equal weight. This leads to some symmetries: numbering the edges e_i, $i = 1, \ldots, 7$ we can take e_4 as a center of symmetry and then observe that $e_{4-k} = e_{4+k}$ for $k = 1, 2, 3$. Recent work [91] has shown that graphs that have these kinds of symmetries, and related ones, lead to GFTs that can be decomposed into the product of two matrices. The first matrix contains a series of "butterfly" operations (averages and differences between pairs of inputs), with low computational cost. The second is block diagonal, with blocks of dimension no greater than $N/2$, so that the overall cost is significantly reduced. While we should not expect most graphs to exhibit such symmetries, this idea suggests that node domain simplifications could be useful if they approximate a graph by one having symmetries.

Approximations via Givens rotations   Instead of simplifying the graph, GFT computation for the original graph can be simplified by breaking it down into a series of elementary operations. The article [92] presents techniques to approximate the GFT, $\mathbf{U}$, as the product of a series of Givens rotations written as matrices of the form

$$\boldsymbol{\Theta}_{i,j,k} = \begin{bmatrix} 1 & \cdots & 0 & \cdots & 0 & \cdots & 0 \\ & & \vdots & & \vdots & & \\ 0 & \cdots & \cos\theta_k & \cdots & -\sin\theta_k & \cdots & 0 \\ & & \vdots & & \vdots & & \\ 0 & \cdots & \sin\theta_k & \cdots & \cos\theta_k & \cdots & 0 \\ & & \vdots & & \vdots & & \\ 0 & \cdots & 0 & \cdots & 0 & \cdots & 1 \end{bmatrix}, \tag{6.6}$$

where all columns are taken from the $N \times N$ identity matrix, except columns i and j which include cosines and sines of the angle θ_k. Note that any $\boldsymbol{\Theta}_{i,j,k}$ is orthogonal by construction, so that a product of $\boldsymbol{\Theta}_{i,j,k}$ matrices will also be orthogonal. The positions (i, j) and the corresponding angles θ_k are chosen to approximate $\mathbf{U}$. The approximation can be improved incrementally by adding one Givens rotation at a time. This can be done until the exact transform has been computed or a desired degree of approximation has been achieved, the overall complexity of the computation being proportional to the number of rotations used. In [92] the authors observed that exact approximation is more challenging when the eigenvalues are close. This is consistent with our observations about Theorem 6.2: when eigenvalues are close to each other, small changes to the graph can change their order. Thus, it is possible to achieve good approximations for the GFT, but with some uncertainty regarding the correct ordering of the eigenvectors. Finally, notice that this is a frequency domain method and thus the simplified approximate GFT may not correspond to the eigenvectors of a valid graph operator.

6.1.4 Topology Selection

A second scenario for graph simplification arises when a specific graph topology is desirable for subsequent graph signal processing. For example, a bipartite graph approximation is needed for the graph filterbanks of Section 5.6.4. This approximation can follow ideas discussed in Section 6.1.2, i.e., removal of edges with smaller weight should be favored. Thus, for bipartition, where only edges across the two sets are preserved, an algorithm for identifying a maximum cut (see Definition 2.11) can be used, so that only edges in the cut are kept. In general, existing graph algorithms (for coloring or finding cuts) can be used in the topology selection process.

Case Study: Bipartite Graphs

As a case study, consider bipartite graph approximation. We consider two approaches, depending on whether single- or multiple-level approximations are used.

Single-level approximation   In a single-level approximation, the goal is to find a single bipartite subgraph (same nodes, subset of edges). Then, filtering operations are performed on the resulting graph. This can be viewed as an approximation where the edges removed by the graph simplification are no longer used for processing. Following our earlier discussion, the goal should be to minimize the weight of the edges being eliminated. Conventional bipartition algorithms can be used, e.g., max-cut techniques based on the maximum eigenvector of $\mathbf{L}$.

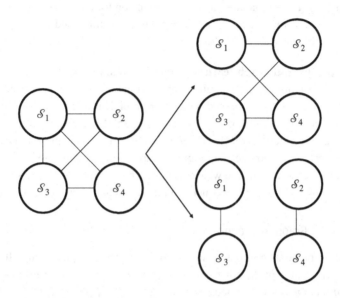

Figure 6.1 Decomposition into multiple bipartite subgraphs. This graph is 4-colorable. After splitting into four sets S_1, S_2, S_3 and S_4, the first level of decomposition is a bipartite subgraph (right, top) that contains all nodes with connections only from $\{S_1, S_3\}$ to $\{S_2, S_4\}$. The second level of decomposition is a bipartite approximation that has two disconnected subgraphs (right, bottom) with connections from S_1 to S_3 and S_2 to S_4.

Multi-level approximation problem As an alternative, multiple bipartite graph approximations can be chosen. This allows all edges to be used for filtering, although they cannot be all used simultaneously. This idea is illustrated in Figure 6.1 for a 4-colorable graph. The first level of decomposition uses only horizontal and diagonal connections, while the second uses vertical connections only. Because in this case all edges are used, the goal may be to obtain more uniform bipartite approximations. As an example, if the number of edges included in the first bipartition were maximized this could lead to successive bipartitions operating on sparse graphs, which could potentially be disconnected.

Multi-level approximation algorithms The designs in Section 5.6.4 were based on the symmetric normalized Laplacian, $\mathcal{L}$. Recall that for a bipartite graph all eigenvalues come in pairs, λ and $2 - \lambda$, and thus there are two filters, a low pass $h(\lambda)$ and a corresponding high pass $h(2 - \lambda)$, which have the same response at $\lambda = 1$. Thus, signals that have frequency $\lambda = 1$ will be present with equal weight in both subbands, i.e., the filterbank has the least selectivity at $\lambda = 1$. This is particularly problematic since the subspace corresponding to $\lambda = 1$ has dimension at least $|N_1 - N_2|$ (Box 5.6), so that for a high-dimension subspace we may have no frequency selectivity. While an unbalanced bipartition may happen to be good in terms of minimizing the number of edges removed, it may be preferable to use alternative bipartite approximation technique [93] with the goal of maintaining a low dimension for the subspace at $\lambda = 1$. This may result in a bipartition where more edges have been eliminated as compared to one obtained through a max-cut algorithm.

Topology and graph learning   In this section we have considered the simplification of an existing graph to obtain an approximation with some desired topology. Topology objectives can be combined with other graph learning scenarios. For example, methods have been developed to learn (estimate) graph Laplacians with desired topologies from covariance matrices [94]. This work shows that good approximations can be achieved for certain (monotone) topologies by approximating the covariance matrix (which can be viewed as a complete weighted graph) by a graph with the desired topology, and then learning the graph weights using methods such as those to be described in Section 6.3.

6.1.5   Graph Reduction

In the previous sections, simplification consisted of preserving all graph nodes but removing some of the edges so as to sparsify the graph. In **graph reduction**, however, a *subset of nodes* is selected and a smaller graph is created with those nodes, while taking into account the connectivity of the original graph. Nodes can be selected using one of the algorithms in Chapter 4. Processing sampled signals on the smaller graph can be advantageous as a way of reducing complexity, or it may be required as part of a multiresolution processing strategy.

Kron reduction   Given a subset of graph nodes, a widely used graph reduction approach consists of two steps: Kron reduction followed by sparsification. Kron reduction (see [95] for a detailed review) takes into account the topology of the original graph in order to obtain the reduced graph. Assume a subset of nodes $\mathcal{V}_1$ has been selected and define $\mathcal{V}_2 = \mathcal{V}_1^c$ as its complement in $\mathcal{V}$. Then, we can rearrange the Laplacian matrix after suitable permutation of the nodes to give

$$\mathbf{L} = \begin{bmatrix} \mathbf{L}_{11} & \mathbf{L}_{12} \\ \mathbf{L}_{21} & \mathbf{L}_{22} \end{bmatrix},$$

where $\mathbf{L}_{12} = \mathbf{L}_{21}^\mathsf{T}$ for the undirected Laplacian. The matrix $\mathbf{L}_{22}$ is invertible when the graph is connected,[1] so that we can create the Laplacian of the Kron-reduced graph using the Schur complement:

$$\mathbf{L}_r = \mathbf{L}_{11} - \mathbf{L}_{12}\mathbf{L}_{22}^{-1}\mathbf{L}_{12}^\mathsf{T}, \tag{6.7}$$

where $\mathbf{L}_r$ can be shown to be a graph Laplacian.

---

[1]   This is true because $\mathbf{L}$ is diagonally dominant, so that any principal submatrix of $\mathbf{L}$ is strictly diagonally dominant, and therefore invertible.

Node by node reduction The Kron reduction can be computed by removing one node at a time and reconnecting the graph. This procedure provides insights about the effect of graph reduction on graph connectivity. Consider any two nodes i and j and assume that a third node k has been removed. Then we will have the following cases.

- If i and j are connected, $w_{ij} \neq 0$, and if only either i or j were originally connected to k then w_{ij} will be unchanged; if both were connected to k then their connection remains but w_{ij} changes.
- If i and j are not connected, $w_{ij} = 0$, then a new edge will now appear between them if and only if both i and j were connected to the removed node k.

More formally, given that k will be removed, permute the Laplacian

$$\mathbf{L} = \begin{bmatrix} \mathbf{L}_{\bar{k}} & \mathbf{l}_k \\ \mathbf{l}_x^\mathsf{T} & d_k \end{bmatrix},$$

where d_k is the degree of node k, and $\mathbf{l}_k$ is a column vector with non-zero entries only for nodes to which k was originally connected. Using (6.7), we have

$$\mathbf{L}_r = \mathbf{L}_{\bar{k}} - \frac{1}{d_k}\mathbf{l}_k\mathbf{l}_k^\mathsf{T},$$

where $\mathbf{l}_k\mathbf{l}_k^\mathsf{T}$ is a rank-1 matrix, with non-zero entries only at nodes that were originally connected to k. This leads to

$$d_{i,r} = d_i - \frac{w_{ik}^2}{d_k} \quad \text{and} \quad w_{ij,r} = w_{ij} + \frac{w_{ik}w_{jk}}{d_k}, \tag{6.8}$$

which shows that the edge weight between i and j will only change if both of them were originally connected to k. This can be illustrated by the following example.

Example 6.2 Consider an unweighted graph where $d_k = 1$ and the only neighbors of node k are i and j, with $w_{ik} = w_{jk} = 1/2$ and $w_{ij} = 0$. Then use (6.8) to find the weights between i and j if k is removed.

Solution
From (6.8) we get

$$d_{i,r} = d_i - (1/4) \quad \text{and} \quad w_{ij,r} = 1/4,$$

which can be interpreted as follows: the weights correspond to the **conductances** of an electric circuit, so that the corresponding resistances would be $r_{ik} = r_{jk} = 2$. Therefore $r_{ij} = 4$, corresponding to the connection of two resistors in series, and thus $w_{ij} = 1/r_{ij} = 1/4$.

Sparsification after reduction From the one-step Kron reduction of (6.8) we can see the local effect of the removal of k. Denote by $\mathcal{N}_k$ the set of neighbors of k in the original graph. All nodes in $\mathcal{N}_k$ are connected to k, but not necessarily connected to each other. In the reduced graph, however, all nodes in N_k are connected to each other, so that N_k becomes a clique (Section 2.2.5) in the reduced graph. This shows that, as more nodes are removed, the graph becomes denser. Thus, even though the Kron reduction is optimal in the sense of preserving the properties of the original graph (e.g., the resistance in Example 6.2), it is common to sparsify a Kron-reduced graph (see Section 6.1.2 and Section 6.2.2).

6.2   Constructing Graphs from Node Similarity

In many applications a notion of node similarity can be defined before a graph is constructed. Examples include: (i) a sensing system, where each sensor node is placed in the environment and its location is known and (ii) a machine learning application, where each node corresponds to a datapoint with which we can associate some attributes or features. In what follows, we call **attributes** these properties associated with nodes.

Let i and j be arbitrary nodes with attributes $\mathbf{y}_i, \mathbf{y}_j \in \mathbb{R}^D$ and define a pairwise distance, $d(i, j) = \|\mathbf{y}_i - \mathbf{y}_j\|$, or some other metric of similarity between attributes. From the definition of the Laplacian quadratic form,

$$\mathbf{x}^\mathsf{T}\mathbf{L}\mathbf{x} = \sum_{ij} w_{ij}(x_i - x_j)^2;$$

the smooth signal $\mathbf{x}$ is such that $(x_i - x_j)^2$ is small when w_{ij} is large. Therefore, selecting a large w_{ij} for nearby nodes ($d(i, j)$ small) leads to graph smoothness (based on w_{ij}) and spatial smoothness (based on $d(i, j)$) being consistent requirements. That is, for a smooth graph signal $\mathbf{x}$ we expect $(x_i - x_j)^2$ to be small when $d(i, j)$ is small. This demonstrates the importance of a good choice of similarity/distance, as it directly affects our definition of graph processing (locality, frequency, etc.) for the particular case in question.

6.2.1   Computing Graph Weights as a Function of Similarity

Gaussian kernel   The attributes $\mathbf{y}_i$, $\mathbf{y}_j$ associated with nodes i and j, respectively, may represent node positions in space (in the sensor case) or features (in the machine learning application). Given a distance $d(i, j) = \|\mathbf{y}_i - \mathbf{y}_j\|$, Gaussian kernel methods define edge weights as

$$w_{ij} = \exp\left(-\frac{d(i, j)^2}{2\sigma^2}\right), \tag{6.9}$$

where the parameter σ is the bandwidth of this Gaussian kernel. Note that if $d(i, j)$ is close to zero then the edge weight is close to 1, while if $d(i, j)$ increases then w_{ij} approaches zero. Thus, unless sparsification is applied (with methods discussed in Section 6.1.2 or to be described in Section 6.2.2) all nodes will be connected, although some weights can be vanishingly small.

Cosine similarity Cosine similarity is often used for data that has discrete attributes. Content-recommendation graphs provide a classical example. In this setting, users rate a certain type of content, e.g., movies, where each item is associated with a node and users give discrete ratings. For simplicity assume only two ratings, $+1$ or -1, are given, corresponding to positive and negative assessments, respectively, while 0 indicates that a particular user has not rated a movie. Then two content items i and j corresponding to two nodes on the graph have respective vectors $\mathbf{y}_i$ and $\mathbf{y}_j$ with entries taking values $+1, -1, 0$ representing the ratings given by all users: the kth entries $y_i(k)$ and $y_j(k)$ represent how user k rated items i and j, respectively. Two movies will be similar if they are given similar ratings by users, and, given $\mathbf{y}_i$ and $\mathbf{y}_j$, the cosine similarity, as in (A.9), is used to define the graph edge weight:

$$w_{ij} = \frac{|\mathbf{y}_i^\mathsf{T} \mathbf{y}_j|}{\|\mathbf{y}_i\| \, \|\mathbf{y}_j\|}. \tag{6.10}$$

Note that maximum similarity is achieved for two movies that generate exactly the same ratings from all users. The same principle can be used to define the similarity between users, by computing the cosine similarity between all their ratings.

6.2.2 Similarity Graph Optimization

Both Gaussian kernels and cosine similarity produce nearly complete graphs, where very few edge weights are exactly zero. Thus, graph sparsification (Section 6.1) can be applied to reduce the number of edges. For example, K-NN can be applied on a graph with weights defined as in (6.9), which requires choosing two parameters: K and σ.

Parameter selection Clearly σ in (6.9) should be chosen after taking into account the distances between neighboring points in the dataset. Defining $d_{\min} = \min_{i,j} \|\mathbf{y}_i - \mathbf{y}_j\|$, the minimum pairwise distance between points in the dataset, if σ is much larger than $d_{\min}$ then many edge weights will be close to 1. Conversely, if we choose σ to be significantly smaller than $d_{\min}$ then most weights will be very small (close to 0). A possible solution is to compute statistics of local distances (e.g., the minimum or average distance for each point in the dataset) and use those to choose a value for σ. For specific tasks, parameter choice (of σ and K) can also be decided using a cross-validation strategy, where some data is held out (i.e., kept back) to choose parameter values that provide the best performance according to an appropriate metric. See [96] for a recent review of methods.

Neighborhood optimization In K-NN, for each i only K pairwise distances, $d(i, j)$, are needed to define a new graph: as long as j is among the closest (or most similar) nodes to i, the corresponding edge weight w_{ij}, computed using (6.9) or (6.10), will be preserved. As a consequence, this approach does not take into account the possibility that two of these chosen nodes, say j and k, might be redundant in some way. To address this issue, alternative techniques have been developed that use $d(i, j)$, $d(i, k)$ **and** $d(j, k)$ to decide whether edges connecting to j, k, or both, should be preserved. Two representative methods are briefly described next.

Local linear embedding (LLE)   In local linear embedding (LLE) [97], for each node i, after K-NN search a matrix $\mathbf{Y}_S$ is formed with columns equal to $\mathbf{y}_j$, the attributes of the K nodes closest to i. Then, LLE computes a vector $\boldsymbol{\theta}$ satisfying

$$\arg\min_{\boldsymbol{\theta}:\ \boldsymbol{\theta}\geq 0} \ \|\mathbf{y}_i - \mathbf{Y}_S\boldsymbol{\theta}\|_2^2, \tag{6.11}$$

whose entries are used as edge weights: the entry in $\boldsymbol{\theta}$ corresponding to $\mathbf{y}_j$ is assigned to w_{ij}. Thus, if $\mathbf{y}_j$ and $\mathbf{y}_k$ are very close, they may be redundant in terms of the linear approximation (6.11), so that w_{ik} or w_{ij} might be zero.

Non-negative kernel (NNK) regression   Non-negative kernel (NNK) graph construction [98] starts with the Gaussian kernel[2] edge weights (6.9) and solves a problem similar to (6.11), but in the kernel space rather than in the space of the $\mathbf{y}_i$ vectors. This type of regression leads to an intuitive geometric interpretation. Consider again two candidates j and k in the neighborhood of i, and let the initial weights be $w_{ij} > w_{ik}$. Then define a hyperplane passing through $\mathbf{y}_j$ and perpendicular to $\mathbf{y}_j - \mathbf{y}_i$. This hyperplane divides the space into two, a region R_{ij} that contains $\mathbf{y}_i$ and its complement $\bar{R}_{ij}$. Then k will be connected to i only if $\mathbf{y}_k \in R_{ij}$. Note that, after finding all connections to i, the only point in the intersection of all R_{ij} regions will be $\mathbf{y}_i$.

Similarity metric reliability   To conclude this section, since graph processing depends on the edge weights, it is important to ask whether these are representative of the "true" similarity between nodes. In particular, a scenario where all edge weights have similar values leads to a complete graph, which does not provide much information about the data. For example, if the attributes $\mathbf{y}_i \in \mathbb{R}^D$ and D is large, distances may not be informative; this is the well-known "curse of dimensionality." For very high dimensions we may observe that many points are at similar distances with respect to each other. As a second example, for discrete attribute data, for which (6.10) is used, the lack of sufficient data may be a problem. In the movie ratings example, if two movies with ratings $\mathbf{y}_i$ and $\mathbf{y}_j$ have not been rated by the same users then $\mathbf{y}_i^\mathsf{T}\mathbf{y}_j = 0$ and therefore $w_{ij} = 0$. If few ratings are available, and not many overlap, then the graph may be very sparsely connected and will not necessarily provide a useful representation of the data.

6.3   Learning Graphs from Signals

In Section 6.2, each node, i, was associated with an attribute vector $\mathbf{y}_i \in \mathbb{R}^D$, allowing us to define distances or similarities between nodes before constructing a graph. In this section, we address the problem of building a graph from example signals. We are given M graph signals forming a **training set**, $\mathcal{X} = \{\mathbf{x}_m \in \mathbb{R}^N,\ m = 1,\dots,M\}$, with individual **training vectors**, $\mathbf{x}_m$. Then, the goal is to select an N-node graph $G(\mathcal{X})$ such that the signals in $\mathcal{X}$ have some desirable graph spectral properties with respect to $G(\mathcal{X})$. For example, the goal may be to choose $G(\mathcal{X})$ to maximize the smoothness of the signals $\mathcal{X}$

---

[2]   More generally, any similarity metric taking values between 0 and 1 can be used.

with respect to the chosen graph. Note that in some cases graphs can be learned from *both* the attributes $\mathbf{y}_i$ and the signal examples $\mathcal{X}$. Furthermore, attribute similarity can be applied to data consisting of signal observations by defining node attributes that are derived from $\mathcal{X}$.

6.3.1 Basic Principles and Intuition: Empirical Covariance

Let us focus on undirected graphs with the graph Laplacian as the fundamental operator. Recall that the eigenvectors of $\mathbf{L}$, the columns of the GFT $\mathbf{U}$, are orthogonal and associated with real graph frequencies (eigenvalues). We wish to construct $G(\mathcal{X})$ in such a way that the signals in $\mathcal{X}$ are similar to the low-frequency eigenvectors of $G(\mathcal{X})$. As a first step consider a simplified scenario where $\mathcal{X}$ contains a single example.

Box 6.3   Matching a single example signal

Assume $\mathcal{X} = \{\mathbf{x}_m\}$. We can use the fact that similarity can be measured using the inner product (Section A.3) and define $P_m(\mathbf{x})$, the projection of $\mathbf{x}$ onto the space spanned by $\mathbf{x}_m$, as

$$P_m(\mathbf{x}) = \frac{1}{\|\mathbf{x}_m\|^2}(\mathbf{x}_m\mathbf{x}_m^{\mathsf{T}})\mathbf{x} = \frac{1}{\|\mathbf{x}_m\|^2}\mathbf{x}_m(\mathbf{x}_m^{\mathsf{T}}\mathbf{x}),$$

where $\mathbf{X}_m = \frac{1}{\|\mathbf{x}_m\|^2}\mathbf{x}_m\mathbf{x}_m^{\mathsf{T}}$ is a rank-1 matrix. Note that maximum similarity is achieved for $\mathbf{x}$ such that $\mathbf{x} = \alpha\mathbf{x}_m$, which maximizes (A.9), i.e.,

$$\cos\theta = \frac{|\mathbf{x}_m^{\mathsf{T}}\mathbf{x}|}{\|\mathbf{x}_m\|\,\|\mathbf{x}\|}.$$

Also note that $\mathbf{X}_m\mathbf{x}_m = \mathbf{x}_m$ and therefore $\mathbf{x}_m$ is an eigenvector of $\mathbf{X}_m$ with eigenvalue 1. Thus, if $\mathcal{X} = \{\mathbf{x}_m\}$ then $\mathbf{I} - \mathbf{X}_m$ has an eigenvector $\mathbf{x}_m$ with eigenvalue 0.

Building on the intuition of Box 6.3, let us consider the general case where $|\mathcal{X}| = M$. In (4.5) we defined the empirical covariance matrix $\mathbf{S}$ and how it relates to the empirical correlation, $\mathbf{R}$. If the signals in $\mathcal{X}$ have mean $\mathbf{0}$, $\mathbf{S} = \mathbf{R}$. Without loss of generality assume that the mean is indeed $\mathbf{0}$, so that the empirical covariance matrix, $\mathbf{S}$ is given by

$$\mathbf{S} = \frac{1}{M}\sum_{m=1}^{M}\mathbf{x}_m\mathbf{x}_m^{\mathsf{T}} = \frac{1}{M}\sum_{m=1}^{M}\|\mathbf{x}_m\|^2\mathbf{X}_m, \tag{6.12}$$

which can be viewed as a weighted average of the rank-1 matrices defined in Box 6.3. Let $\mathbf{x}$ be a vector with norm $\|\mathbf{x}\| = 1$; then $\mathbf{x}^{\mathsf{T}}\mathbf{X}_m\mathbf{x}$ is maximum if $\mathbf{x} = \frac{1}{\|\mathbf{x}_m\|}\mathbf{x}_m$. Following this idea, if $\mathbf{x}$ maximizes $\mathbf{x}^{\mathsf{T}}\mathbf{S}\mathbf{x}$ then $\mathbf{x} = \mathbf{s}_1$, the eigenvector of $\mathbf{S}$ with maximum eigenvalue, and $\mathbf{s}_1$ is maximally aligned with all $\mathbf{x}_m \in \mathcal{X}$.

Graph construction from S   Assume we wish to find a graph $G(\mathcal{X})$ such that signals in $\mathcal{X}$ are maximally smooth. For a given graph G and for $\mathbf{Z} = \mathbf{L}$ the smoothest signal is the first eigenvector of $\mathbf{L}$, which minimizes $\mathbf{x}^{\mathsf{T}}\mathbf{L}\mathbf{x}$. Assuming that $\mathbf{S}$ is invertible, note that $\mathbf{s}_1$ maximizes $\mathbf{x}^{\mathsf{T}}\mathbf{S}\mathbf{x}$ and minimizes $\mathbf{x}^{\mathsf{T}}\mathbf{S}^{-1}\mathbf{x}$. Then, since $\mathbf{s}_1$ has maximum similarity

to the signals in $\mathcal{X}$, choosing an $\mathbf{L}$ that approximates $\mathbf{S}^{-1}$ will fulfill the design requirement of finding a graph where the vectors in $\mathcal{X}$ are maximally smooth. Direct application
of this idea is not possible in general because $\mathbf{S}$ may not be invertible, and, even if it is
invertible, (i) the inverse is unlikely to be sparse and will not be in the form of a graph
Laplacian and (ii) the estimated edge weights from $\mathbf{S}$ may not be reliable (e.g., if M, the
number of observed signals in $\mathcal{X}$, is much smaller than $N^2/2$, the maximum number of
unique edge weights to be estimated).

To address these challenges, methods proposed in the literature estimate $G(\mathcal{X})$ from
$\mathbf{S}$ using two main types of approaches: (i) **node domain techniques** (Section 6.3.3)
find a graph operator with suitable sparsity that approximates the inverse of $\mathbf{S}$, and derive the GFT from this operator and (ii) **frequency domain techniques** (Section 6.3.4)
use the eigenvectors of $\mathbf{S}$ to define the GFT and then select the eigenvalues to achieve
different graph operators. Before describing these methods, we review several classical
approaches that solve a similar problem and learn operators corresponding to graphs
that are more general (e.g., having negative edge weights) than those typically used in
GSP applications.

6.3.2   Learning Unrestricted Operators from Data

Note that while sparsity is helpful for model interpretation (see also Box 6.4 below),
statistical models for graph data are not necessarily sparse. A model is said to be sparse
if its **precision** matrix, i.e., the inverse of its covariance matrix, is sparse and represents a
graph. To determine whether the observed data can be characterized by a sparse model,
we can try to recover the precision matrix, $\mathbf{Q}$, from the sample covariance matrix, $\mathbf{S}$. A
naive approach to doing this would be to attempt to compute $\mathbf{S}^{-1}$. However, the latter
estimator does not exist if $M < N$. An alternative approach is covariance selection [99],
where an edge set $\mathcal{E}$ is selected first and then the following problem is solved:

$$\min_{\mathbf{Q} \geq 0, \, \mathbf{Q}_{ij} = 0, \, ij \notin \mathcal{E}} (\mathrm{tr}(\mathbf{QS}) - \log \det(\mathbf{Q})), \tag{6.13}$$

where $\mathbf{Q}$ is required to be positive definite. A solution to (6.13) may exist for certain
types of graph topologies $\mathcal{E}$ even if $M < N$ [100], but it requires knowledge of the
true edge set, which is infeasible in practice. A more practical estimator is the graphical
lasso [101], which favors a sparse estimate by applying an ℓ_1 **regularization** term[3] to
the entries of $\mathbf{Q}$:

$$\min_{\mathbf{Q} \geq 0} (\mathrm{tr}(\mathbf{QS}) - \log \det(\mathbf{Q}) + \gamma \|\mathbf{Q}\|_1), \tag{6.14}$$

where γ controls the relative strength of the ℓ_1 regularization. This estimator always
exists for $\gamma > 0$, does not require that the edge set $\mathcal{E}$ be identified *a priori* and has other
desirable statistical properties.

The connection between this problem and finding the maximum likelihood of model
parameters for a Gauss–Markov random field is described below. Note that the graphical

---

[3] When regularization is used, new terms are added to a problem formulation with the goal of selecting
solutions that meet specific criteria, e.g., sparsity.

lasso approach and the solution to (6.13) lead to positive semidefinite solutions and that in the case of the graphical lasso these solutions are sparse but there is no guarantee that the graph weights will be positive.

Node attributes from signals Finally, it is worth mentioning techniques based on local regression [102], which can be formulated by taking the viewpoint of node similarity (Section 6.2). Given observations in X we associate with node i an attribute vector containing all M scalar observations at the node. Thus, in the notation of Section 6.2, the attribute vector at node i is

$$\mathbf{y}_i = [x_{1i}, \ldots, x_{mi}, \ldots, x_{Mi}]^\mathsf{T} \tag{6.15}$$

where x_{mi} is the entry of the training vector $\mathbf{x}_m$ corresponding to node i. With this notation, one can identify weights by, for example, solving a problem similar to local linear embedding, (6.11). With appropriate regularization (e.g., ℓ_1 as in [102]) this local embedding can be sparse, and the edge weights can be constrained to be positive, but the resulting graph will be directed. Note that this local regression formulation can be interpreted in terms of the one-hop local prediction of Section 2.4.1. The regression in [102], and related techniques to be described below, can be seen as selecting a graph operator that provides optimal prediction, that is, one for which the error in predicting $\mathbf{y}_i$ from one-hop nodes in the graph is minimized.

6.3.3 Node Domain Techniques

Smoothness-based techniques Assume we are given a candidate graph G with corresponding Laplacian $\mathbf{L}$ and GFT $\mathbf{U}$. Then, we can compute the GFT of each signal in X as

$$\tilde{\mathbf{x}}_i = \mathbf{U}^\mathsf{T} \mathbf{x}_i.$$

Since this can be done for any G, we can compare different candidate graphs by choosing a criterion to quantify how well they represent the data in X. For example, the approach in [103] assumes that each signal $\mathbf{x}_i \in X$ is approximated by $\mathbf{c}_i$, chosen such that $\mathbf{c}_i^\mathsf{T} \mathbf{L} \mathbf{c}_i$ is small, where $\mathbf{L}$ depends on the graph of interest. Since

$$\mathbf{c}_i^\mathsf{T} \mathbf{L} \mathbf{c}_i = \sum_k \lambda_k \tilde{c}_{i,k}^2, \tag{6.16}$$

a constraint on $\mathbf{c}_i^\mathsf{T} \mathbf{L} \mathbf{c}_i$ is a penalty on the higher frequencies of $\mathbf{c}_i$ (corresponding to larger λ_k), and thus is equivalent to imposing a smoothness constraint. For a given constraint such as (6.16), defining two matrices $\mathbf{X}$ and $\mathbf{C}$, with ith columns $\mathbf{x}_i$ and $\mathbf{c}_i$, respectively, an optimization problem can be formulated as

$$\min_{\mathbf{C}, \mathbf{L}} \left(\underbrace{\|\mathbf{X} - \mathbf{C}\|_F^2}_{(A)} + \underbrace{\alpha \operatorname{tr}(\mathbf{C}^\mathsf{T} \mathbf{L} \mathbf{C})}_{(B)} + \underbrace{\beta \|\mathbf{L}\|_F^2}_{(C)} \right) \tag{6.17}$$

under constraints that ensure that $\mathbf{L}$ is a valid graph Laplacian matrix:

$$\text{tr}(\mathbf{L}) = N, \ \mathbf{L1} = \mathbf{0}, \ l_{ij} \leq 0, \forall i \neq j,$$

where $\| \cdot \|_F$ denotes the Frobenius norm of a matrix (i.e., the square root of the sum of the absolute squares of its entries) and $\text{tr}(\cdot)$ is the trace. The first term in (6.17), (A), quantifies how well each $\mathbf{x}_i$ is approximated by the corresponding $\mathbf{c}_i$. The term (B) quantifies the smoothness of the resulting set of $\mathbf{c}_i$ vectors on the chosen graph with Laplacian $\mathbf{L}$, while (C) penalizes the choice of large entries for $\mathbf{L}$. To obtain the term (B) note that $\mathbf{C}^\mathsf{T}\mathbf{LC}$ is an $M \times M$ matrix, where the i, j entry is $\mathbf{c}_i^\mathsf{T}\mathbf{Lc}_j$ and thus its trace is $\sum_i \mathbf{c}_i^\mathsf{T}\mathbf{Lc}_i$. The non-negative scalar parameters α and β control the relative weight given to each term. An interesting property of this approach is that it makes explicit the smoothness criterion applied to select the graph. Thus it would be possible to replace the term (B) by one that reflects a different smoothness criterion.

An alternative approach [104] defines attribute vectors $\mathbf{y}_i$ (6.15) for each node, as proposed in [102], so that the smoothness term can be rewritten as

$$\text{tr}(\mathbf{C}^\mathsf{T}\mathbf{LC}) = \text{tr}(\mathbf{WY}) = \|\mathbf{W} \circ \mathbf{Y}\|_1,$$

where $\circ$ denotes the entry-wise product and $\mathbf{Y}$ is an $N \times N$ matrix whose i, j entry is the squared distance $\|\mathbf{y}_i - \mathbf{y}_j\|^2$; $\mathbf{W}$ is a matrix of positive weights:

$$\text{diag}(\mathbf{W}) = \mathbf{0}, \ W_{ij} = W_{ji} \geq 0.$$

This leads to the following formulation of the problem:

$$\min_{\mathbf{W}} \left(\|\mathbf{W} \circ \mathbf{Y}\|_1 - \alpha \mathbf{1}^\mathsf{T} \log(\mathbf{W1}) + \frac{\beta}{2}\|\mathbf{W}\|_2 \right), \tag{6.18}$$

where the first term in the minimization favors a sparse graph, while the purpose of the log term is to ensure that none of the nodes is disconnected. The resulting graph is undirected and has positive weights and no self-loops.

Gauss–Markov random fields (GMRFs)   In order to develop an alternative formulation, let us focus on the term (B) in (6.17), but write it for the data matrix $\mathbf{X}$, i.e., $\text{tr}(\mathbf{X}^\mathsf{T}\mathbf{LX})$. First note that, using the cyclic property of the trace operator, we have

$$\text{tr}(\mathbf{X}^\mathsf{T}\mathbf{LX}) = \text{tr}(\mathbf{LXX}^\mathsf{T}),$$

where $\mathbf{XX}^\mathsf{T}$ is an $N \times N$ matrix and we can define, $\mathbf{S}$, the **empirical covariance matrix**, by

$$\mathbf{S} = \frac{1}{M}\mathbf{XX}^\mathsf{T} = \frac{1}{M}\sum_i \mathbf{x}_i \mathbf{x}_i^\mathsf{T}, \tag{6.19}$$

under the assumption that the mean has been removed. Then, the term (B) for a given $\mathbf{X}$ can be written as $\text{tr}(\mathbf{LS})$, up to a constant normalization factor.

Next, for a given graph with Laplacian $\mathbf{L}$, define $\tilde{\mathbf{L}} = \mathbf{L} + \delta\mathbf{I}$, for some scalar δ, so that now $\det(\tilde{\mathbf{L}}) > 0$, and we can write a Gaussian multivariate model as follows:[4]

---

[4] If $(\lambda_i, \mathbf{u}_i)$ is an eigenpair of $\mathbf{L}$ then $(\lambda_i + \delta, \mathbf{u}_i)$ is an eigenpair of $\tilde{\mathbf{L}}$.

$$p(\mathbf{x}|\tilde{\mathbf{L}}) = \frac{\det(\tilde{\mathbf{L}})^{1/2}}{(2\pi)^{N/2}} \exp\left(-\frac{1}{2}\mathbf{x}^{\mathsf{T}}\tilde{\mathbf{L}}\mathbf{x}\right), \tag{6.20}$$

where $\tilde{\mathbf{L}}$ plays the role of inverse covariance, or precision, matrix for this Gauss–Markov random field (GMRF) model. If our goal is to obtain the best model for a given set of data samples $\mathbf{x}_i$ then it makes sense to formulate a maximum likelihood problem for the available data. The log-likelihood for the observed data is (ignoring constant terms)

$$\log p(\mathbf{x}|\tilde{\mathbf{L}}) \propto \log \det(\tilde{\mathbf{L}}) - \sum_i \mathbf{x}_i^{\mathsf{T}}\tilde{\mathbf{L}}\mathbf{x}_i$$

and therefore, using the result (6.19) and choosing the model that minimizes the negative log-likelihood, given the empirical covariance $\mathbf{S}$, would lead to solving

$$\min_{\tilde{\mathbf{L}} \geq 0,\ \mathbf{L}\mathbf{1} = 0,\ l_{ij} \leq 0 \text{ for } i \neq j} \left(\mathrm{tr}(\tilde{\mathbf{L}}\mathbf{S}) - \log \det(\tilde{\mathbf{L}})\right). \tag{6.21}$$

If $\mathbf{S}$ is invertible and there are no Laplacian constraints, then the optimal solution to this problem would be $\tilde{\mathbf{L}} = \mathbf{S}^{-1}$. That is, the maximum likelihood estimate of the precision matrix is simply the inverse of the empirical covariance. An advantage of incorporating Laplacian constraints is that the solution to (6.21) always exists as long as $M > 2$ [105]. This is in contrast with the stricter conditions that apply to its covariance estimation counterpart (6.13).

The model (6.20) provides a probabilistic interpretation of the eigenpairs associated with $\tilde{\mathbf{L}}$. Observe that only the exponential term depends on the signal $\mathbf{x}$. Clearly a higher probability can be achieved when the term $\mathbf{x}^{\mathsf{T}}\tilde{\mathbf{L}}\mathbf{x}$ is small. In Section 3.5, we discussed in detail how the eigenvectors of $\mathbf{L}$ can be obtained via successive minimizations. Thus $(\delta, \mathbf{1})$ is the eigenpair corresponding to minimum $\mathbf{x}^{\mathsf{T}}\tilde{\mathbf{L}}\mathbf{x}$, and successive eigenvectors minimize $\mathbf{x}^{\mathsf{T}}\tilde{\mathbf{L}}\mathbf{x}$ while being orthogonal to the previously found eigenvectors. Then, the most likely signals are obtained along the direction $\mathbf{u}_1 = \mathbf{1}$.

Box 6.4 Precision matrix interpretation

In a GMRF model, the connectivity induced by the precision matrix $\tilde{\mathbf{L}}$ has a well-defined interpretation. Denote by x_k and x_l the kth and lth entries of a random vector $\mathbf{x}$ distributed according to the model (6.20). Denote by $\bar{\mathbf{x}}_{kl}$ the vector containing all entries of $\mathbf{x}$ except x_k and x_l. Then, if $\tilde{\mathbf{L}}$ is such that there is no edge between k and l, i.e., $\tilde{\mathbf{L}}_{kl} = \tilde{\mathbf{L}}_{lk} = 0$, then x_k and x_l are conditionally independent:

$$p(x_k, x_l|\bar{\mathbf{x}}_{kl}) = p(x_k|\bar{\mathbf{x}}_{kl})p(x_l|\bar{\mathbf{x}}_{kl}).$$

Based on this idea, several approaches have been proposed that modify the cost function of (6.21) to incorporate additional criteria, such as for example a sparsity-favoring term [106, 107]:

$$\min_{\mathbf{L} \geq 0,\ \mathbf{L}\mathbf{1} = 0,\ l_{ij} \leq 0 \text{ for } l \neq j} \left(\mathrm{tr}(\mathbf{L}\mathbf{S}) - \log \det(\mathbf{L}) + \alpha\|\mathbf{L}\|_1\right). \tag{6.22}$$

6.3.4   Frequency Domain Methods

Frequency domain methods, such as those discussed in [108], use the eigenvectors of the empirical covariance matrix S as a starting point to obtain the graph operator. These approaches start by making the assumption that the observations are realizations of stationary graph processes. A stationary graph process is defined here (see also Section 4.1.4) as a filtered version of an independent identically distributed white noise signal, where the filter F is a polynomial of the graph operator Z (refer to Chapter 2). Under these conditions it is easy to show that the eigenvectors of S are equal to the eigenvectors of Z (since the input random signal is i.i.d., its covariance is I). Therefore in [108] the eigenvectors of S were first estimated, and assumed to be equal to the eigenvectors of Z, and a set of eigenvalues for Z was obtained such that the resulting operator has some desirable properties. The authors also developed techniques that relax the requirement of using the eigenvectors of S, as these may not be reliable.

6.3.5   Discussion

As mentioned at the beginning of this chapter, graph selection is essentially a modeling task. Work in the literature often evaluates algorithm performance by treating it as a detection task, where the goal is to recover a graph that is somehow linked to the signals in X. As an example, given the precision matrix for a GMRF model (6.21), random data can be generated and the goal is to estimate the precision matrix from the data. While this kind of detection formulation provides a useful tool to assess performance, in practice it is a fairly artificial scenario. In a more realistic setting, there may not be a "correct" underlying graph to be identified. Instead, graph-learning performance may have to be assessed based on achievable results, and whether they can be interpreted, for a given specific application.

Qualitatively, we can compare frequency and node domain techniques on the basis of the amount of data available (how large is the number of edges to be estimated relative to the size of the training set) and on the importance of prior assumptions (a knowledge of the desirable graph properties). Frequency domain techniques that use a GFT derived from S are more appropriate when large amounts of data are available. The reason is that these methods use as a GFT the spectrum of S, which may not be reliable if M is small. One advantage of these methods is that they do not rely significantly on assumptions about graph structure. In contrast, node domain techniques do have to make assumptions about the graph structure (sparsity, specific topology), and yield different results under each choice of priors. However, these assumptions serve as a form of regularization, and give increased reliability in the presence of noise or for M small.

Chapter at a Glance

Given that GSP tools, filters, frequency analysis, sampling and transforms depend on the graph, choosing a good graph is a fundamental step to achieving good results for a specific application. We can think of graph learning as being a modeling task, where one has to balance potentially conflicting requirements, namely, (i) to provide an appropriate model for the data and (ii) to facilitate the use of graph processing tools. Methods for graph simplification (Section 6.1) are used when the graph is given, but there is a need to adapt them so as to facilitate processing. Graphs can also be constructed from attributes (Section 6.2) or from signal data (Section 6.3), in which case those properties required to improve processing are incorporated into the optimization as additional constraints.

Further Reading

There have been several publications describing the problem of learning graphs from data from the graph signal processing perspective. Summaries of recent work can be found in [109, 110].

GraSP Examples

In Section B.7.1 an example of graph learning is presented. More generally, there is Matlab code available for many techniques described in this chapter, and GraSP makes it easy to integrate third party tools as shown in Section B.7.1.

7 Applications

Graph signal processing methods have been proposed for a broad range of applications. This is both an opportunity and a challenge. Diverse applications provide different perspectives to illustrate GSP concepts. However, an in-depth discussion of a large set of different applications would not be practical, while a superficial coverage of many examples might not help the reader achieve a better understanding of either the tools or the applications. Since this book is aimed at students and practitioners interested in the basics of GSP, and who may already have deep expertise in some of these application domains, we decided to present a broad range of representative applications but to emphasize a graph signal perspective, evaluating some of its potential benefits and challenges. Thus, we have not reviewed extensively the state of the art in these areas or compare graph-based approaches with others to be found in the literature. In contrast, we have explained how graphs can be associated with specific applications, provided a rationale for the graphs chosen and described some representative results using GSP.

In this chapter, we start by presenting our general approach (Section 7.1), and include an overview of application areas and a high-level methodology for application of GSP tools. Then we discuss a series of representative applications in physical, biological and social networks (Section 7.2) and in sensor networks (Section 7.3) and images (Section 7.4), and we conclude with an overview of applications in machine learning (Section 7.5).

7.1 Overview, Application Domains and Methodology

In describing GSP theory we have emphasized the basic ideas without attempting to cover exhaustively all contributions to the recent state of the art. Similarly, we have not attempted to be exhaustive in describing applications of GSP. Instead, we selected representative examples in a series of domains, and used them to illustrate the connection between application requirements and GSP tools. Our aim is not to show that GSP methods have advantages. In fact, it is a rare occurrence when a new set of tools can immediately solve existing problems. Indeed, progress within a specific application domain frequently comes from combinations of new and existing methods, building on available domain knowledge.

Thus, our goal here is to provide examples of how GSP tools are being *used*, at the time of writing of this book. A more detailed description of applications, and of alternative methods that have been used in those applications, goes beyond the scope of the book. Recent overview papers and book chapters by application domain experts are appropriate resources for those wishing to go deeper into specific application domains.

7.1.1 Application Domains

A major challenge in presenting a diverse range of applications is to choose a reasonable grouping of these areas, so as to make it easier for the reader to identify common characteristics. We attempt to do so now (see also Section 1.4).

Physical, biological and social networks   In some applications involving physical, biological and social networks, we expect that a graph (which may be observable) has had a direct influence on the data to be processed. Examples of physical networks include transportation networks (road or rail), communication networks (the Internet) and infrastructure networks (the smart grid or a water distribution infrastructure). In any of these cases, measurements made on the graph are likely to be a result of behavior that can be explained by the graph's connectivity. For example, traffic intensity measurements at a road intersection are certainly related to prior measurements at nearby intersections, since vehicles physically traverse the network. Unlike physical networks, biological networks, such as those explaining brain behavior or the interactions between genes, are not easily observable, and indeed identifying them from data might be the principal goal of data analysis. However, in common with physical networks, the observed data is generated via interactions on the graph. For example, neural activity on the brain reflects the connection between neurons. Finally, there are also graphs underlying the phenomena observed in online social network platforms, including news and rumor propagation. A few examples will be discussed in Section 7.2.

Sensing   In some sensing applications (Section 7.3) observations of a physical phenomenon (temperature, air quality) can be made by deploying sensors anywhere in the environment. While a graph can be constructed on the basis of relative sensor positions, the physical phenomenon being observed exists independently of the presence of sensors and their positions. Therefore, a graph-based model can help analyze data, but in general it does not provide an accurate description of the underlying phenomenon leading to the observed data. In imaging and point-cloud sensing applications (Section 7.4), where pixel information in 2D or 3D space corresponds to an actual scene, sensors may be uniformly placed (e.g., a 2D array of sensors on a camera). In this case, graph-based approaches can be used to take into account the image structure (e.g., differences in pixel intensities) as part of the processing.

Machine learning   Machine learning applications (Section 7.5) may correspond to both of the cases just discussed. In some situations we are interested in tools to analyze a dataset, where each data point is an observation and where what may explain the location of the points (e.g., the existence of a smooth manifold containing the data) is not directly observable and is independent of the observations. One major difference with respect to other cases considered (e.g., images and sensor networks) is that the dimension of the ambient space is much higher. In other situations, we may be processing information on an existing graph, e.g., by running a neural network on road data or a point cloud in order to classify its behavior.

7.1.2 An Application Checklist

Our ability to apply graph signal processing concepts in various application domains is at an early stage. Here, rather than attempting to define a formal methodology, we propose a series of ideas that can be part of a checklist for applying graph signal processing methods in various application domains.

Know your graph   A point that has been made repeatedly throughout this book is that all GSP tools can be adapted to the characteristics of a specific graph. Thus, whether we select a graph based on data or whether we are given a graph (and will perhaps simplify it), choosing a graph means making modeling assumptions (Chapter 6) and we should be aware of their implications. Typical changes we can make, such as sparsifying or converting from a directed graph to an undirected graph, may facilitate computations but may also change our assumptions about the data. Elementary graph signals, the GFT basis vectors, depend on graph topology, so that any changes to the topology can lead to significant changes to our analysis.

How do we determine whether a graph is suitable? Consider two cases.

- If a graph is known and data are available, we can use the graph GFT to analyze the frequency content of the data and ask whether the observed frequencies match our expectations. If the available data is not well localized in the graph frequencies, that may mean that some tasks (e.g., denoising or anomaly detection) may not be easy to accomplish. Thus, the existing graph might be modified on the basis of the data (e.g., by learning new weights for the given topology).
- Even if no example data is available, we can still analyze the GFT basis vectors, for example using the visualization tools described in this book. We often observe a transition from more localized to less localized basis functions (lower to higher frequencies), and may wish to treat those two ranges of frequencies separately. Since localization properties depend on graph topology, the choice of graph will also affect the extent to which graph signal processing will be localized.

Test your solutions   We have studied the problems of defining graph signal sampling (Chapter 4), representations (Chapter 5) and graph learning (Chapter 6). It is tempting to assess the various available methods on the basis of their end-to-end performance, essentially treating the graph signal processing tools as one more component of a complex system. However, since the tools we have presented lend themselves to some degree of interpretation, it would be preferable to exploit this property in order to achieve more robust results.

- For the sampling methods of Chapter 4 it can be useful to observe the actual location or index of the nodes chosen by each candidate algorithm before deciding on which one to use. Since sampling adapts to irregular graph characteristics (Figure 4.3) a sampling rate may have to be chosen for each graph. For example, if a graph has a small number of clusters and some GFT bases are localized, the sampling rate may have to be adapted to ensure uniform reconstruction quality across clusters.

- Since the transformations in Chapter 5 may be exactly or approximately polynomial on the graph, it is useful to relate the polynomial degree to the diameter of the graph. This can help determine to what extent the processing reflects local data characteristics.

- Finally, the methods to learn graphs from data discussed in Chapter 6 can provide a range of solutions, with specific topologies, different levels of sparsity, and so on; the solution that is chosen should facilitate interpretation for the application at hand.

7.2 Physical, Biological and Social Networks

We first describe several scenarios where data is known to have been produced by a network, which can either be observed directly (as will be the case for transportation and infrastructure networks) or is known to exist but has to be inferred (brain networks). In these scenarios we expect to see a strong dependence between the data and the network.

7.2.1 Transportation Networks: Road Networks

Many sensors are available to measure traffic on a road network, including built-in inductive-loop detectors, video cameras and data from mobile phone applications. The resulting data can be used to optimize routes, adapt road pricing levels and minimize congestion. Sensors placed at fixed positions (loops or cameras) capture time series of traffic intensity. Graph signal processing methods are being used to combine time-based and space-based predictions for traffic forecasting, by taking into account the relative locations of sensors and using knowledge of the road network.

As a representative example, the authors of [111] incorporated spatial prediction into the model by considering diffusion on a graph obtained from the road network. This method uses graph weights that are a negative exponential function of the distance, as in (6.9). Because these graphs are naturally directed (as illustrated by Example 1.3) this work used directed polynomial filters with the directed random graph Laplacian as the graph operator. That is, the filter coefficients were learned from the data and are polynomials of $\mathbf{D}_{out}^{-1}\mathbf{A}$, where $\mathbf{D}_{out}$ is the diagonal matrix of out-degrees (Definition 2.4). Another representative example is [112], where the road network graph was partitioned to identify clusters that exhibit stationarity, making it easier to predict traffic behavior within each cluster.

7.2.2   Infrastructure Networks

Infrastructure systems, such as the electric grid or a water distribution system, form networks. The electric grid can be viewed as a large-scale circuit and can be represented by a graph where nodes correspond to electrical junctions (buses) connected to loads or generators, while edges correspond to transmission lines between buses. The weights associated with each of the edges represent the electric admittance. The DC and AC models lead to real and complex admittance values, respectively, with corresponding real and complex graph Laplacians. In [113], a graph signal is defined as a vector of voltages, where each entry is the voltage observed at the corresponding node at a given time. The authors demonstrate that standard loads lead to smooth graph signals (with greater smoothness for the real than for the imaginary part of the signal). On this basis, GSP methods can be used for anomaly detection, such as a malicious injection of incorrect data into the measurements. Such attacks can be detected by noting that they lead to higher-frequency observed graph signals, as compared to signals observed under normal conditions.

In some cases graphs constructed from data are used. To tackle the problem of load disaggregation and appliance monitoring, the authors of [114] used non-intrusive monitoring by observing variations in the total load corresponding to a house. A similarity graph is constructed where there are N measurements, with the measurement at node i corresponding to a (possibly zero) change in total load Δp_i. The edge weights between nodes are computed using (6.9), with the distance between nodes i and j equal to $|\Delta p_i - \Delta p_j|$. The graph signal depends on the actual distribution of power across appliances. For each appliance the authors define a switching state variable, which can take values $\{1, -1, 0\}$ depending on whether the appliance was turned on or off (1), remained in the same state (-1) or no information was available (0). Knowing the Δp values associated with different appliances, and using training data where exact appliance usage was known, the authors showed that it was possible to estimate the load changes at each appliance (the disaggregated load) with regularization based on the Laplacian quadratic form (3.9). The main underlying assumption is that two nodes with similar values of Δp are likely to correspond to similar underlying appliance activity. As a simple example, if there is a single appliance for which the transition value is exactly $\Delta p = a$, then every observation where the state is a will be expected to be closely connected to all others with the same value a, and they are assigned to the state where only that one appliance is turned on or off.

7.2.3   Brain Networks

Magnetic resonance imaging (MRI) is widely used to understand the human brain. With diffusion MRI it is possible to discover structural connections (fiber tracts in the white matter), while functional MRI (fMRI) can be used to discover how different parts of the brain are activated as various tasks are being performed. A summary of recent work applying GSP in this area can be found in [115].

Graphs are defined with each node representing a brain region, where the volume and location of these regions depends on prior knowledge of brain function and on the resolution of imaging devices. The graph edges are based on (i) structural connections depending on the underlying brain anatomy and (ii) functional connections, which also depend on brain structure but are associated with specific tasks. The edge weights are estimates of the degree of coupling between the regions and are based on either structure (larger weights corresponding to a greater estimated number of fiber tracts) or activity (larger weights associated with greater correlation), where activity graphs can be learned using the methods of Section 6.3.

The graph signals consist of measurements of brain activity (e.g., from fMRI data) for each brain region. For a given choice of graph, if a graph signal has most of its energy in the low frequencies, this can be seen as evidence that structurally connected regions are being activated together. In contrast, greater energy in the high frequencies indicates that non-connected regions can be active simultaneously. Recent work shows that there is a correlation between the amount of high-frequency energy for a given graph signal and the difficulty of the corresponding task for a subject [116, 115]. Other work [117, 118] uses graph-based representations for dimensionality reduction in the context of classification for both fMRI and other brain signal modalities. For example, the authors of [118] considered several graph constructions (data-driven, structural and hybrid) and showed that various types of dimensionality reduction (e.g., the sampling of graph nodes, as in Chapter 4 or the selection of low or high graph frequencies) can achieve good performance, in some cases without requiring prior knowledge of the data.

7.2.4   Social Networks

Online social networks are increasingly important communication tools, where users are connected by directed (follower) or undirected (friendship) edges. With respect to other applications considered in this chapter, social networks are characterized by several properties: (i) these are networks for which the graph structure is known, (ii) very irregular degree distributions are common: some nodes have very few connections and others have millions and (iii) the overall size of these networks can be very large.

Since each node in the social network graph corresponds to a user, the graph signals represent different types of information observed or inferred about users. As an example, a recommendation system suggests items they may like (e.g., movies) to users on the basis of their known preferences and those expressed by other users. Approaches have been proposed where the similarities between users (based on their known preferences) are combined by using a graph with their social connections. The underlying assumption is that users connected in the social graph are more likely to have similar interests and preferences [119]. Other applications where social network connectivity is used include the detection of fake users, the modeling of news propagation and the identification of influential users.

7.3   Distributed Sensor Networks

A sensor network consists of a series of sensing devices that are placed in the environment and communicate with each other, with the data they gather being ultimately transmitted to be processed. The scale and purpose of such deployments can vary significantly: sensors can monitor temperature within a building, they can be embedded to monitor the structural health of a bridge, they can be placed along a pipeline, etc. Other types of sensing can range from temperature and air quality to images and video.

There are several reasons why a GSP perspective can be a natural choice for sensor network data: (i) measurements are made at multiple locations that are not necessarily spaced in a regular manner in the environment, (ii) sensors are often battery powered, which requires their power consumption to be managed so as to increase their lifetime and (iii) measurements may be unsynchronized, subject to noise or unreliable owing to sensor malfunction. In relation to (i) and (ii) we first discuss GSP approaches that can help to optimize a distributed sensing system. Then, we provide examples of GSP sensor data analysis.

7.3.1   Data Transport

In a distributed sensing system, active sensors collect data and then relay it to a central node (or sink) from where it will be accessible for processing and analysis. Data could be relayed from sensor to sensor without further processing, but the data captured by neighboring sensors tends to be correlated so the system can be made more efficient by in-network processing. This idea is illustrated by the following example.

Example 7.1   In-network processing along a path graph   Assume a sensor captures data to be relayed to a destination node over N sensor-to-sensor links, starting at node 1 (the source) and ending at node $N + 1$ (the destination). In this example the **routing graph** is a directed version of a path graph (Figure 1.8). If a node can only relay information to its immediate neighbor on the routing graph, then data from node i will be sent to node $i + 1$, then to $i + 2$ and so on.

In-network processing can be useful if data exhibit spatial correlation (neighboring nodes make similar observations). More efficient communication can be achieved by transmitting differences. Sensor 1 can send measurement x_1 to sensor 2, which can compute $e_2 = x_2 - x_1$ and forward both e_2 and x_1 to node 3. Node k receives data forwarded by node $k - 1$, i.e., $x_{k-1}, e_{k-1}, \ldots, e_1, x_1$, where x_{k-1} is forwarded along with the other data in order to avoid having to reconstruct x_{k-1} from all the measurements. Note that the difference between successive nodes can be written in vector form as $(\mathbf{A} - \mathbf{I})\mathbf{x}$, a polynomial filter of the directed adjacency matrix. Better compression can be achieved by quantizing each residual signal and then using each reconstructed value for prediction.

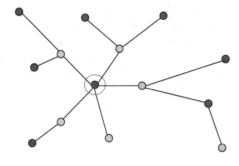

Figure 7.1 In this simple example several sensing nodes collaborate by relaying information until it reaches the sink node (circled). It can be seen that the nodes form a tree, where the nodes at each level of the tree correspond to one of two colors (shown with a different shading). Thus, nodes that are an even or odd number of hops away from the sink are placed in different sets.

The basic concepts used in the above example can be applied in more complex routing scenarios (e.g., a tree instead of a path) and more sophisticated transformation and compression can be used (e.g., distributed wavelets). Note also that for simplicity we have assumed that each sample is transported as soon as it is captured, but in practice other approaches are possible. For example, multiple samples may be captured and stored at each sensor over a period of time before any transmission occurs. In this case, it would be possible to use both temporal and spatial transformations.

Selecting the routing graph   Given the sensor positions (graph nodes) and given a sink node, there are many algorithms to optimize the routing and processing graph. In Example 7.1, in-network processing can be viewed as a combination of distributed graph filtering and quantization. Since we are measuring information in space, connecting neighboring nodes can lead to a more efficient representation. Experimentally it can be easily verified that random samples obtained from a spatially smooth continuous field, connected with a k-NN similarity graph (Section 6.2), lead to smooth graph signals, such that most of the energy is concentrated in the lowest frequencies [120]. Selecting graphs that connect neighboring nodes is also useful owing to the distributed nature of the system, since the cost of communication between sensors will increase with distance. While in general the communication graph and the data analysis graph are not necessarily the same, it can be useful to select them jointly.

In-network processing using lifting   Distributed transforms can be developed based on lifting (Section 5.4.2), where the transform operates on a bipartite graph and is guaranteed to be invertible. An in-network lifting transform requires the nodes to be separated into two sets, even and odd (i.e., a bipartition), so that the information in the even nodes can be used to process the odd nodes, and vice versa. Several strategies have been proposed to choose a bipartition of the sensor nodes, making it possible to apply lifting. When data is routed through a tree, the authors of [121] used the natural bipartition, where the two sets are formed by the nodes at odd and even levels of the tree, respectively. This idea is illustrated in Figure 7.1. Other proposed examples include that discussed in [122], which considers selecting a series of nodes and routing to them. A key difference in this case is that the sensors are selected first in order to optimize the representation, and the routing strategy is defined next. This can be described as a transform-then-route strategy in that the transform is designed first. Clearly this can lead to better transforms but at the cost of potentially higher transport costs. A more complete description of such compression over networks can be found in [123].

7.3.2 Sensor Data Analysis

Conventional processing methods   Several approaches can be used to analyze sensor measurements from a frequency perspective. First, starting with some original irregularly sampled data, interpolation tools can be used to estimate values on a finer, regular grid, on which conventional 2D signal processing techniques can be used. Second, the vector of measurements can be processed without taking into account the relative locations of the nodes. For example, on the basis of the vector of measurements, we can use principal component analysis (PCA), starting from the empirical covariance of the observed data. Projection onto these principal components provides a spectral representation based on data statistics.

Graph selection   A GSP approach can take into account the relative positions of the nodes. Since sensor measurements often exhibit spatial correlation, either as samples of some continuous phenomenon (e.g., temperature) or because they are related to each other (e.g., traffic at neighboring intersections in a road network), a distance-based graph (Section 6.2) can be expected to lead to smooth graph signals. Alternatively, a data-driven graph learning approach (Section 6.3) leads to GFTs that are closely linked to principal component analysis (PCA). Indeed, if we construct the graph Laplacian of a graph as an approximation to the inverse empirical covariance of the observed data (Section 6.3) then the corresponding GFT can be viewed as a regularized version of the eigenvectors of the empirical covariance (which would be used directly in PCA). Smoothness properties of signals on the chosen graph can be used for sensor selection and anomaly detection, among other applications.

Sensor selection   In some scenarios it is preferable not to have all sensor nodes active simultaneously; this gives a way to save power and increase the time the network can be active without requiring maintenance. Then selecting a subset of sensors to be activated can be viewed as a graph signal sampling problem and addressed with the methods described in Chapter 4, where the goal is to identify a subset of sensors that is most informative [56]. While in Chapter 4 the focus was to select the best subset of nodes, for more efficient power consumption a more practical approach would be to identify a series of non-overlapping subsets of nodes. Then, instead of having all sensors active or activating a fixed subset, it is possible to rotate through each of those subsets, so that the power consumption is more evenly distributed across sensors.

Anomaly detection   To define anomalies we need to specify first what constitutes normal behavior. In the sensor setting, from the previous discussion a common approach is to consider normal signals to be smooth. When a graph is learned from data (Section 6.3) we make the assumption that lower graph frequency signals are more likely ((6.3.3)). Then, an anomaly can be detected from the presence of higher frequencies [124]. A more sophisticated approach may start by modeling the observed data. Given a graph based on sensor positions and a series of data observations, a first goal may be to determine whether the data has similar characteristics in different parts of the graph (Section 4.1.5). If this analysis shows that the data is non-stationary (it does have different properties in different parts of the graph), then localized approaches for anomaly detection may be preferable.

7.4 Image and Video Processing

In a graph approach to image and video processing, each pixel corresponds to a node and the edge weights between nodes are chosen to capture information about pixel position and pixel similarity. The graph signal associated with each node is the intensity or color information at the corresponding pixel. Notice that if the selected graph connects neighboring pixels, this will lead to a regular grid graph with a 4-connected or 8-connected topology (Section 1.3.2). In this section we first study image data representations derived from regular grid graphs (Section 7.4.1). Next we consider alternative definitions for graphs that can better capture pixel similarity (Section 7.4.2). We then explore a series of applications where graphs have been used, including filtering (Section 7.4.3), compression (Section 7.4.4), restoration (Section 7.4.5) and segmentation (Section 7.4.6). Our goal is to provide an overview of how graph signal processing is used, and we refer readers to a recent overview paper [125] for a more complete description of work in this area.

7.4.1 Transforms on Regular Grid Graphs

In conventional digital images, pixels are located on a regular grid in space so that the distances between neighboring pixels along the horizontal and vertical directions are all equal. Because of this, a straightforward graph choice would be a regular grid graph, with all edge weights equal to one. Indeed, as discussed next, two of the most widely used transforms in image and video processing can be interpreted as graph-based representations on path or grid graphs.

Discrete cosine transform (DCT)   The discrete cosine transform (DCT) is the most widely used transform for image and video coding. The combinatorial Laplacian $\mathbf{L}$ for the graph of Figure 1.8 with all edge weights equal to 1 is as follows:

$$
\mathbf{L} =
\begin{bmatrix}
1 & -1 & 0 & 0 & 0 & 0 & 0 & 0 \\
-1 & 2 & -1 & 0 & 0 & 0 & 0 & 0 \\
0 & -1 & 2 & -1 & 0 & 0 & 0 & 0 \\
0 & 0 & -1 & 2 & -1 & 0 & 0 & 0 \\
0 & 0 & 0 & -1 & 2 & -1 & 0 & 0 \\
0 & 0 & 0 & 0 & -1 & 2 & -1 & 0 \\
0 & 0 & 0 & 0 & 0 & -1 & 2 & -1 \\
0 & 0 & 0 & 0 & 0 & 0 & -1 & 1
\end{bmatrix},
\tag{7.1}
$$

and its corresponding GFT, $\mathbf{U}_{DCT}$, is the discrete cosine transform (DCT) [1] shown in Figure 7.2 for $N = 8$. A path graph with N nodes can be viewed as representing N neighboring pixels along a line. For each block of eight neighboring pixels the corresponding graph signal is the vector of pixel intensities. If all edge weights are equal to 1 then this indicates maximum and equal similarity between all pairs of neighboring pixels. A more general, probabilistic interpretation of edge weights is discussed in Box 7.1.

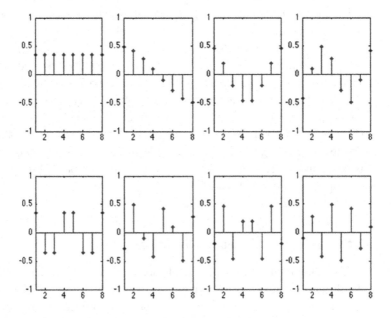

Figure 7.2 Discrete cosine transform (DCT) basis for $N = 8$. Each vector has eight entries and there are eight vectors forming an orthogonal basis for $\mathbb{R}^N$. These basis vectors are ordered (left to right, top to bottom) by their frequency. We can see that as the frequency increases, so does the number of zero crossings (sign changes from one sample to the next). The first basis vector has no zero crossings, while there are seven zero crossings in the highest-frequency basis vector.

Box 7.1 Probabilistic interpretation of DCT

A slight modification of the graph Laplacian, $\tilde{\mathbf{L}} = \mathbf{L} + \delta\mathbf{I}$, can be viewed as the precision matrix in a Gauss–Markov random field, where the likelihood can be written as in (6.20) and the connectivity of $\tilde{\mathbf{L}}$ represents conditional independence properties (Box 6.4). On this basis, since a pixel connects only to its two immediate neighbors in the graph of Figure 1.8, in the corresponding model pixels are conditionally independent from those that are not immediate neighbors. This models leads to the following observations.

- All edges between pixels connected in the DCT graph have weights equal to 1, indicating maximum similarity. Minimal similarity between two pixels is reached if the corresponding edge weight is set to 0. Setting any of the edge weights in the path graph to zero leads to two disconnected graphs, making the two subgraphs statistically independent.
- Note also that by construction the lowest-frequency eigenpair of $\mathbf{L}$ is $(0, \mathbf{1})$. As discussed in the paragraph before Box 6.4, the lowest-frequency eigenvector can be seen as the most likely signal in the corresponding GMRF. Thus, the popularity of the DCT for encoding natural images can be explained by the fact that these tend to be "smooth," and therefore an all-constant (or DC) signal is a good approximation for many signals.

Two-dimensional DCT and one-dimensional DCT   We have defined the 1D DCT on a path graph. This can be easily extended to 2D grids. Separable transforms can be defined for signals on 2D graph grids using transforms for 1D path graphs. In matrix form, if $\mathbf{X}$ is a block of $N \times N$ pixels and $\mathbf{U}_{\mathrm{DCT}}$ is the DCT matrix then the transformed block is

$$\mathbf{Y} = \mathbf{U}_{\mathrm{DCT}}^{\mathsf{T}} \mathbf{X} \mathbf{U}_{\mathrm{DCT}}, \tag{7.2}$$

where the left transform operation applies the DCT matrix to the columns of $\mathbf{X}$, while the right operation does the same thing for the rows of $\mathbf{X}$. This is equivalent to representing each block of pixels, a vector in a space of dimension $\mathbb{R}^{N \times N}$, in terms of a set of $N \times N$ basis vectors, each obtained as the outer product of two basis vectors from the 1D DCT. The 2D basis vector formed from the column basis vector with index i and row basis vector with index j is

$$\mathbf{U}_{i,j} = \mathbf{u}_i \mathbf{u}_j^{\mathsf{T}}. \tag{7.3}$$

Denote by vect$(\mathbf{X})$ a column vector of size N^2 obtained from the $N \times N$ matrix $\mathbf{X}$. Any of the multiple possible vectorizations can be chosen, as long as the same vectorization is used for all matrices representing $N \times N$ image blocks and for the transform. For example, vect$(\mathbf{X})$ can be obtained by concatenation of the columns $\mathbf{X}$ following their original order. For a given vect$(\cdot)$ we can rewrite the separable transform of (7.2) as

$$\mathrm{vect}(\mathbf{Y}) = \mathbf{U}_{\mathrm{S\text{-}DCT}}^{\mathsf{T}} \mathrm{vect}(\mathbf{X}), \tag{7.4}$$

where $\mathbf{U}_{\text{S-DCT}}$ is an $N^2 \times N^2$ matrix for which each column is one of the N^2 vectorized basis vectors from (7.3) vect($\mathbf{U}_{i,j}$). It can also be shown that the 2D $N \times N$ graph can be obtained as the Kronecker product of two line graphs, so that consequently its algebraic representations can be obtained in the same way, which leads to $\mathbf{U}_{\text{S-DCT}}$.

Note that, more generally, graphs that are obtained as Kronecker products of two graphs have GFTs that can be constructed in the form of (7.4). If graph G is formed as the Kronecker product of graphs G_1 and G_2, with respective sizes N_1 and N_2 and GFTs $\mathbf{U}$ and $\mathbf{V}$, then the GFT for G can be constructed with $N_1 \times N_2$ elementary basis vectors vect($\mathbf{u}_i \mathbf{v}_j^\top$) for $i \in \{1, \ldots, N_1\}$ and $j \in \{1, \ldots, N_2\}$ [10]. As an example, in human activity analysis a spatio-temporal graph can be obtained by combining via a Kronecker product (i) the spatial skeleton graph connecting joints in the human body and (ii) a temporal path graph, that helps track motion over time [126].

Blocks and subgraphs   A typical mechanism for adapting transforms to image content is to select different block sizes so as to best capture differences in image properties. If we view this from a graph perspective, this idea can be interpreted as creating two subgraphs from the original graph, so that the data in each subgraph is represented independently. Refer to Section 5.4.3 for an example of filterbank design based on this idea.

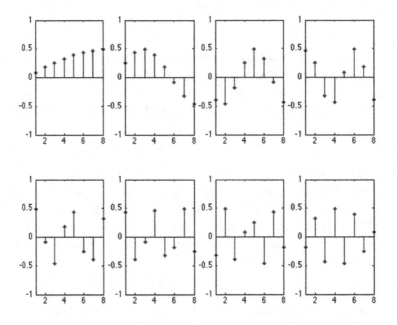

Figure 7.3 Asymmetric discrete sine transform (ADST) basis for vectors of dimension 8. As compared with the DCT basis of Figure 7.2, notice that the value for the leftmost node becomes smaller, in particular for the first basis vector.

Asymmetric discrete sine transform (ADST)   The asymmetric discrete sine transform (ADST) is typically used to encode **intra-prediction residuals**. In intra-prediction, images are split into non-overlapping blocks (as in JPEG), but pixels from neighboring blocks are used to predict the current block. As an example, if $\mathbf{x}$ is a vector of pixel values and if the pixel used for prediction has value x_0 then the intra-prediction residual to be encoded will be $\mathbf{x} - x_0\mathbf{1}$ instead of $\mathbf{x}$. This fundamentally changes the statistics of the vector to be encoded, reflecting the empirical observation that the residual energy at nodes closer to the predictor tends to be lower. The ADST (Figure 7.2) is the GFT of a line graph with generalized graph Laplacian

$$
\mathbf{L}' =
\begin{bmatrix}
2 & -1 & 0 & 0 & 0 & 0 & 0 & 0 \\
-1 & 2 & -1 & 0 & 0 & 0 & 0 & 0 \\
0 & -1 & 2 & -1 & 0 & 0 & 0 & 0 \\
0 & 0 & -1 & 2 & -1 & 0 & 0 & 0 \\
0 & 0 & 0 & -1 & 2 & -1 & 0 & 0 \\
0 & 0 & 0 & 0 & -1 & 2 & -1 & 0 \\
0 & 0 & 0 & 0 & 0 & -1 & 2 & -1 \\
0 & 0 & 0 & 0 & 0 & 0 & -1 & 1
\end{bmatrix}.
\tag{7.5}
$$

Note that the only difference with respect to (7.1) is that the first diagonal term has changed from 1 to 2. This can be interpreted as adding a "self-loop" to the first node on the left of the path graph, that is,

$$\mathbf{L}' = \mathbf{L} + \mathrm{diag}(1, 0, \ldots, 0),$$

where $\mathbf{L}$ corresponds to the DCT in (7.1). From Section 3.2.2, introducing a self-loop creates an additional penalty in the variation associated with the corresponding node. Thus, the node with a self-loop will tend to have a lower absolute value than other nodes. This is illustrated by a comparison of Figure 7.2 and Figure 7.3, where we can see that the lowest-frequency eigenvector (i.e., the highest-probability signal) changes from being $\mathbf{1}$ in the DCT (no self-loop) to a non-constant value for ADST (self-loop added). Thus, signals that can be well represented by ADST are more likely to have lower energy in the first node on the path. This explains why ADST is well suited for intra-prediction: predicting using the pixel immediately to the left is likely to be more successful (and thus lead to smaller residual values) for pixels close to the left side of the block.

Generalization   The connection between DCTs and graphs established in [1] was generalized to all trigonometric transforms, i.e., all types of DCT and DST, in [127]. This work and related contributions by the same authors [24, 25] were fundamental to establishing an algebraic perspective for GSP [15].

7.4.2 Graph Selection

In Section 7.4.1 we saw that two popular transforms for images and video, such as DCT and ADST, correspond to graph models for pixels. Next we use the ideas of Chapter 6 to construct graphs for image data.

Data-driven methods   To derive a graph model using a statistical method, we start by collecting image data: a set of vectors $\mathcal{X}$ containing samples $\mathbf{x}_m \in \mathbb{R}^N$, representing image patches. For example we can collect 8×8 pixel blocks and vectorize them into size $N = 64$ vectors, as described in the previous section. Next, an empirical covariance matrix is computed, as in (6.12), and then a graph Laplacian is learned as an approximation to the inverse of the covariance (Section 6.3). As it is approximately the inverse of the covariance (or precision) matrix, the connections captured by the graph Laplacian can be interpreted in terms of conditional independence (see Box 6.4 and Box 7.1). The DCT corresponds to the particular case where (i) two non-neighboring pixels are statistically independent, given all the other pixels and (ii) the correlation is identical between any pair of neighboring pixels. These statistical methods lead to different transforms when using training data specific to the application (e.g., corresponding to different intra-prediction modes) or when the graph is learned under different constraints (e.g., allowing non-separable transforms for 2D image data). Refer to [128] for examples of transform designs obtained with these approaches. We explore in more detail graph-based transform designs specifically for image and video compression in Section 7.4.4.

Similarity-based methods   The similarity-based methods of Section 6.2 connect a set of N nodes (in our case N pixels) by defining the Gram matrix of all pairwise distances between pixels. Therefore the key step is to define a meaningful distance between pixels. Since pixels are located on a regular grid, a distance based solely on position will lead us back to a regular grid. Instead, let $x_i = x(i_1, i_2)$ and $x_j = x(j_1, j_2)$ be the signal values (intensities) of two pixels at respective positions (i_1, i_2) and (j_1, j_2); then, in addition to a distance in space

$$d_s(i, j) = \sqrt{(i_1 - j_1)^2 + (i_2 - j_2)^2}, \qquad (7.6)$$

we define an intensity distance

$$d_p(i, j) = |x(i_1, i_2) - x(j_1, j_2)|. \qquad (7.7)$$

By combining these two distances we are viewing pixels as points in 3D space. Then, similarly to the idea behind (6.9), we can define a Gaussian kernel distance as a function of $d_s(i, j)$ and $d_p(i, j)$ like that used to define bilateral filters [3]:

$$w_{i,j} = \exp\left(-\frac{d_s(i, j)^2}{2\sigma_s^2}\right)\exp\left(-\frac{d_p(i, j)^2}{2\sigma_p^2}\right), \qquad (7.8)$$

where σ_s and σ_p control the relative importance of the spatial and intensity distances, respectively. This measure of the similarity between pixels (or alternative measures) allows us to create an $N \times N$ matrix with positive weights, which can be viewed as the adjacency matrix of a graph. All other GSP tools can then be derived from this matrix.

Discussion   The choice between these two types of methods, which make different assumptions about the image data, will typically be application dependent. Data-driven methods are optimized for particular sets of signals (similar to $\mathcal{X}$) and can be used for any signal in that class. In applications such as compression, where it is necessary to record or transmit which graph was used, the overhead can be maintained reasonably low (all that needs transmitting is the class of signals to which a particular input belongs). In contrast, the overhead would be significantly larger if weights such as those in (7.8) are used, since these would be different for each set of pixels. A limitation of data-driven approaches is that the results depend on how well the data is grouped into classes. Furthermore, the quality of the model will depend on the number of signal examples used for training, i.e., $|\mathcal{X}|$.

7.4.3   Filtering

Bilateral filters   The graph weights of (7.8) are derived from bilateral filters [3], which were designed for applications such as denoising, where the goal is to achieve smoothing without blurring image contours. If $w_{i,j}$ quantifies the similarity between pixels i and j then a smoothing filter can estimate the intensity $\hat{x}(i_1, i_2)$ using a weighted average of neighboring pixel intensities:

$$\hat{x}(i_1, i_2) = \frac{1}{\sum_{j \in N_i} w_{i,j}} \sum_{j \in N_i} w_{i,j} x_j, \tag{7.9}$$

where N_i is a set of pixels close to i and including i, and $w_{i,i} = 1$. Because the filter weights depend on pixel intensities, this filter is non-linear and input-dependent, so that j has little effect on i (i.e., $w_{i,j} \approx 0$) if $d_p(i, j)$ is large. The rationale for this is that significant differences in intensity between pixels are unlikely to be due to noise. The parameters σ_d and σ_p control the relative importance given to neighboring pixels (since the weight $w_{i,i}$ is always 1) and their relative values determine whether position or intensity is more important. The choice of values for σ_d and σ_p may also be a function of the application. If pixel intensities are noisy then the intensity distance (7.7) is not very reliable, so that a smaller σ_p may be preferable.

Graph interpretation and extensions   Note that, writing $\mathbf{x}$ as the vector of pixel intensities, denoting by $\mathbf{W}$ the matrix of filter weights $w_{i,j}$ from (7.8) and setting $\mathbf{D} = \mathbf{W1}$, we can rewrite the bilateral filter of (7.9), with weights given by $\mathbf{W}$, as

$$\tilde{\mathbf{x}} = \mathbf{D}^{-1}\mathbf{W}\mathbf{x}, \tag{7.10}$$

which can be seen as a one-hop graph filtering based on the random walk Laplacian graph operator.[1] Other popular image filtering approaches, such as using non-local means [129], can also be viewed from a graph perspective. Non-local filtering involves identifying patches that are similar within an image and using weights obtained from the matching patch to define filter coefficients. Thus, if in the matching patch pixels i, j correspond to pixels i', j', respectively, i.e., $d_s(i, j) = d_s(i', j')$, then in (7.8) $d_p(i, j)$ is

---

[1] The filtering operation written in graph form in (7.10) is not meant to lead to a practical implementation for typical images with millions of pixels. The pixel-wise localized filtering operations originally proposed in the literature are much better for implementation, but a graph point of view can be helpful to understand the properties of these filters.

replaced by $d_p(i', j')$. Kernel-based methods (Section 6.2.1) for image processing [130] can be viewed from a graph perspective, which provides a general framework for image filtering [33].

Filter normalization   The normalized operator in (7.10) is a non-symmetric matrix $\mathbf{D}^{-1}\mathbf{W}$. Alternative approaches for normalization include approximating $\mathbf{D}^{-1}\mathbf{W}$ by a doubly stochastic matrix $\mathbf{W}'$ (such that the entries of each row and each column add to 1) as described in [34, 33]:

$$\mathbf{W}' = \mathbf{D}_l\mathbf{D}^{-1}\mathbf{W}\mathbf{D}_r,$$

where $\mathbf{D}_l$ and $\mathbf{D}_r$ are diagonal matrices that can be computed with the Sinkhorn algorithm [35]. See Section 3.4.4 for a discussion of this type of normalization.

Filter design   The weights (7.8) define a (signal-dependent) graph, while the normalization in (7.10) leads to a fundamental operator $\mathbf{Z} = \mathbf{D}^{-1}\mathbf{W}$, so that graph filtering can be written $\mathbf{Z}\mathbf{x}$ for input $\mathbf{x}$. This idea can be extended by choosing as a filter any polynomial of $\mathbf{Z}$. Note that, in the original graph, node i is connected to all nodes in a neighborhood $\mathcal{N}_i$. Typically $\mathcal{N}_i$ is chosen so that all pixels within the 5×5 or 7×7 regions around i are included. Thus, even a small polynomial degree k will lead to filters $p(\mathbf{Z})$ having a large footprint within an image. As an alternative, we can define graphs with sparser connectivity (e.g., with $\mathcal{N}_i$ a 3×3 neighborhood) while using a larger polynomial degree k [131].

7.4.4   Graph-Based Image and Video Compression

Separable block-based transforms, such as DCT (Figure 7.2) and ADST (Figure 7.3), implemented as in (7.2), are GFTs of 2D (grid) graphs obtained as Kronecker products of 1D path graphs (Section 7.4.1). Separable transforms can be combined (e.g., ADST for rows and DCT for columns), but cannot provide basis vectors with non-horizontal or non-vertical orientations (i.e., directional transforms). Also, DCT and ADST are associated to specific graphs, which may not provide the best representation for a particular type of signal. In what follows, we review methods for learning graphs, and corresponding GFTs, in order to achieve better directionality and statistical fit than existing methods. Since a GFT is associated with any graph, the problem of transform design becomes a problem of graph construction, for which the methods in Chapter 6 are the starting point.

Directional transforms: motivation   Separable transforms can only produce basis vectors with vertical or horizontal orientation. To understand the need for directional transforms, consider the example of Figure 7.4, which shows the variance in residual pixel values across multiple video blocks predicted using a specific intra-prediction mode. In **intra-prediction**, pixels in a block are predicted by using pixels only from blocks that precede the current block in the block-scan order, that is, the blocks immediately above and to the right of the current block. In Figure 7.4 we can observe that the

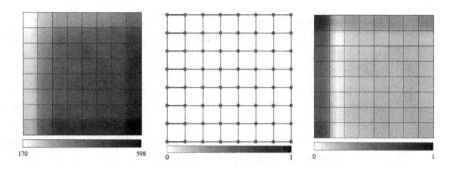

Figure 7.4 Variance of residual pixel value (left), where in each square (pixel) the shading is proportional to the variance of the residual pixel value (darker means greater variance). Notice that higher variance occurs away from the predictors (in the left and top neighboring blocks). The learned graph edge weights (center). The learned graph self-loop weights (right).

error increases away from those predictors, i.e., away from the left and top edges. This is to be expected, since prediction is likely to be more accurate (i.e., with less error) for pixels that are closer to the predictors. The energy pattern of Figure 7.4 suggests that basis vectors with increasing values away from the left and top edges would be preferable, like the first basis vector of ADST (Figure 7.3) but in 2D. We next explore graph-based techniques leading to basis vectors with directionality.

Steerable transforms It can be shown that in separable DCT transforms there are multiple basis vectors (eigenvectors of the grid graph) sharing the same eigenvalue. As discussed in Chapter 2, a higher-multiplicity eigenvalue leads to a subspace of dimension greater than 1 where all vectors have the same graph frequency. Thus, we can select different sets of basis vectors within each of these subspaces. Using this observation, recent work proposes steerable DCT transforms, where within each of those subspaces a rotation angle can be chosen for the basis vectors [132]. This rotation can be selected using the characteristics of the data and is shown to lead to improved performance.

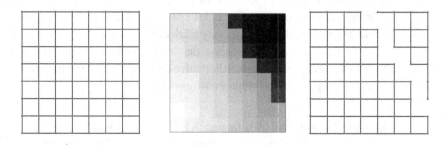

Figure 7.5 An illustration of graph construction for a given 8×8 residual block signal where $w_c = 1$ and $w_e = 0.1$, and the corresponding graphs: original graph (left), residue (middle), adapted graph (right).

Transforms from deterministic contour maps   Similarity-based graph construction (Section 7.4.2) provide an alternative approach to achieve directionality. A graph with the bilateral filter weights of (7.8) would not be suitable for compression, since these weights are a function of pixel intensities and additional bits would have to be used to transmit them to the decoder. A more practical approach is to identify significant contours within each block and to use only those contours for graph construction. For example, as shown in Figure 7.5, the encoder can start with a grid graph with all equal weights. If no significant contours are identified then the original grid graph is used, corresponding to the separable DCT. If a significant contour exists, side information is transmitted to the decoder. The graph edges linking two edges that are on different sides of a strong contour can then be given a lower edge weight, ϵ, as discussed in Box 7.2.

Box 7.2   Graph selection and compression

Owing to the overhead required to transmit a graph chosen to design a transform, it is important for the contour selection to identify significant contours while limiting how much information needs to be communicated. As a simple example, we can start with a regular grid graph and use a threshold Δ: if two neighboring pixels with intensities x_i and x_j are similar, i.e., $|x_i - x_j| \le \Delta$, then the edge weight is unchanged, $w_{ij} = 1$; otherwise we set the edge weight $w_{ij} = \epsilon$, with $\epsilon < 1$, and call these **weak edges**.

The position of significant contours varies from block to block and has to be encoded and transmitted to the decoder. This can be done in different ways. For example, since the weight of each edge in the initial grid graph can only take two values, transmitting the location of modified edges can be viewed as a problem of encoding binary information, and efficient encoders for this type of data have been proposed [133]. Ultimately, the choice of transform (non-separable DCT versus the GFT associated with a graph containing weak edges) will depend on the trade-off between their corresponding rate–distortion performances [134].

Representative approaches based on this type of graph construction for depth images can be found in [135, 136, 137]. Depth images are nearly piecewise constant, so that efficient encoding of contours can lead to significant compression gains. Moreover, the contours are typically sharp, as they mark differences in depth between objects. In contrast, for natural images we observe that image contours are not as sharp, in part due to the use of anti-aliasing filters. Alternative approaches have been proposed where lifting is used to implement the graph transform [138, 139].

Statistically optimal transforms: KLT   From Figure 7.4 we can see that pixels in the corresponding intra-prediction residuals have different energies. A classical approach to finding an optimal transform for a set of example data vectors is the Karhunen–Loève transform (KLT). To construct the KLT the first step is to collect representative data in order to construct a training set $\mathcal{X}$, and then to obtain the empirical covariance matrix using (6.12). The KLT is the matrix of eigenvectors of the covariance matrix.

While this approach can be applied to intra-residual data, such as that represented by Figure 7.4, there are two main drawbacks.

First, the number of examples used for training may be insufficient and thus the design may be unreliable. For example, a non-separable KLT for 8×8 pixel blocks is based on computing a 64×64 covariance matrix, which essentially means estimating more than 2000 parameters. To do this reliably is likely to require thousands of example blocks. Second, the transforms themselves may be hard to interpret. Under the GMRF model (Section 6.3), the sparsity of the inverse covariance or precision matrix can be interpreted in terms of the conditional independence between pixels (Box 6.4). But the inverse of an arbitrary empirical covariance matrix is unlikely to be sparse.

Techniques given in Section 6.3 can be used to overcome these challenges. As we shall see, learning inverse covariance matrices in the form of graph Laplacians leads to solutions that are more robust (fewer parameters need to be learned) and easier to interpret (sparser). For example, while $N^2/2$ values need to be estimated for a block of dimensions $N \times N$ (the number of free variables in the covariance matrix), the number of parameters to be learned can be of the order of $2N$ for a graph with sparsity similar to that of the grid graph.

Graph-based KLT alternatives   Data-optimized transforms, which can serve as alternatives to the KLT, can be designed using training data and then incorporated into a codec (coder–decoder). In some cases the encoder can use signaling bits (overhead) to indicate its choice among multiple transforms, to be used in addition to DCT and ADST. In other cases, the transform can be mode-dependent, i.e., the same transform is used for all intra-residuals produced by a given intra-prediction mode. For example, all blocks that use the intra-prediction mode of Figure 7.4 will use the same transform, derived from that residual data. Examples of designs based on these ideas can be found in [140, 141]. Graph learning approaches such as those in Section 6.3 can be used. Next we focus on how to choose (i) the training set X and (ii) the desired graph properties to be used for learning.

Training set selection   Intuitively, the choice of training set should balance two conflicting requirements. The blocks used for training should have similar characteristics, which suggests defining a large number of training sets, each specialized to blocks with some specific and similar characteristics. But there should be a sufficient number of blocks within each class, so that the overhead is not high (there are not too many classes) and there is enough data for training (a large number of example blocks per class). A typical design strategy is to define a class on the basis of example data obtained from modes in an existing codec, e.g., one class for each intra-prediction mode and one for each inter-prediction mode. Refer to [128] for examples of this approach.

Graph properties   Graph choice is important for interpretability, accuracy and complexity. Designing graphs with some desired properties typically comes at the cost of deviating from the original (unconstrained) representation provided by the KLT. For example, in (6.22) if a larger weight is given to sparsity in the cost function, this will result in more differences between the eigenvectors of the learned graph Laplacian and those of the empirical covariance matrix. However, note that, when insufficient data is available, a model with fewer parameters (e.g., greater sparsity) may in fact provide better performance. Better interpretability may also be achieved by introducing graph topology constraints. For example, since we can readily interpret transforms associated with line or grid graphs with uniform weights (DCTs), learning weighted line or grid graphs from data can lead to easy interpretation. In a learned line graph the relative edge weights provide a measure of relative correlation between pixels, with larger weights corresponding to greater correlation.

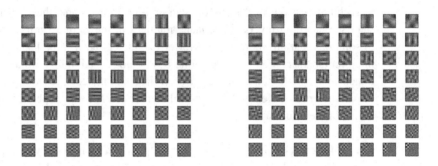

Figure 7.6 Comparison of separable (left) and non-separable (right) 2D data-driven transforms for 8×8 blocks. These two sets of 2D basis vectors were learned from exactly the same training data, which illustrates how different sets of constraints lead to very different learned transforms.

Computation complexity is also related to the choice of graph constraints. It will be easier to apply lifting approaches to define a transform (e.g., [139, 138]) for a sparse graph. Furthermore, if graphs have symmetry properties, the corresponding transforms have faster implementation [142]. Finally, we can choose to learn non-separable transforms to better approximate directional properties, or instead we can use separable approaches for lower complexity [140]. Figure 7.6 provides an example to illustrate this idea. The two sets of basis vectors were obtained from the same data, corresponding to a specific intra-prediction mode. In the separable design of Figure 7.6 (left), the data were used to learn two separate path graphs, one corresponding to rows and the other to columns. In the non-separable design of Figure 7.6 (right), a single grid graph was learned. The additional degrees of freedom made possible by the non-separable design allow the basis functions to have diagonal orientation. A recent paper demonstrates how these techniques can be integrated into a state-of-the-art video codec [143].

7.4.5 Image Restoration

In a restoration problem, we observe $\mathbf{y}$, a corrupted version of an image $\mathbf{x}$ affected by a combination of noise and some, possibly linear, transformation:

$$\mathbf{y} = \mathbf{Hx} + \mathbf{n},$$

where $\mathbf{n}$ is the noise term and $\mathbf{H}$ is the transformation. Note that $\mathbf{y}$ and $\mathbf{x}$ need not have the same dimension. This problem, i.e., finding $\mathbf{x}$ given $\mathbf{y}$, is underdetermined: there exist an infinite number of choices for $\mathbf{x}$. Graphs can be used to define image *priors* for image restoration. A prior is a set of assumptions, a model, to describe our knowledge about the images that we wish to reconstruct. Priors are needed in order to find a solution to inverse problems. Defining a graph prior requires first selecting a graph.

Graph selection   Since by definition we do not have access to $\mathbf{x}$, we need to construct a graph from either $\mathbf{y}$ or intermediate solutions in order to approximate $\mathbf{x}$. Consider a denoising application (i.e., $\mathbf{H} = \mathbf{I}$). Then we can define a graph with the bilateral weights of (7.8) but based on $\mathbf{y}$. Since the pixels in $\mathbf{y}$ are noisy versions of $\mathbf{x}$ we could also first compute an approximate denoised version $\mathbf{x}_1$ of $\mathbf{x}$ and use the weights obtained from $\mathbf{x}_1$ in the denoising process. Another alternative, given that the information in $\mathbf{y}$ is corrupted by noise, is to introduce quantization in the weight computation, for example:

$$w_{i,j} = \exp\left(-\frac{d_s(i,j)^2}{2\sigma_s^2}\right)\exp\left(-\frac{Q_p(d_p(i,j))^2}{2\sigma_p^2}\right), \tag{7.11}$$

using the definitions of (7.6) and (7.7) and where Q_p denotes the quantizer. To justify this, consider that two identical pixels in the noise-free image are likely to be different owing to the noise in $\mathbf{y}$, with a difference that will depend on the noise variance. Thus, the greater the noise variance, the greater the step size of Q_p.

Graph priors   Graph priors translate image properties intro graph frequency domain properties. Once a graph has been selected, several graph prior selection strategies can be used. These associate a cost function on the basis of the chosen graph G with each candidate reconstructed signal, $\mathbf{z}$. A frequency prior uses the GFT $\mathbf{U}$ obtained from the fundamental graph operator and computes $\tilde{\mathbf{z}} = \mathbf{U}^\mathsf{T}\mathbf{z}$. Possible priors include frequency localization,

$$\tilde{z}_i = 0, \quad \forall i > k,$$

or sparsity, which favor $\|\tilde{\mathbf{z}}\|_0$ or $\|\tilde{\mathbf{z}}\|_1$ being small. Use of ℓ_1 as a regularization term (Section 6.3.2) leads to well-known soft thresholding solutions in the graph transform domain. An alternative prior uses the Laplacian quadratic form [144],

$$L(\mathbf{z}) = \mathbf{z}^\mathsf{T}\mathbf{L}\mathbf{z} = \sum_i \lambda_i(\tilde{z}_i)^2,$$

where the penalty on a given frequency is the corresponding eigenvalue of the graph Laplacian. For the denoising case, this prior leads to a solution

$$\hat{\mathbf{x}} = \min_{\mathbf{z}} \|\mathbf{x} - \mathbf{z}\| + \gamma \mathbf{z}^\mathsf{T}\mathbf{L}\mathbf{z},$$

where $\gamma > 0$ controls how much importance is given to the prior and, as for the other priors, a smooth reconstructed image is favored, where smoothness is defined on the basis of the chosen graph G.

7.4.6   Image Segmentation

For **image segmentation**, an image-dependent graph is defined, such as the one that would be obtained with the weights from (7.8). To simplify the discussion, assume that only connections in a 3×3 neighborhood are used. Given this, note that if j is within the 3×3 neighborhood of i, the distance between the two pixels can only be 1 or $\sqrt{2}$. Thus the value of $w_{i,j}$ is dominated by the difference in intensity between pixels. Noting that image segmentation aims at identifying regions with similar intensities, this problem can be formulated as one of identifying clusters of nodes in the graph, or equivalently finding cuts [145, 146]. The problem of finding clusters using graph eigenvectors will be discussed in Section 7.5.2. In image segmentation, a cluster of nodes corresponds to identifying regions with similar intensities. Note that such a spectral clustering will have significant advantages over a naive node domain approach, which could be based on simply removing those edges whose weight is small, e.g., below some threshold. First, such a removal would not be guaranteed to be a true segmentation, since regions may not become completely separated. Second, textured areas would be problematic, since both large and small edge weights may be found and a one-hop approach is unlikely to group them together into a single region.

7.5   Machine Learning

Given data in the form of N pairs $(\mathbf{x}_i, y_i)$, where $\mathbf{x}_i \in \mathbb{R}^M$, a machine learning task involves learning a function f to predict $y = f(\mathbf{x})$ where $\mathbf{x} \in \mathbb{R}^M$ is an arbitrary vector. In the case when $y \in \mathbb{R}$ we have a regression problem, while if y takes only discrete values (**labels**) then this will lead to a classification task. In addition to this supervised problem, we also consider unsupervised (or semi-supervised) scenarios, where no y_i (or only some y_i) is available. A GSP perspective for machine learning problems starts by building a similarity graph (Section 6.2), where any two points $\mathbf{x}_i$ and $\mathbf{x}_j$ can be connected by an edge with weight that is a function of the distance $d(\mathbf{x}_i, \mathbf{x}_j)$, based for example on the Gaussian kernel (6.9). Given a similarity graph, $\mathbf{y}$ can be viewed as a graph signal.

   We will first discuss how to choose a graph (Section 7.5.1) and then consider unsupervised and semi-supervised learning, in Section 7.5.2 and Section 7.5.3, respectively. Graph signal processing methods are also useful when the data points, $\mathbf{x}_i$, are themselves defined on graphs. The classification of 3D point clouds, containing pixels associated with points in 3D space, would be an example. We provide an overview of graph neural networks (Section 7.5.4) to illustrate how GSP tools can be used in these scenarios.

7.5.1   Graphs and Graph Signals in Learning Applications

In the above setting, for the $(\mathbf{x}_i, y_i)$ pair, $\mathbf{x}_i$ will correspond to the ith node in the graph, while y_i will be used to construct one or more graph signals. In the regression case the graph signal $\mathbf{y} \in \mathbb{R}^N$ represents outputs of the function $f(\cdot)$ to be estimated, with y_i the *i*th scalar entry, and $y_i = f(\mathbf{x}_i)$. Alternatively, in a C-class classification problem, we can define C vectors, each representing the **membership function**, $\mathbf{y}_c$, for the corresponding class $c = 1, \ldots, C$, where $\mathbf{y}_c$ is such that $y_{c,i} = 1$ if $\mathbf{x}_i$ is in class c and $y_{c,i} = 0$ otherwise. Here each $\mathbf{x}_i$ is a *feature vector*, i.e., a representation of the data point that is assumed to be suitable for the task. For image classification we may extract features from the image, see e.g., [147], and use those instead of the original pixel values. In a neural network, $\mathbf{x}_i$ may be the output of the penultimate layer for the *i*th input. The dimension of this feature space, M, will in general be different from that of the original space where the data was defined (e.g., the size of the image in numbers of pixels).

Graph construction and interpretation   Given a distance $d(\mathbf{x}_i, \mathbf{x}_j)$ between two data points $\mathbf{x}_i$ and $\mathbf{x}_j$, the methods described in Section 6.2 can be used to construct a similarity graph.[2] This graph captures information about the relative positions of data points in the M-dimensional feature space and can provide useful insights. The most important one is that given two nodes i and j connected by an edge of weight w_{ij} we would expect that if w_{ij} is close to 1 then y_i and y_j are likely to be equal. Intuitively, if the feature space and the similarity metric are well defined, they should capture information that is relevant to the classification task. Thus, if neighboring data points often have different labels, this could be an indication that the choice of feature representation is not adequate or that the classification task is very challenging.

Label signal smoothness and task complexity   Given a similarity graph with Laplacian $\mathbf{L}$, $\mathbf{y}_c$ can also be viewed as a **label graph signal** or **label signal**, and we can define its smoothness as (see Remark 3.1)

$$\Delta_{\mathbf{L}}(\mathbf{y}_c) = \mathbf{y}_c^\mathsf{T}\mathbf{L}\mathbf{y}_c = \sum_{i \sim j} w_{ij}(y_{c,i} - y_{c,j})^2, \qquad (7.12)$$

where we can see that $y_{c,i} - y_{c,j} \neq 0$ if and only if one of the two points belongs to class c and the other does not. Thus, if $\Delta_{\mathbf{L}}(\mathbf{y}_c) = 0$ then there are no connections on this graph between points in class c and points in other classes. If this occurs, then we can conclude that class c is easy to separate from the others. More generally, $\Delta_{\mathbf{L}}(\mathbf{y}_c)$ is equal to the sum of the weights of all the edges connecting a point in class c and a point in any other class. Since these similarity weights are a decreasing function of distance, the larger the values that $\Delta_{\mathbf{L}}(\mathbf{y}_c)$ takes, the closer points in different classes will be and the more difficult the classification problem. Denoting by $\tilde{\mathbf{y}}_c$ the GFT of $\mathbf{y}_c$,

$$\tilde{\mathbf{y}}_c = \mathbf{U}^\mathsf{T}\mathbf{y}_c$$

and, using Remark 3.2, we can conclude that if $\Delta_{\mathbf{L}}(\mathbf{y}_c)$ is small then most of the energy of $\mathbf{y}_c$ will be in the low frequencies. The reason is that

$$\Delta_{\mathbf{L}}(\mathbf{y}_c) = \sum_k \lambda_k \tilde{y}_{c,k}^2$$

---

[2] Instead of working with this complete unweighted graph, in practice the graph is simplified using techniques such as K-nearest-neighbor (Section 6.1).

and, since the λ_k are non-negative and increase with k, it follows that, for $\Delta_{\mathbf{L}}(\mathbf{y}_c)$ to be small, the $\tilde{y}_{c,k}$ should have smaller values for larger k.

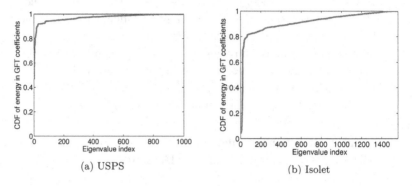

(a) USPS   (b) Isolet

Figure 7.7 Cumulative energy of the GFTs (7.13) of two different datasets used as testbeds for machine learning algorithms [148].

Empirical label signal smoothness   This property is illustrated empirically in the example in Figure 7.7 [148], which shows the cumulative energy contained in the GFT frequencies (ordered by increasing frequency). The average cumulative energy E_K up to index K is

$$E_K = \frac{1}{C} \sum_{c=1}^{C} \sum_{k=1}^{K} \frac{\tilde{y}_{c,k}^2}{\|\mathbf{y}_c\|^2}, \tag{7.13}$$

where the cumulative energy is averaged over all label signals. From the orthogonality and energy preservation properties of the GFT (see Figure 7.7) and the normalization (dividing by $\|\mathbf{y}_c\|^2 = 1$) we have that $E_K \leq 1$ and $E_N = 1$. The results in Figure 7.7 clearly show that the energy content of these label graph signals is biased towards the low frequencies. Furthermore, comparing the two datasets, it can be observed experimentally that those tasks for which the label signal is smoother are easier, in the sense of leading to a lower classification error rate [148].

Application to neural networks   The basic idea that the smoothness of the label type of graph signal relates to task complexity can be used to analyze the performance of a neural network [149]. In a neural network, the ith data point has a feature representation following each network layer. Thus, if $\mathbf{x}_i^{(k)}$ is the representation at the output of layer k, we can construct a graph for each layer. Then denoting by G_k the graph constructed with the outputs of layer k, it can be observed experimentally that the label signal (which is the same for every layer) becomes smoother at deeper (larger k) layers [149]. Thus, defining as $\mathbf{L}_k$ and $\mathbf{L}_\ell$ the graph Laplacians for G_k and G_ℓ, corresponding to layers k and ℓ, it can be observed empirically that in general, if $k < \ell$, we have

$$\Delta_{\mathbf{L}_k}(\mathbf{y}_c) > \Delta_{\mathbf{L}_\ell}(\mathbf{y}_c),$$

which shows that classes are increasingly better separated in the deeper layers.

Discussion   While the energy of $\mathbf{y}_c$ is concentrated in the low frequencies, this signal is not exactly bandlimited. As shown by Figure 7.7 the $\mathbf{y}_c$ have some energy in the high frequencies, so other models, e.g., piecewise constant models (Section 4.1.3), may be more appropriate. Furthermore, the selection of the graph itself is an open question, as some of the choices made, e.g., edge weights or K-nearest-neighbor sparsification methods, may not be theoretically justified. In particular, since the starting point of the graph definition is a distance metric, it is also important to investigate the robustness of this metric when operating in a high-dimensional space.

7.5.2   Unsupervised Learning and Clustering

In the previous section we made a link between labels associated with data points and the smoothness of the corresponding label graph signal on the similarity graph. We now consider an unsupervised scenario, where data are available but we have no labels associated with the data, i.e., we do not have a y_i corresponding to $\mathbf{x}_i$. We show how the similarity graph can be used to identify data clusters.[3] This can viewed as a form of unsupervised learning, in the sense that we may expect each cluster to be assigned a different label. Graph spectral methods have been frequently used to speed up clustering.

Problem formulation   Given a set X of vectors $\mathbf{x}_i \in \mathbb{R}^M$, we wish to group these vectors into clusters, where the jth cluster is a set $C_j \subset X$ and we have that $C_1 \cup C_2 \cdots =$ X with $C_j \cap C_\ell = \emptyset$ for $j \neq \ell$. The centroid of C_j is the average of the vectors in the set

$$\mathbf{c}_j = \frac{1}{|C_j|} \sum_{\mathbf{x}_i \in C_j} \mathbf{x}_i. \tag{7.14}$$

Qualitatively, a cluster contains vectors from X that are closer to each other than to vectors in other clusters, and since $\mathbf{c}_j$ represents the average of points C_j, we can assign $\mathbf{x}$ to C_j if $\mathbf{x}$ is closer to $\mathbf{c}_j$ than to any other centroid $\mathbf{c}_\ell$. This observation leads to the k-means clustering algorithm described in Box 7.3.

Box 7.3 k-Means clustering

For the set X of vectors $\mathbf{x}_i \in \mathbb{R}^M$, find k clusters, with $\mathbf{c}_j$ corresponding to the centroid of C_j, by alternating between the following steps.

- **Nearest neighbor step**: assign each $\mathbf{x}_i \in X$ to cluster C_j such that the distance $\|\mathbf{x}_i - \mathbf{c}_j\|^2$ is minimized.
- **Centroid step**: compute a new centroid $\mathbf{c}_j$ for each of the current C_j using (7.14).

Similarity graph and clustering   To understand how a similarity graph can be helpful for clustering, consider a problem where we wish to identify two clusters. The

---

[3] Clustering problems can be encountered in the design of vector quantizers or classifiers.

Rayleigh quotient or normalized Laplacian quadratic form, (3.39), for an arbitrary non-zero vector $\mathbf{y}$ such that $\mathbf{y}^\mathsf{T}\mathbf{1} = 0$ can be written as

$$R(\mathbf{y}) = \frac{\Delta(\mathbf{y})}{\mathbf{y}^\mathsf{T}\mathbf{y}} = \frac{\mathbf{y}^\mathsf{T}\mathbf{L}\mathbf{y}}{\mathbf{y}^\mathsf{T}\mathbf{y}}. \tag{7.15}$$

From the definition of the eigenvectors of $\mathbf{L}$ we know that $R(\mathbf{y})$ is minimized by $\mathbf{u}_2$, the second eigenvector of $\mathbf{L}$, also known as the Fiedler eigenvector. Then, we can use the sign pattern of $\mathbf{u}_2$ to induce a clustering of the graph nodes; namely, if $u_2(i) > 0$ then $\mathbf{x}_i$ is assigned to C_1 and otherwise $\mathbf{x}_i$ is assigned to C_2. Thus, $\mathbf{y}$ defined as $\mathbf{y} = \text{sign}(\mathbf{u}_2)$ can be seen as a cluster indicator signal derived from the graph: $\mathbf{y}$ takes the values $+1$ and -1 for points belonging to C_1 and C_2, respectively. Using (3.39) we can rewrite (7.15):

$$R(\mathbf{y}) = \frac{\mathbf{y}^\mathsf{T}\mathbf{L}\mathbf{y}}{\mathbf{y}^\mathsf{T}\mathbf{y}} = 2N\frac{\sum_{i\sim j} w_{ij}(y(i) - y(j))^2}{\sum_{i=1}^{N}\sum_{j=1}^{N}(y(i) - y(j))^2} = 2N\frac{A(\mathbf{y})}{B(\mathbf{y})}.$$

We can simplify $A(\mathbf{y})$ and approximate $B(\mathbf{y})$ as follows:

$$A(\mathbf{y}) = 4\sum_{i\in C_1, i'\in C_2} w_{i,i'}, \quad B(\mathbf{y}) \approx 4\frac{N^2}{2},$$

where $A(\mathbf{y})$ is a function of the sum of weights of the edges linking the two clusters. We would like to minimize $A(\mathbf{y})$, which is smaller if distances across clusters are large. The approximation for $B(\mathbf{y})$ is obtained by assuming that the graph is complete (N^2 edges) and the two clusters have approximately equal size. Then, only connections across clusters (about half of all connections) have a non-zero cost, equal to 4. Overall we have that $R(\mathbf{y})$ is proportional to the cost of the "cut," i.e., the sum of the weights of the edges linking C_1 and C_2:

$$R(\mathbf{y}) \propto \sum_{i\in C_1, i'\in C_2} w_{i,i'}.$$

Note that $\mathbf{u}_2$ minimizes $R(\mathbf{u}_2)$ but the choice $\mathbf{y} = \text{sign}(\mathbf{u}_2)$ does not lead necessarily to an optimal cut (min-cut). That is, we may be able to find $\mathbf{y}' \neq \mathbf{y}$ such that $R(\mathbf{y}') < R(\mathbf{y})$. In conclusion, $R(\cdot)$ is a good target metric for clustering, but because $\mathbf{u}_2$ is not restricted to take integer values ($+1$ and -1), $\mathbf{y} = \text{sign}(\mathbf{u}_2)$ is not guaranteed to be optimal.

Spectral clustering   In the two-cluster example, we computed $\mathbf{u}_2$ and then assigned to each vector $\mathbf{x}_i$ a scalar $u_2(i)$, the ith entry of $\mathbf{u}_2$. This can be seen as an embedding, which maps points in $\mathbb{R}^M$ to points on the real line $\mathbb{R}$. Clearly, we should expect a significant loss of information about the original data geometry, since we are going from dimension M to dimension 1. Spectral clustering addresses this problem by mapping each input vector $\mathbf{x}_i$ onto a space of dimension L, where $M > L > 1$. To do so, first compute L eigenvectors of $\mathbf{L}$: $\mathbf{u}_2, \mathbf{u}_3, \ldots, \mathbf{u}_{L+1}$ (where since $\mathbf{u}_1 = \mathbf{1}$ the first eigenvector is not informative). Then, assign to each node i the corresponding L entries, i.e., define a vector $\mathbf{f}_i = [u_2(i), u_3(i), \ldots, u_{L+1}(i)]^\mathsf{T}$. Finally apply the K-means clustering algorithm of Box 7.3 to $\mathbf{f}_i \in \mathbb{R}^L$ instead of to $\mathbf{x}_i \in \mathbb{R}^M$. This leads to dimensionality reduction and a simpler clustering problem.

As a simple example, consider the case where $L = 2$ and $K = 4$ and use $\mathbf{u}_2$ and $\mathbf{u}_3$

to create a 2D embedding. Since $\mathbf{u}_i^\top \mathbf{1} = 0$, the eigenvectors $\mathbf{u}_2$ and $\mathbf{u}_3$ will be expected to have roughly equal numbers of positive and negative entries.[4] Since $\mathbf{u}_2^\top \mathbf{u}_3 = 0$ it follows that $\mathbf{u}_2$ and $\mathbf{u}_3$ cannot have the same sign pattern. Thus, for any $\mathbf{x}_i$, the pair $(\text{sign}(u_2(i)), \text{sign}(u_3(i)))$ defines to which of the four quadrants of the 2D plane $\mathbf{x}_i$ is assigned. This can be viewed as a very rough clustering that improves the clustering achieved with $\mathbf{u}_2$ in the $L = 1$, $K = 2$ case.

7.5.3 Semi-supervised Learning

Motivation Collecting raw unlabeled data, i.e., the $\mathbf{x}_i$ vectors, can be relatively easy, while labeling it, i.e., finding the corresponding y_i, is often challenging, involves human effort and is prone to errors. For example, images can be found on the Internet, but these are frequently not labeled, or labels that can be associated with them (e.g., text on the web page containing those images) may not be reliable. At the same time it has been shown that using both labeled and unlabeled data for training a classification system can lead to significant benefits over using only labeled data, leading to the concept of semi-supervised learning [150].

Active learning and sampling set selection We focus first on the problem of active semi-supervised learning, where unlabeled data has been collected but we can only spend a limited amount of resources to label some of it. In both scenarios, we will need to predict unknown labels from known labels, i.e., perform some form of label propagation. From Section 7.5.1 we expect the (unobserved) label signal $\mathbf{y}$ to be smooth on the similarity graph, with the degree of smoothness depending on the difficulty of the classification task. A solution to the active learning problem can be found as follows [148]: (i) construct a similarity graph using all N available data points, $\mathbf{x}_i$, (ii) use one of the sampling algorithms in Section 4.4 to select some target $N_1 < N$ points forming a set $\mathcal{S}_1$, (iii) label all selected data points in $\mathcal{S}_1$ and (iv) use the label information y_j for all $\mathbf{x}_j \in \mathcal{S}_1$ to estimate the labels of unlabeled data (see below). Semi-supervised learning algorithms make use of both observed and estimated labels to train a model.

Label propagation After a subset $\mathcal{S}_1$ has been identified (using active learning or other techniques such as random sampling) and the data in $\mathcal{S}_1$ have been labeled, the next step is label propagation, which can be viewed as an interpolation process similar to those discussed in Section 4.3. Without loss of generality, assume that data points $1, \ldots, N_1$ have been labeled, while points $N_1 + 1, \ldots, N$ are unlabeled. With the same ordering, define a vector of observed labels $\mathbf{y}_l \in \mathbb{R}_1^N$ and a vector of unobserved labels $\mathbf{y}_u \in \mathbb{R}^{N-N_1}$, where $\mathbf{y}_u = \mathbf{0}$. Then the goal is to compute a label signal $\hat{\mathbf{y}} \in \mathbb{R}^N$ containing estimated labels for all the entries in $\mathbf{y}_u$. While an exactly bandlimited model (Section 4.1.2) may not be a valid assumption for $\mathbf{y}$, we can at least assume $\mathbf{y}$ to be smooth on the graph (Section 7.5.1). Thus, a graph-based regularizer with penalty $\hat{\mathbf{y}}^\top \mathbf{L} \hat{\mathbf{y}}$, where $\mathbf{L}$ is the Laplacian for the similarity graph, is often used in semi-supervised learning

---

[4] In general this is a good approximation for low-frequency eigenvectors, which tend not to be localized.

[150]. Multiplying a graph signal $\mathbf{y}$ by $\mathbf{L}$ is a high pass filtering operation, so that the penalty term $\hat{\mathbf{y}}^{\mathsf{T}}\mathbf{L}\hat{\mathbf{y}}$ will increase if the signal has higher energy content in the high frequencies.

This idea can be further extended by using other high pass filters, such as an ideal high pass filter, which penalizes signals only if they have energy above a certain graph frequency. To do so, assume that $\hat{\mathbf{y}}$ is expected to have energy only in the first K frequencies of the GFT $\mathbf{U}$, where in general $K \leq N_1$. Then, denoting by $\mathbf{U}_{\bar{K}}$ the $N \times (N-K)$ matrix formed by the last $N - K$ columns of $\mathbf{U}$, we can write the penalty term as

$$\hat{\mathbf{y}}^{\mathsf{T}}\mathbf{U}_{\bar{K}}\mathbf{U}_{\bar{K}}^{\mathsf{T}}\hat{\mathbf{y}}.$$

With the chosen penalty term, reconstruction techniques such as those of Section 4.3.4 and Section 4.3.5 can be used.

7.5.4 Graph Convolutional Neural Networks

In this section our goal is not to provide an overview of the recent state of the art in this fast changing field. Instead, we highlight how key ideas in GSP are used for machine learning on graph datasets.

Motivating scenario   We consider a classification scenario where we have a number of elementary objects, each associated with a different graph and graph signal, and where the goal is to assign a label to each of these. As a concrete example, consider the problem of classifying point-cloud image data, where each object to be classified contains a set of points in 3D space, with corresponding attributes (e.g., color). In this scenario each of the objects has a different number of points. In this case a graph would have to be constructed (as in Section 7.3 and Section 7.4), but the same ideas can be applied if the graph is given.

Convolutional neural networks   Graph neural networks were developed with the goal of extending highly successful convolutional neural networks (CNNs) to graph data. A simple CNN can be constructed using a series of basic operations.

- Linear time- or space-invariant *filtering* (using convolution) with filter coefficients learned during the training process can be used. In a given layer, multiple convolutions can be performed in parallel. Using convolutions has the advantage of limiting the number of parameters to be learned (which is essentially equal to the filter length) and also favors the invariance of classification to shifts. Intuitively we can think of each filter as defining a "pattern," with the inner product in the filtering operation matching shifted versions of this pattern to the input signal (Section A.3.2).
- An element-wise non-linearity of the filter outputs is used, the most popular choice being the rectified linear unit (ReLU), which is simply a threshold that sets to zero any negative inputs and leaves unchanged the positive inputs.
- Pooling operations are used to enable the classification of objects at multiple scales. Pooling is essentially a downsampling operation. However, rather than using a fixed

downsampling pattern as in Section 5.6, other strategies such as max-pooling (selecting the largest among multiple values) are used. Max-pooling is motivated by providing local invariance, i.e., if a pattern matches an input, the shift that provides the maximum match is selected. Note that for regular domain signals (e.g., images) the output of the pooling operation is again an object of the same type (i.e., smaller images).

Graph Convolutional Neural Network Design

The most widely used graph neural network designs are developed by extending to graphs the operations described above. Note that non-linearities are applied entry-wise and thus they can be directly used for graph data. Thus, we consider filtering and pooling in what follows.

Filtering   As discussed in Chapter 2 and Chapter 3, graph filters that are polynomials of the graph operator $\mathbf{Z}$ also have a polynomial response in the frequency domain. Thus, a polynomial $p(\mathbf{Z})$ corresponds to a frequency response $p(\lambda)$. By restricting learned filters to be polynomials we will have at least some knowledge of the frequency response without having to compute the GFT. If $\mathbf{Z}$ is a normalized operator, e.g., the normalized Laplacian, then the range of frequencies is known although the exact discrete frequencies are not. That is, given $p(\lambda)$ and the range of graph frequencies, we know the response at any λ but we do not know the specific discrete λ_i values for each specific graph. Note also that we cannot define directly a convolution operation in the node domain for graph signals. However, a polynomial filter of degree K provides a different, local (K-hop) template for processing data around each graph node. Thus by carefully choosing the degree of the polynomial (ensuring that it is small relative to the radius of the graph) we can achieve local processing, similar to that obtained for CNNs applied to regular domain signals.

Learned spectral filters   Initially proposed techniques for graph neural networks focused on designing filters in the spectral domain [151], but this method was found to have the disadvantage of requiring a GFT to be computed. This approach reduced the GFT computation overhead by requiring only a subset of the GFT basis vectors (the first k out of N) to be computed. While this would be more efficient than computing the whole GFT and provides similar performance, since the lower-frequency basis vectors are the ones representing the smooth structure of the graph, this overhead is incurred for every input in the training set and for every test input.

Learned polynomial filters   This limitation of spectral methods explains the popularity of techniques such as those proposed in [152] or [153]. The learning algorithm in [152] optimizes the parameters of a Chebyshev polynomial (Section 5.5.2). The approach in [153] further simplifies the filtering operation by using only one-hop operators, for which learning a single parameter is needed. More recent approaches are exploring further the idea of designing polynomial filters of a selected node domain graph operator.

Pooling   In the context of CNNs for conventional signals such as images, pooling consists of selecting a subset of pixel outputs and constructing a new, lower resolution, signal with those outputs. Typically, non-linear approaches such as max-pooling are preferred. Taking images as an example, for a set of pixels (e.g., a 2×2 block) max-pooling selects the maximum output value in the block. This maps an $N \times N$ image to a new $N/2 \times N/2$ image containing the maximum value. The size of the original block (and thus the amount of downsampling) as well as the pooling strategy (maximum, average, etc.) can all be selected and are part of the neural network architecture. Note that the subsampled output is an image with the same connectivity (i.e., it consists of pixels positioned on a regular grid) but with a different size.

To extend pooling ideas to graphs presents significant challenges. Images can be viewed as graphs with a rectangular grid topology, on which it is easy to partition the pixels into regular blocks. In contrast, for arbitrary graphs there are many possible ways of partitioning, leading to different pooling strategies. Also, when an image has been downsampled, its topology is preserved, i.e., the downsampled image is still defined on a grid. The same is not true for a graph. Assume we define a strategy to select key nodes on a graph (which could be done using some of the sampling strategies in Chapter 4). Then, a pooling approach could be used to select from the filtered signal the values to associate with each of those key nodes. Next we would need to reconnect the graph.

This could be done using the ideas in Section 6.1.5, most notably the Kron reduction. However, the difficulty with these methods is that they can make graphs significantly denser, especially if applied multiple times in succession. In practice this means that while pooling is possible for graph signals, it may require the choice of additional design parameters and may be sensitive to the choice of parameters that is made.

Challenges   Notice that one fundamental challenge in designing classifiers for graph-based data is that both the graph topology and the graph signals themselves can change substantially. Thus, it would be interesting to study how stable the choice of filter parameters is to the characteristics of the graph. One of the fundamental messages in this book has been to show that the frequency characteristics of a graph change significantly as a function of its topology. Thus, for the practical application of graph neural networks it is likely to be necessary to develop some constraints on the type of graphs that can be handled by a given classifier. Just as for images we may know that a classifier is best suited for images of a certain resolution, we may have to restrict the set of graphs for which a certain classifier is used as a function of their size and topology.

Recent work [154] has addressed the stability problem more formally and has developed interesting insights. This work identifies a trade-off between stability and frequency selectivity, by noting that small changes in topology lead to small changes in frequency (see Theorem 6.1). Therefore while the same filter $p(\lambda)$ may be used for two graphs, one a perturbed version of the other, the actual frequency response at the discrete frequencies will be different. How different it is depends on how flat the original response was. This leads to the following trade-off: flat responses (less frequency selectivity) are more stable to graph changes but by definition they may also be less effective for classification.

Chapter at a Glance

The main focus of this chapter has been to provide a series of representative examples of graph signal processing applications. These were grouped into: (i) applications where graphs exist and signals may be a function of the graph (Section 7.2), (ii) distributed applications, where graph processing is not necessarily centralized (Section 7.3), (iii) imaging applications (Section 7.4), demonstrating that graph representations can be useful even for signals for which conventional signal processing tools exist and (iv) machine learning applications (Section 7.5), where we show that graph representations can help us understand large-scale, high-dimensional datasets.

While any choice of applications is subjective, and indeed related to the author's own research interests, the main goal has been to emphasize ideas presented in earlier chapters. In particular, the choice of a graph and related tools (graph operator, filters, downsamplers) is the fundamental first step to applying GSP in any application domain. Therefore, the first step in applying GSP tools is to develop a detailed understanding of how a graph model captures important characteristics of this application (see also Section 7.1). A GSP perspective does not provide a solution to a problem; it provides a way to think about the problem in order to find a solution.

Further Reading

As stated in the introduction to the chapter, our goal was not to survey all possible GSP applications, let alone introduce the state of the art and competing methods in each of those application domains. In addition to the references cited in this chapter, the reader can refer to references in overview papers such as [9, 10, 11], where several applications are discussed, as well as to overview papers in specific areas such as imaging [125] and learning from graph data [155, 156].

Appendix A Linear Algebra and Signal Representations

We have two main goals in this chapter. First, while we assume most readers will have some familiarity with the topic, this chapter provides a summary of key concepts in elementary linear algebra, which can be reviewed before reading Chapter 2 and Chapter 3. Second, we emphasize the importance of a linear algebra perspective for (graph) signal processing, allowing us to present various problems such as signal representation or signal approximation from this perspective, which will be used throughout this book.

We work with sets of signals that share some common properties, so that these sets can be mathematically modeled as vector spaces or subspaces (Section A.1). Signals in a given class can be represented as a linear combination of elementary signals which form a basis or an overcomplete representation (Section A.2). This will allow us to define the properties of any signal using the properties of the basis vectors. Approximate (inexact) representations, where only a small set of elementary signals is used, are often of practical interest. In order to quantify how good a given approximation is, we define the notion of the distance between signals (Section A.3). We next revisit the concept of bases and introduce orthogonal bases, which provide signal representations with desirable properties in terms of distance (Section A.4). Finally, we return to overcomplete sets and discuss techniques to obtain signal representations using them (Section A.5).

A.1 Signal Spaces and Subspaces

A basic introduction to signal processing will usually cover a series of *transforms*, such as the Fourier transform, the discrete-time Fourier transform (DTFT) and the discrete Fourier transform (DFT). All these transforms are invertible, each providing a unique representation for signals in specific spaces, namely, the space of continuous-time finite-energy signals, $L_2(\mathbb{R})$, the space of discrete-time finite-energy signals, $l_2(\mathbb{Z})$, and $\mathbb{C}^N$, respectively. All these transformations can be viewed as projections of input signals onto an orthogonal set of basis vectors for the corresponding space.

Elementary signal operations   In this book we consider finite-dimensional signals, i.e., signals are vectors defined on graphs with a finite number of nodes, N. Each signal vector consists of the values at each node in the graph. We generally consider real signals[1] so that the space of signals of interest is $\mathbb{R}^N$. In $\mathbb{R}^N$ we can define the sum of two

---

[1] The extension to complex signals would be straightforward.

graph signal vectors and the multiplication of a graph signal vector by a scalar in the usual way as respectively the entry-wise sum of two vectors and the entry-wise multiplication of a vector by a scalar. Then, given two vectors $\mathbf{x}, \mathbf{y} \in \mathbb{R}^N$ and a scalar $\alpha \in \mathbb{R}$, the ith entry of $\mathbf{x} + \mathbf{y} \in \mathbb{R}^N$ will be $x(i) + y(i)$ and the ith entry of $\alpha\mathbf{x} \in \mathbb{R}^N$ will be $\alpha x(i)$.

Thus, using standard definitions, vector sums and scalar multiplication operate independently at each node of the graph and do not depend on the graph topology. These vector operations are easy to interpret in terms of graph data. For example, consider a dataset of temperature measurements, where each vector $\mathbf{x}_t$ corresponds to the observations of node temperature values at a given time, t, and each entry in the vector corresponds to one specific measurement (at a given sensor). The average over N_t measurements (the number of nodes) would naturally be

$$\bar{\mathbf{x}} = \frac{1}{N_t} \sum_t \mathbf{x}_t.$$

Sets and subspaces   We are interested in defining sets of signals that share some properties and in particular vector spaces, i.e., sets that are closed under addition and scalar multiplication. Let $S \subset \mathbb{R}^N$ be a subset of vectors in $\mathbb{R}^N$.

DEFINITION A.1   (SUBSPACE) $S \subset \mathbb{R}^N$ is a subspace of $\mathbb{R}^N$ if it is a vector space: $\mathbf{0} \in S$ and S is closed under addition and scalar multiplication, that is, for $\mathbf{x}, \mathbf{y} \in S$ we have $\mathbf{x} + \mathbf{y} \in S$ and $\alpha\mathbf{x} \in S$, for any $\alpha \in \mathbb{R}$.

Note that the zero vector, denoted $\mathbf{0}$, belongs to any subspace of $\mathbb{R}^N$. Thus, if we take any two arbitrary subspaces S_1 and S_2 we will always have that $\{\mathbf{0}\} \subset S_1 \cap S_2$. As we shall see next, all signals in a subspace can be written as a linear combination of a set of elementary signals that form a basis (Section A.2). Given $S \subset \mathbb{R}^N$, not all vectors in $\mathbb{R}^N$ belong to S. Thus, we later address the question of how to approximate any vector in $\mathbb{R}^N$ by a vector in S (Section A.4.3).

A.2   Bases

Non-trivial vector subspaces contain an infinite number of vectors. This can be easily seen: if a subspace contains $\mathbf{x} \neq \mathbf{0}$ then, by definition, $\alpha\mathbf{x}$ belongs to the subspace for any $\alpha \in \mathbb{R}$. Since α can take an infinite number of values, the number of vectors in the space is also infinite. In finite-dimensional spaces, such as $\mathbb{R}^N$, any of the infinite vectors in a subspace can be represented as a unique linear combination of a *finite* number of vectors which form a basis. This idea is reviewed in what follows.

A.2.1   Span

Denote by $S = \{\mathbf{u}_1, \mathbf{u}_2, \ldots, \mathbf{u}_k\}$ a set of vectors in $\mathbb{R}^N$. Then we define the *span* of this set as a set containing all vectors that can be obtained as linear combinations of $\{\mathbf{u}_1, \mathbf{u}_2, \ldots, \mathbf{u}_k\}$.

> **DEFINITION A.2** (SPAN) Given a set of vectors $S = \{\mathbf{u}_1, \mathbf{u}_2, \ldots, \mathbf{u}_k\}$,
>
> $$\text{span}(S) = \left\{ \mathbf{x} \in \mathbb{R}^N \mid \exists \alpha_i \in \mathbb{R}, \ \mathbf{x} = \sum_{i=1}^{k} \alpha_i \mathbf{u}_i \right\}.$$

In matrix form, defining $\mathbf{U}_k$ as:

$$\mathbf{U}_k = \begin{bmatrix} \vdots & \vdots & & \vdots \\ \mathbf{u}_1 & \mathbf{u}_2 & \cdots & \mathbf{u}_k \\ \vdots & \vdots & & \vdots \end{bmatrix} \tag{A.1}$$

the span of $\{\mathbf{u}_1, \mathbf{u}_2, \ldots, \mathbf{u}_k\}$ is the set of $\mathbf{x} \in \mathbb{R}^N$ for which there exists $\mathbf{a} \in \mathbb{R}^k$ such that

$$\mathbf{x} = \mathbf{U}_k \mathbf{a}.$$

> **PROPOSITION A.1** (SPANS ARE SUBSPACES) $\text{span}(S)$ is a subspace of $\mathbb{R}^N$.

Proof Any two vectors in the span can be written as linear combinations of the vectors $\mathbf{u}_1, \mathbf{u}_2, \ldots, \mathbf{u}_k$; therefore their sum is also a linear combination of $\mathbf{u}_1, \mathbf{u}_2, \ldots, \mathbf{u}_k$, or, in matrix form,

$$\mathbf{x}_1 + \mathbf{x}_2 = \mathbf{U}_k \mathbf{a}_1 + \mathbf{U}_k \mathbf{a}_2 = \mathbf{U}_k (\mathbf{a}_1 + \mathbf{a}_2).$$

A similar argument can be given for scalar multiplication, so that the span is closed under addition and scalar multiplication, and therefore is a subspace (Definition A.1). □

A.2.2 Linear Independence

For any $\mathbf{x} \in \text{span}(S)$, we can write by definition $\mathbf{x} = \mathbf{U}_k \mathbf{a}$, a linear combination of the vectors in S. Is this representation unique? For it to be unique the vectors in S must be *linearly independent*.

> **DEFINITION A.3** (LINEAR INDEPENDENCE) A set of vectors $\{\mathbf{u}_1, \mathbf{u}_2, \ldots, \mathbf{u}_k\}$ is linearly independent if we have that
>
> $$\sum_{i=1}^{k} \alpha_i \mathbf{u}_i = 0 \iff \alpha_i = 0, \forall i = 1, 2, \ldots, k.$$

If the vectors are linearly independent then one cannot express one as a linear function of the others. If the set is linearly dependent, on the other hand, there exists a set of α_i with at least one $\alpha_j \neq 0$ such that

$$\sum_{i=1}^{k} \alpha_i \mathbf{u}_i = 0,$$

and then we can write

$$\mathbf{u}_j = -\frac{1}{\alpha_j} \sum_{i=1, i \neq j}^{k} \alpha_i \mathbf{u}_i. \tag{A.2}$$

In matrix form, if we can find a non-zero vector $\mathbf{b} \in \mathbb{R}^k$ such that $\mathbf{U}_k \mathbf{b} = 0$ (linear dependence), then it is easy to see that, if $\mathbf{x} = \mathbf{U}_k \mathbf{a}$, we have also

$$\mathbf{x} = \mathbf{U}_k \mathbf{a} = \mathbf{U}_k (\mathbf{a} + \beta \mathbf{b}),$$

for any scalar β and so there exist an infinite number of ways of representing $\mathbf{x}$ as a function of the columns of $\mathbf{U}_k$. For any matrix $\mathbf{U}_k$ the set of vectors $\mathbf{a}$ such that $\mathbf{U}_k \mathbf{a} = 0$ forms the *null space* of $\mathbf{U}_k$. From Definition A.3, the columns of $\mathbf{U}_k$ are linearly independent if and only if the null space of $\mathbf{U}_k$ is equal to the zero vector, $\mathbf{0}$.

A.2.3   Bases

While the concepts discussed here extend to infinite-dimensional spaces, we focus our discussion on finite-dimensional spaces, as these will be directly applicable to representing signals on (finite) graphs. If (A.2) holds, it is clear that we can remove $\mathbf{u}_j$ and still have the same ability to represent vectors. That is, the span does not change if $\mathbf{u}_j$ is removed:

$$\text{span}(\{\mathbf{u}_1, \mathbf{u}_2, \dots, \mathbf{u}_k\}) = \text{span}(\{\mathbf{u}_1, \mathbf{u}_2, \dots, \mathbf{u}_k\} \setminus \{\mathbf{u}_j\}).$$

Thus, since we can obtain signals in a subspace through linear combinations of vectors, the linear independence of $\{\mathbf{u}_1, \mathbf{u}_2, \dots, \mathbf{u}_k\}$ guarantees that every vector belonging to $\text{span}(\{\mathbf{u}_1, \mathbf{u}_2, \dots, \mathbf{u}_k\})$ can be represented in a unique way as a linear combination of these vectors. More formally, we can define the concept of a basis.

> DEFINITION A.4   (BASIS AND DIMENSION)   A set of vectors $\mathcal{B} = \{\mathbf{u}_1, \mathbf{u}_2, \dots, \mathbf{u}_k\}$ is a **basis** for a vector space S if (i) $\mathbf{u}_1, \mathbf{u}_2, \dots, \mathbf{u}_k$ are linearly independent and (ii) $\text{span}(\mathcal{B}) = S$. The number of vectors in a basis is the **dimension** of the space.

A set of vectors $\mathbf{u}_1, \mathbf{u}_2, \dots, \mathbf{u}_N$ forming a basis for $\mathbb{R}^N$ can be represented as an $N \times N$ matrix $\mathbf{U}$, where the ith column of $\mathbf{U}$ is $\mathbf{u}_i$, as shown in (A.1). Notice that multiplying a vector $\mathbf{a}$ by $\mathbf{U}$ can be written as

$$\mathbf{U}\mathbf{a} = a_1 \mathbf{u}_1 + a_2 \mathbf{u}_2 + \cdots + a_N \mathbf{u}_N$$

that is, the result of multiplication can be written as a linear combination of the columns. If $\mathbf{U}$ has linearly independent columns then its null space will contain only the zero vector and it will be invertible.

In this case, if a vector $\mathbf{x}$ is written as a linear combination of the basis vectors,

$$\mathbf{x} = \sum_{i=1}^{N} \tilde{x}_i \mathbf{u}_i = \mathbf{U}\tilde{\mathbf{x}},$$

then we can easily find $\tilde{\mathbf{x}}$:

$$\tilde{\mathbf{x}} = \mathbf{U}^{-1}\mathbf{x}. \tag{A.3}$$

This representation is unique and can be computed directly from the vector $\mathbf{x}$ if $\mathbf{U}^{-1}$ exists, but, as we will see in Section A.5, there are cases where it is useful to represent signals using sets that do not form a basis.

A.3 Inner Product, Distance and Similarity

A.3.1 Definitions

DEFINITION A.5   (INNER PRODUCT, NORM AND DISTANCE)   For $\mathbf{x}, \mathbf{y} \in \mathbb{C}^N$, the inner product is defined as

$$\langle \mathbf{x}, \mathbf{y} \rangle = \mathbf{y}^H \mathbf{x} = \sum_{v_i} x(v_i) y^*(v_i) \tag{A.4}$$

where the asterisk $*$ denotes conjugation and the superscript H denotes the conjugate transpose operation; thus $\mathbf{y}^H = (\mathbf{y}^*)^T$. The ℓ_2 norm of a vector is $\|\mathbf{x}\|$, where

$$\|\mathbf{x}\|^2 = \langle \mathbf{x}, \mathbf{x} \rangle.$$

Then the distance between two vectors $\mathbf{x}$ and $\mathbf{y}$ can be defined as

$$d(\mathbf{x}, \mathbf{y}) = \|\mathbf{x} - \mathbf{y}\|.$$

Note that if the vectors are real-valued, i.e., $\mathbf{x}, \mathbf{y} \in \mathbb{R}^N$, we can define $\langle \mathbf{x}, \mathbf{y} \rangle = \mathbf{y}^T \mathbf{x} = \mathbf{x}^T \mathbf{y} = \sum_{v_i} x(v_i) y(v_i)$. In this book, we deal mostly with real-valued signals. With an inner product we can introduce the concept of orthogonality:

DEFINITION A.6   (ORTHOGONALITY)   Vectors $\mathbf{x}$ and $\mathbf{y}$ are orthogonal if and only if

$$\langle \mathbf{x}, \mathbf{y} \rangle = 0. \tag{A.5}$$

Observe that if S_1 and S_2 are vector subspaces then $S_1 \cap S_2$ and $S_1 + S_2$ are also subspaces, where we define $S_1 + S_2 = \{\mathbf{x} + \mathbf{y} \mid \mathbf{x} \in S_1$ and $\mathbf{y} \in S_2\}$. If $S_1 \cap S_2 = \{\mathbf{0}\}$ then we have that any vector in $S_3 = S_1 + S_2$ can be expressed in a unique way as a sum of two vectors, one in S_1 and one in S_2. The subspace S_3 is then the direct sum of S_1 and S_2, denoted as $S_3 = S_1 \oplus S_2$.

A particular case of interest occurs when S_1 and S_2 are *orthogonal subspaces*, that is, for any $\mathbf{x} \in S_1$ and $\mathbf{y} \in S_2$ we have that $\langle \mathbf{x}, \mathbf{y} \rangle = 0$, the *orthogonal direct sum* of two subspaces. Lastly, we can define the *orthogonal complement* of a subspace.

DEFINITION A.7   (ORTHOGONAL COMPLEMENT)   The orthogonal complement $S^\perp$ of $S \subset \mathbb{R}^N$ is defined as

$$S^\perp = \left\{\mathbf{x} \in \mathbb{R}^N \,\middle|\, \forall \mathbf{y} \in S, \mathbf{x} \perp \mathbf{y}\right\}. \tag{A.6}$$

> PROPOSITION A.2   (PROPERTIES OF ORTHOGONAL COMPLEMENT)   (i) $S \cap S^\perp = \{\mathbf{0}\}$ and
> (ii) $S^\perp$ is a subspace of $\mathbb{R}^N$.

Proof   Item (i) clearly holds since $\langle \mathbf{x}, \mathbf{x} \rangle = 0$ if and only if $\mathbf{x} = \mathbf{0}$. As for (ii), let $\mathbf{x}_1$ and $\mathbf{x}_2$ be vectors in $S^\perp$: then using the linearity properties of the inner product we have that $\langle \mathbf{y}, \alpha \mathbf{x}_1 \rangle = \alpha \langle \mathbf{y}, \mathbf{x}_1 \rangle = 0$ and $\langle \mathbf{y}, \mathbf{x}_1 + \mathbf{x}_2 \rangle = \langle \mathbf{y}, \mathbf{x}_1 \rangle + \langle \mathbf{y}, \mathbf{x}_2 \rangle = 0$ for any $\mathbf{y} \in S$, which shows that any linear combination of $\mathbf{x}_1$ and $\mathbf{x}_2$ in $S^\perp$ belongs to $S^\perp$ and thus $S^\perp$ is a subspace of $\mathbb{R}^N$.   □

If S is a subspace then $S \oplus S^\perp = \mathbb{R}^N$. Denote by $\mathbf{U}_S$ a matrix with columns that form a basis for S; then if $\mathbf{y} \in S^\perp$ it will be orthogonal to any vector in S, and thus will be orthogonal to the basis vectors, so that

$$\mathbf{U}_S^\mathsf{T} \mathbf{y} = \mathbf{0}, \tag{A.7}$$

where $\mathbf{U}_S^\mathsf{T}$ is the transpose of $\mathbf{U}_S$.

A.3.2   Interpretation: Signal Similarity

The definitions of inner product and distance in the previous subsection allow us to define norm-preserving (i.e., orthogonal) transforms. The inner product allows us to quantify the similarity between signals. The *Cauchy–Schwarz inequality*,

$$|\langle \mathbf{x}, \mathbf{y} \rangle| \le \|\mathbf{x}\| \, \|\mathbf{y}\|, \tag{A.8}$$

allows us to define θ as the angle between the spaces spanned by two vectors $\mathbf{x}$ and $\mathbf{y}$, so that

$$\cos \theta = \frac{|\langle \mathbf{x}, \mathbf{y} \rangle|}{\|\mathbf{x}\| \, \|\mathbf{y}\|}; \tag{A.9}$$

we can observe that the absolute value of the inner product is *minimized* when the vectors are orthogonal. There is an equality in (A.8), and therefore $\cos \theta$ is *maximized* when the vectors are aligned, i.e., if $\mathbf{y} = \alpha \mathbf{x}$ for some non-zero α then we have

$$|\langle \mathbf{x}, \mathbf{y} \rangle| = \|\mathbf{x}\| \, \|\mathbf{y}\|. \tag{A.10}$$

This leads to the following important remark (see also Figure A.1).

> *Remark* A.1   (INNER PRODUCT AND SIMILARITY)   For any $\mathbf{x}, \mathbf{y}$ the absolute value (or magnitude in the complex case) of the inner product, $|\langle \mathbf{x}, \mathbf{y} \rangle|$, is a measure of the similarity between vectors:
>
> $$0 \le |\langle \mathbf{x}, \mathbf{y} \rangle| \le \|\mathbf{x}\| \, \|\mathbf{y}\|. \tag{A.11}$$
>
> Maximum similarity occurs when the two vectors are aligned, while two orthogonal vectors have minimum similarity.

The visualization in Figure A.1 can be applied in N-dimensional spaces ($N > 2$). Given two vectors, $\mathbf{x}$ and $\mathbf{y}$ in $\mathbb{R}^N$, the subspace span$(\mathbf{x}, \mathbf{y})$ has dimension 2 if the vectors are linearly independent. Thus the corresponding plane containing $\mathbf{x}$ and $\mathbf{y}$ can be represented as in Figure A.1.

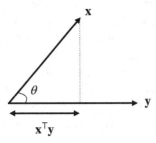

Figure A.1 Inner product example in $\mathbb{R}^2$. Note that the absolute value of the inner product is maximized when the vectors $\mathbf{x}$ and $\mathbf{y}$ are aligned.

Matrix vector multiplication and inner products Let $\mathbf{U}$ be an $N \times N$ matrix, with $\bar{\mathbf{u}}_i^\mathsf{T}$ denoting its ith row, that is,

$$
\mathbf{U} = \begin{bmatrix} \cdots & \bar{\mathbf{u}}_1^\mathsf{T} & \cdots \\ & \vdots & \\ \cdots & \bar{\mathbf{u}}_N^\mathsf{T} & \cdots \end{bmatrix}; \tag{A.12}
$$

then, for any $\mathbf{x} \in \mathbb{R}^N$ we can write

$$
\mathbf{U}\mathbf{x} = \begin{bmatrix} \bar{\mathbf{u}}_1^\mathsf{T}\mathbf{x} \\ \vdots \\ \bar{\mathbf{u}}_n^\mathsf{T}\mathbf{x} \end{bmatrix} = \sum_{k=1}^{N} x_k \mathbf{u}_k.
$$

Thus, $\mathbf{U}\mathbf{x}$ computes (i) a linear combination of the columns of $\mathbf{U}$ and and (ii) inner products with the rows of $\mathbf{U}$. The latter view of this operation can provide useful insights in the context of our discussion of signal similarity (Box A.1).

Box A.1 Matrix vector multiplication as a measure of similarity

Assume we have a matrix $\mathbf{U}$ with normalized rows, that is, $\mathbf{U}$ can be written as in (A.12) with $\bar{\mathbf{u}}_i^\mathsf{T}\bar{\mathbf{u}}_i = 1$ for all i. Then, using (A.9) and defining

$$
\cos\theta_i = \frac{|\langle \bar{\mathbf{u}}_i, \mathbf{x} \rangle|}{\|\mathbf{x}\|},
$$

the entry-wise absolute value of $\mathbf{U}\mathbf{x}$ can be written as

$$
\begin{bmatrix} |\bar{\mathbf{u}}_1^\mathsf{T}\mathbf{x}| \\ \vdots \\ |\bar{\mathbf{u}}_n^\mathsf{T}\mathbf{x}| \end{bmatrix} = \|\mathbf{x}\| \begin{bmatrix} \cos\theta_1 \\ \vdots \\ \cos\theta_N \end{bmatrix}.
$$

Therefore, finding the entry in $\mathbf{U}\mathbf{x}$ with the largest absolute value tells us which of the rows of $\mathbf{U}$ is most similar to $\mathbf{x}$.

The fact that $\mathbf{U}\mathbf{x}$ measures the similarity to the rows of $\mathbf{U}$ will be important in the context of signal representations (Section A.4).

Norm, orthogonality and graph structure   When $\mathbf{x}$ and $\mathbf{y}$ are signals on a graph G, Definition A.5 and Definition A.6 do not take into consideration the structure of G. That is, we can compute a distance $d(\mathbf{x}, \mathbf{y}) = \|\mathbf{x} - \mathbf{y}\|$ without using any information about the underlying graph. Similarly, the norm of a signal $\mathbf{x}$ is the same whether it is associated with G_1 or G_2, two different graphs with same number of nodes. Thus, when designing an orthogonal basis for signals on a specific graph, orthogonality itself does not depend on the graph structure. Section 3.1 describes how sets of basis vectors can be derived by taking into account the graph topology, while in Section 3.4.5 we explore alternative norm definitions.

A.4   Orthogonal and Biorthogonal Bases

A.4.1   Biorthogonal Bases

Let $\mathbf{U}$ be an $N \times N$ invertible matrix, and define

$$
\mathbf{U}^{-1} = \begin{bmatrix} \tilde{\mathbf{u}}_1^{\mathsf{T}} \\ \vdots \\ \tilde{\mathbf{u}}_n^{\mathsf{T}} \end{bmatrix} = \tilde{\mathbf{U}}^{\mathsf{T}},
$$

where $\tilde{\mathbf{u}}_i^{\mathsf{T}}$ is the ith row of $\mathbf{U}^{-1}$ and therefore the ith column of $\tilde{\mathbf{U}}$. This leads to several observations. First, by definition of the inverse we have that

$$
\mathbf{U}^{-1}\mathbf{U} = \tilde{\mathbf{U}}^{\mathsf{T}}\mathbf{U} = \mathbf{I}. \tag{A.13}
$$

Each entry of $\tilde{\mathbf{U}}^{\mathsf{T}}\mathbf{U}$ is equal to an inner product $\tilde{\mathbf{u}}_i^{\mathsf{T}}\mathbf{u}_j$, involving a row from $\mathbf{U}^{-1}$ and a column from $\mathbf{U}$. Thus, from (A.13) we have that

$$
\tilde{\mathbf{u}}_i^{\mathsf{T}}\mathbf{u}_j = \delta(i - j).
$$

This can be interpreted by saying that the rows of $\mathbf{U}^{-1}$ are orthogonal to the columns of $\mathbf{U}$. Also, from our definition $\tilde{\mathbf{U}} = (\mathbf{U}^{-1})^{\mathsf{T}}$, and we note that

$$
\mathbf{I} = (\mathbf{U}\mathbf{U}^{-1})^{\mathsf{T}} = (\mathbf{U}^{-1})^{\mathsf{T}}\mathbf{U}^{\mathsf{T}} = \tilde{\mathbf{U}}\mathbf{U}^{\mathsf{T}}.
$$

In summary we have two matrices $\mathbf{U}$ and $\tilde{\mathbf{U}}$ each with linearly independent columns, such that the columns of $\mathbf{U}$ are orthogonal to those of $\tilde{\mathbf{U}}$. This leads to the definition of a biorthogonal basis.

DEFINITION A.8   (BIORTHOGONAL BASES)   The set of vectors $\mathbf{u}_1, \ldots, \mathbf{u}_N$ forms a **biorthogonal basis** (or simply a **basis**) for $\mathbb{R}^N$ if there exists a dual basis $\tilde{\mathbf{u}}_1, \ldots, \tilde{\mathbf{u}}_N$ such that

$$
\langle \tilde{\mathbf{u}}_i, \mathbf{u}_j \rangle = \delta(i - j).
$$

The columns of $\mathbf{U}$ are the basis vectors, while the rows of $\mathbf{U}^{-1}$ (the columns of $\tilde{\mathbf{U}}$) are the dual basis vectors. Note that a biorthogonal basis is simply a basis, but the dual basis concept is useful for infinite-dimensional spaces, where the dual is not found via a matrix inversion. Note also that the roles of a basis and its dual are interchangeable. That is, the columns of $\mathbf{U}$ are the dual basis for the columns of $\tilde{\mathbf{U}}$. We can write the representation for a signal $\mathbf{x}$ as

$$\mathbf{x} = \sum_{i=1}^{N} (\tilde{\mathbf{u}}_i^\mathsf{T} \mathbf{x}) \mathbf{u}_i.$$

A.4.2 Orthogonal Bases

If the vectors in the basis are orthogonal to each other, i.e., $\mathbf{u}_i^\mathsf{T} \mathbf{u}_j = 0$ if $i \neq j$, then we have an **orthogonal basis**. An **orthonormal basis** is an orthogonal basis where the vectors have norm equal to 1, that is, $\mathbf{u}_i^\mathsf{T} \mathbf{u}_i = 1$. Thus, orthogonal bases are biorthogonal bases such that $\mathbf{U}^{-1} = \mathbf{U}^\mathsf{T}$. In the particular case where the dual basis is the same as the original basis, i.e., $\tilde{\mathbf{u}}_i = \mathbf{u}_i$ for all i, then we have an orthogonal basis, or equivalently

$$\mathbf{U}^{-1} = \mathbf{U}^\mathsf{T},$$

and therefore

$$\mathbf{x} = \sum_{i=1}^{N} (\mathbf{u}_i^\mathsf{T} \mathbf{x}) \mathbf{u}_i = \sum_{i=1}^{N} \alpha_i \mathbf{u}_i. \tag{A.14}$$

Therefore, in the orthogonal case we can measure the similarity between an input vector and the basis functions $\mathbf{u}_i^\mathsf{T} \mathbf{x}$, so that, from Box A.1, the weights corresponding to each basis vector $\mathbf{u}_i$ are based on its similarity to $\mathbf{x}$.

One important property of orthogonal representations is norm preservation, given by the **Parseval relation**, where for $\mathbf{x}$ written as in (A.14), we have

$$\sum_{i=1}^{N} |\alpha_i|^2 = \sum_{i=1}^{N} |x_i|^2, \tag{A.15}$$

or, letting $\tilde{\mathbf{x}} = \mathbf{U}\mathbf{x}$,

$$\|\tilde{\mathbf{x}}\|^2 = \mathbf{x}^\mathsf{T} \mathbf{U}^\mathsf{T} \mathbf{U} \mathbf{x} = \|\mathbf{x}\|^2.$$

In contrast, if $\mathbf{U}$ is invertible but not orthogonal, we would have just

$$\|\tilde{\mathbf{x}}\|^2 = \mathbf{x}^\mathsf{T} \mathbf{U}^\mathsf{T} \mathbf{U} \mathbf{x}.$$

Since $\mathbf{U}$ is invertible and $\mathbf{U}^\mathsf{T}\mathbf{U}$ is symmetric, we have that $\mathbf{U}^\mathsf{T}\mathbf{U}$ will have real and strictly positive eigenvalues. The maximum eigenvalue (corresponding to the matrix norm) will define the maximum gain in the norm of the "transformed" signal $\tilde{\mathbf{x}}$. In the orthogonal case $\mathbf{U}^\mathsf{T}\mathbf{U} = \mathbf{I}$, so that all the eigenvalues are equal to 1 and thus the norm does not change.

A.4.3 Least Squares Approximation

We now develop the basic ideas behind the least squares approximation, which is needed in various parts of this book (in particular in Chapter 4). Assume we have a subspace

S of dimension M for which a basis is available. Denote by $\mathbf{U}_S$ the $N \times M$ matrix with column vectors forming a basis for S and let $\mathbf{x}$ be an arbitrary vector in $\mathbb{R}^N$. Since S is a subspace ($M < N$), in general $\mathbf{x}$ does not belong to S.

Our goal is to find the vector $\mathbf{x}_S \in S$ that is closest to S. Mathematically, we are looking for $\mathbf{x}_S \in S$ such that $\|\mathbf{x} - \mathbf{x}_S\|$ is minimized. First, note that if $\mathbf{x} \in S$ then the solution is obviously $\mathbf{x}_S = \mathbf{x}$, with zero error. Next, assume that $\mathbf{x}$ is not in S; then we can state the following important result.

PROPOSITION A.3   (ORTHOGONALITY OF ERROR) If $\mathbf{x}_S \in S$ is the closest vector to $\mathbf{x}$ in S, then $\mathbf{x} - \mathbf{x}_S$ is orthogonal to any vector in S, that is, $\forall \mathbf{y} \in S$, $\mathbf{y}^\mathsf{T}(\mathbf{x} - \mathbf{x}_S) = 0$. Thus, orthogonal projection onto a subspace minimizes the mean square approximation error.

Proof   This result can be proven by contradiction. Assume that $\mathbf{x}_S$ is such that $\mathbf{y}^\mathsf{T}(\mathbf{x} - \mathbf{x}_S) = 0$ for any $\mathbf{y} \in S$. Then assume that there is a second vector $\mathbf{x}'_S \in S$ that is closer to $\mathbf{x}$ than $\mathbf{x}_S$. Because $\mathbf{x}_S$ and $\mathbf{x}'_S$ are both in S it follows that their difference $\mathbf{d} = \mathbf{x}_S - \mathbf{x}'_S$ is also in S, since S is a vector subspace. Thus, we can write

$$\|\mathbf{x} - \mathbf{x}'_S\|^2 = \|(\mathbf{x} - \mathbf{x}_S) + (\mathbf{x}_S - \mathbf{x}'_S)\|^2 = \|(\mathbf{x} - \mathbf{x}_S)\|^2 + \|\mathbf{d}\|^2 \geq \|(\mathbf{x} - \mathbf{x}_S)\|^2,$$

where the second equality comes from the fact that $\mathbf{x} - \mathbf{x}_S$ is orthogonal to all vectors in S (including $\mathbf{d}$) and from the Pythagorean theorem. Thus the distance $\|\mathbf{x} - \mathbf{x}'_S\|$ can only be minimal if $\|\mathbf{d}\| = 0$, implying that $\mathbf{d} = 0$ and proving that $\mathbf{x}_S$ is indeed the optimal solution.   □

Next we can define a linear operator to obtain the orthogonal projection of any vector $\mathbf{x}$ onto a subspace S. We denote this operator as $P_S(\mathbf{x})$.

PROPOSITION A.4   (ORTHOGONAL PROJECTION)   With $\mathbf{U}_S$ a basis for a subspace S, the orthogonal projection of $\mathbf{x}$ onto S is

$$P_S(\mathbf{x}) = \mathbf{U}_S (\mathbf{U}_S^\mathsf{T} \mathbf{U}_S)^{-1} \mathbf{U}_S^\mathsf{T} \mathbf{x}. \tag{A.16}$$

Proof   Given that $\mathbf{U}_S$ forms a basis, any vector in S can be written as $\mathbf{y} = \mathbf{U}_S \mathbf{a}$ where $\mathbf{a} \in \mathbb{R}^M$. Also, a vector $\mathbf{z}$ orthogonal to S will be such that $\mathbf{U}_S^\mathsf{T} \mathbf{z} = \mathbf{0}$, as seen in (A.7). Then from Proposition A.3 the orthogonal projection is such that $\mathbf{U}_S^\mathsf{T}(\mathbf{x} - P(\mathbf{x})) = \mathbf{0}$ and, using the fact that this projection is in S, we have

$$\mathbf{U}_S^\mathsf{T}(\mathbf{x} - \mathbf{U}_S \mathbf{a}) = \mathbf{0}$$

from which we obtain that

$$\mathbf{a} = (\mathbf{U}_S^\mathsf{T} \mathbf{U}_S)^{-1} \mathbf{U}^\mathsf{T} \mathbf{x}$$

and the proposition follows. Note that $\mathbf{U}_S$ is a basis for S, so its columns are linearly independent and thus $\mathbf{U}_S^\mathsf{T} \mathbf{U}_S$ is $M \times M$, has rank M and is invertible.   □

Observe that if $\mathbf{x} \in S^\perp$ then $P_S(\mathbf{x}) = \mathbf{0}$, while for $\mathbf{x} \in S$ we have $P(\mathbf{x}) = \mathbf{x}$. Also, for any $\mathbf{x}$, we can see that $P_S(P_S(\mathbf{x})) = P_S(\mathbf{x})$. Also note that if the basis for S, $\mathbf{V}_S$, is orthonormal then (A.16) can be written simply as

$$P_S(\mathbf{x}) = \mathbf{V}_S \mathbf{V}_S^\mathsf{T} \mathbf{x}, \tag{A.17}$$

since $\mathbf{V}_S^\mathsf{T} \mathbf{V}_S = \mathbf{I}$ given that the columns of $\mathbf{V}_S$ are orthogonal.

A.4.4 Gram–Schmidt Orthogonalization

The results in the previous section can be used to obtain an orthonormal basis $\mathbf{V}_S$ for a space, S, for which we already have a biorthogonal basis $\mathbf{U}_S$. Assume first that we have a subspace of S, S_k, of dimension k, for which $\mathbf{V}_k$ is a matrix whose columns form an orthonormal basis for $\mathbf{V}_k$. Then we have $\mathbf{V}_k^\mathsf{T} \mathbf{V}_k = \mathbf{I}_k$ and, from (A.17), the orthogonal projection of any $\mathbf{x} \in S$ onto S_k is

$$P_{S_k}(\mathbf{x}) = \mathbf{V}_k \mathbf{V}_k^\mathsf{T} \mathbf{x} = \sum_{i=1}^{k} (\mathbf{v}_i^\mathsf{T} \mathbf{x}) \mathbf{v}_i.$$

From Proposition A.3 the approximation error is orthogonal, that is,

$$\mathbf{V}_k^\mathsf{T} (\mathbf{x} - \mathbf{V}_k \mathbf{V}_k^\mathsf{T} \mathbf{x}) = \mathbf{0}.$$

This orthogonality property can be used to iteratively build an orthogonal basis for S using the Gram–Schmidt orthogonalization procedure.

ALGORITHM A.1   (GRAM–SCHMIDT ORTHOGONALIZATION)   Given $\mathbf{U}_S$, whose columns form a basis for S, find $\mathbf{V}_S$, an orthogonal basis for S.

Step 1: Let $\mathbf{v}_1 = \mathbf{u}_1 / \|\mathbf{u}_1\|$, where we simply apply a normalization such that $\|\mathbf{v}_1\| = 1$.

Step i: We have selected $\mathbf{v}_1, \ldots, \mathbf{v}_{i-1}$ using $\mathbf{u}_1, \ldots, \mathbf{u}_{i-1}$, forming $\mathbf{V}_{i-1}$. In general $\mathbf{u}_i$ is not orthogonal to the span of $\mathbf{V}_{i-1}$ but we know that projecting onto a subspace leads to an error that is orthogonal to the subspace. Thus we choose

$$\mathbf{v}_i = \frac{\mathbf{u}_i - \mathbf{V}_{i-1} \mathbf{V}_{i-1}^\mathsf{T} \mathbf{u}_i}{\|\mathbf{u}_i - \mathbf{V}_{i-1} \mathbf{V}_{i-1}^\mathsf{T} \mathbf{u}_i\|},$$

where the numerator is the error vector and the denominator is a normalization term. The error cannot be zero since $\mathbf{U}_k$ has k linear independent columns, and so $\mathbf{u}_i$ is linearly independent of the columns of $\mathbf{V}_{i-1}$.

Note that Algorithm A.1 can be applied to any ordering of the columns of $\mathbf{U}_S$, each resulting (in general) in a different orthogonal basis.

A.5   Dictionaries and Overcomplete Representations

We now describe overcomplete representations, based on a set of vectors, or **atoms**, forming a **dictionary**, that spans $\mathbb{R}^N$ but whose members are no longer linearly independent. In this case, any vector can be represented in an infinite number of different

ways. While this lack of uniqueness appears to be inefficient, having multiple representations available makes it possible to choose the one with the most desirable properties,
at the cost of increased complexity due to searching for the best representation.

A.5.1 Signal Representations Are Not Unique

An overcomplete dictionary is a set of M vectors (atoms) $\mathbf{v}_i \in \mathbb{R}^N$, which can be written
as a rectangular $N \times M$ matrix $\mathbf{V}$, where $M > N$ and the column vectors are linearly
dependent. Then there exists a non-zero vector $\mathbf{a} \in \mathbb{R}^M$ such that

$$\mathbf{Va} = \mathbf{0}.$$

The rank of $\mathbf{V}$ has to be N so that it is possible to represent any vector in $\mathbb{R}^N$ using $\mathbf{V}$.
Then for any given $\mathbf{x} \in \mathbb{R}^N$ we can find an $\tilde{\mathbf{x}} \in \mathbb{R}^M$ such that

$$\mathbf{x} = \mathbf{V}\tilde{\mathbf{x}}.$$

We can also represent $\mathbf{x}$ using $\tilde{\mathbf{x}}' = \tilde{\mathbf{x}} + \alpha\mathbf{a}$ since $\mathbf{Va} = \mathbf{0}$ and therefore

$$\mathbf{V}\tilde{\mathbf{x}}' = \mathbf{V}(\tilde{\mathbf{x}} + \alpha\mathbf{a}) = \mathbf{V}\tilde{\mathbf{x}} = \mathbf{x}.$$

Since the number of representations is infinite, the key question is how to select the
representation that is best for the specific application.

A.5.2 Minimum ℓ_2 Norm Representation

Since rank$(\mathbf{V}) = N$, the $N \times N$ matrix $\mathbf{VV}^\mathsf{T}$ also has rank N and is invertible, and we can
write

$$\mathbf{x} = (\mathbf{VV}^\mathsf{T})(\mathbf{VV}^\mathsf{T})^{-1}\mathbf{x}.$$

Therefore one possible choice for $\tilde{\mathbf{x}}$ is

$$\tilde{\mathbf{x}} = \mathbf{V}^\mathsf{T}(\mathbf{VV}^\mathsf{T})^{-1}\mathbf{x}, \tag{A.18}$$

where we define

$$\tilde{\mathbf{V}}^\mathsf{T} = \mathbf{V}^\mathsf{T}(\mathbf{VV}^\mathsf{T})^{-1}$$

to be an $M \times N$ "tall" matrix, in which the ith row vector can be written as $\tilde{\mathbf{v}}_i^\mathsf{T}$.

Frames The dictionary $\mathbf{V}$ is a **frame**, $\tilde{\mathbf{V}}$ is the **dual frame** and we can write the signal
representation (keeping in mind that it is not unique) for any $\mathbf{x}$ as

$$\mathbf{x} = \begin{bmatrix} \vdots & \vdots & & \vdots \\ \mathbf{v}_1 & \mathbf{v}_2 & \ldots\ldots & \mathbf{v}_M \\ \vdots & \vdots & & \vdots \end{bmatrix} \begin{bmatrix} \tilde{\mathbf{v}}_1^\mathsf{T} \\ \tilde{\mathbf{v}}_2^\mathsf{T} \\ \vdots \\ \vdots \\ \tilde{\mathbf{v}}_M^\mathsf{T} \end{bmatrix} \mathbf{x},$$

so that, similarly to the biorthogonal basis representation, we have

$$\mathbf{x} = \sum_{i=1}^{M} (\tilde{\mathbf{v}}_i^\mathsf{T} \mathbf{x}) \mathbf{v}_i.$$

Note that the norm of the new representation $\tilde{\mathbf{x}}$ is not the same as the norm of the original vector $\mathbf{x}$:

$$\tilde{\mathbf{x}}^\mathsf{T} \tilde{\mathbf{x}} = \mathbf{x}^\mathsf{T} ((\mathbf{V}\mathbf{V}^\mathsf{T})^{-1})^\mathsf{T} \mathbf{V}\mathbf{V}^\mathsf{T} (\mathbf{V}\mathbf{V}^\mathsf{T})^{-1} \mathbf{x} = \mathbf{x}^\mathsf{T} ((\mathbf{V}\mathbf{V}^\mathsf{T})^{-1})^\mathsf{T} \mathbf{x} = \mathbf{x}^\mathsf{T} (\mathbf{V}\mathbf{V}^\mathsf{T})^{-1} \mathbf{x}. \qquad (A.19)$$

While this solution is straightforward to compute it is not likely to be the most useful. In many applications, where the goal is to represent efficiently a vector $\mathbf{x}$, it is preferable to select a representation that has a small number of non-zero coefficients, i.e., it is **sparse**.

Tight frames   To illustrate this, consider a particular case where the rows of $\mathbf{V}$ are orthogonal and have the same norm α, that is,

$$\mathbf{V}\mathbf{V}^\mathsf{T} = \alpha \mathbf{I}_N \quad \text{and} \quad (\mathbf{V}\mathbf{V}^\mathsf{T})^{-1} = \frac{1}{\alpha} \mathbf{I}_N.$$

This is called a **tight frame** representation, allowing us to rewrite (A.18) as

$$\tilde{\mathbf{x}} = \mathbf{V}^\mathsf{T} (\mathbf{V}\mathbf{V}^\mathsf{T})^{-1} \mathbf{x} = \frac{1}{\alpha} \mathbf{V}^\mathsf{T} \mathbf{x} \qquad (A.20)$$

which shows the analogy with the orthogonal representation, since the coefficients for the representation are obtained by projecting onto the transpose $\mathbf{V}^\mathsf{T}$ of the representation vectors $\mathbf{V}$. In this case (A.19) can be rewritten as

$$\|\tilde{\mathbf{x}}\|^2 = \tilde{\mathbf{x}}^\mathsf{T} \tilde{\mathbf{x}} = \mathbf{x}^\mathsf{T} (\mathbf{V}\mathbf{V}^\mathsf{T})^{-1} \mathbf{x} = \frac{1}{\alpha} \|\mathbf{x}\|^2. \qquad (A.21)$$

Example A.1   A tight frame for $\mathbb{R}^2$ can be constructed using two orthogonal basis vectors. Let $\mathbf{R}_{\theta_1}$ and $\mathbf{R}_{\theta_2}$ be 2×2 rotation matrices for two angles $\theta_1 \neq \theta_2$. Then define

$$\mathbf{V} = [\mathbf{R}_{\theta_1} \, \mathbf{R}_{\theta_2}] \quad \text{and} \quad \mathbf{V}^\mathsf{T} = \begin{bmatrix} \mathbf{R}_{\theta_1}^\mathsf{T} \\ \mathbf{R}_{\theta_2}^\mathsf{T} \end{bmatrix}$$

and clearly $\mathbf{V}\mathbf{V}^\mathsf{T} = \alpha \mathbf{I}_N$.

In this book there are two different scenarios where overcomplete representations play a role (Chapter 5): (i) when designing a basis with some specific properties is not possible and (ii) when critically sampled basis representations do exist but more compact representations are preferred (e.g., sparse representations).

A.5.3   Matching Pursuits and Orthogonal Matching Pursuits

Sparse representations   For a given overcomplete set there is an infinite number of *exact* representations for any signal $\mathbf{x}$. We are also interested in *approximate* representations, where we are given $\mathbf{x}$ and an overcomplete set $\mathbf{V}$, and we wish to find $\tilde{\mathbf{x}}$ such that the ℓ_2 norm of the approximation error $\|\mathbf{x} - \mathbf{V}\tilde{\mathbf{x}}\|_2$ is small.

A representation for an input $\mathbf{x}$ is sparse if the number of non-zero values in $\tilde{\mathbf{x}}$ is small, e.g., less than N. Formally, finding such a representation would mean looking for (among all the possible representations of $\mathbf{x}$) one that minimizes the number of non-zero values:

$$\tilde{\mathbf{x}}_0 = \min_{\tilde{\mathbf{x}} \text{ s.t.} \|\mathbf{x}-\mathbf{V}\tilde{\mathbf{x}}\|_2 \le \epsilon} \|\tilde{\mathbf{x}}\|_0 \tag{A.22}$$

where $\|\tilde{\mathbf{x}}\|_0$ is the ℓ_0 norm of $\tilde{\mathbf{x}}$. This is a combinatorial problem for which a typical alternative is to find the best approximation in terms of the ℓ_1 norm:

$$\tilde{\mathbf{x}}_1 = \min_{\tilde{\mathbf{x}} \text{ s.t.} \|\mathbf{x}-\mathbf{V}\tilde{\mathbf{x}}\|_2 \le \epsilon} \|\tilde{\mathbf{x}}\|_1. \tag{A.23}$$

Matching pursuits   Matching pursuits (MP) consists of a series of iterations, where one column of $\mathbf{V}$ is selected at each step, until a stopping criterion is met. This is a *greedy* procedure: once a column is chosen it cannot be removed in subsequent steps. The MP procedure computes the inner product between the input vector (and then successive approximation residues) and each column of $\mathbf{V}$ and selects the one that has the highest absolute value.

- Step 1: Compute $\alpha_k = \mathbf{v}_k^\mathsf{T}\mathbf{x}/\|\mathbf{v}_k\|$ for all k and choose $k_1 = \arg\max_k |\alpha_k|$. The maximum is chosen because it corresponds to maximum similarity (Remark A.1). Denote $\hat{\mathbf{x}}_1 = \alpha_{k_1}\mathbf{v}_{k_1}/\|\mathbf{v}_{k_1}\|$ and compute $\mathbf{e}_1 = \mathbf{x} - \hat{\mathbf{x}}_1$. Note that this is equivalent to selecting the orthogonal projection of $\mathbf{x}$ onto the subspace spanned by $\mathbf{v}_{k-1}$ and thus the approximation is optimal (the error is minimized).
- Step i: At this point we have $\mathbf{e}_{i-1}$, the error from the previous iteration. Then we find $\alpha_k = \mathbf{v}_k^\mathsf{T}\mathbf{e}_{i-1}/\|\mathbf{v}_k\|$ for all $k \ne k_1, k_2, \ldots, k_{i-1}$ and select $k_i = \arg\max_k |\alpha_k|$; then $\hat{\mathbf{x}}_i = \hat{\mathbf{x}}_{i-1} + \alpha_{k_i}\mathbf{v}_{k_i}/\|\mathbf{v}_{k_i}\|$ and $\mathbf{e}_i = \mathbf{x} - \hat{\mathbf{x}}_i$. Note that the selection of columns of $\mathbf{V}$ is greedy (see Section 4.4.3) and α_{k_i} is fixed when $\mathbf{v}_{k-i}$ is selected. Also note that this is no longer an optimal approximation since the columns of $\mathbf{V}$ are not orthogonal.
- The iteration concludes at iteration J if a desired number of columns has been chosen or if the norm of the error is small enough.

The vector $\tilde{\mathbf{x}}$ obtained from matching pursuits will be such that $\tilde{x}(k) = \alpha_k$ for $k = k_1, k_2, \ldots, k_J$ and $\tilde{x}(k) = 0$ otherwise. The representation is sparse in the sense that only J out of M vectors have a non-zero weight. Matching pursuits is suboptimal not only because of the greedy nature of the procedure, but also because the representation for a given set of vectors, $\mathbf{v}_{k_1}, \mathbf{v}_{k_2}, \ldots, \mathbf{v}_{k_i}$, is not optimal. In particular notice that the α_k are chosen independently along each direction $\mathbf{v}_k$. However, this is only an optimal approximation if the vectors selected, $\mathbf{v}_{k_1}, \mathbf{v}_{k_2}, \ldots, \mathbf{v}_{k_i}$, are orthogonal (the successive approximation property).

Orthogonal matching pursuits Orthogonal matching pursuits (OMP) follows a similar greedy vector selection procedure, but recomputes the projection coefficients at each iteration so as to achieve the best approximation to the input.

To do so, at step i assume that $\mathbf{v}_{k_1}, \mathbf{v}_{k_2}, \ldots, \mathbf{v}_{k_i}$ have been selected. Denote by $\mathbf{V}_i$ the matrix whose i columns contain the selected vectors. Then we would like to find $\mathbf{a}_i$ such that the error in approximating $\mathbf{x}$ by $\mathbf{V}_i \mathbf{a}_i$ is minimized. As discussed earlier this can be accomplished by performing an orthogonal projection onto the columns of $\mathbf{V}_i$. The columns are not orthogonal in general (but can be shown to be linearly independent, since they provide an increasingly better approximation to $\mathbf{x}$); we can find the approximation by solving a least squares problem. Using Proposition A.4 we select $\mathbf{a}_i$ as

$$\mathbf{a}_i = (\mathbf{V}_i^\mathsf{T} \mathbf{V}_i)^{-1} \mathbf{V}_i^\mathsf{T} \mathbf{x},$$

which guarantees that the approximation error is minimized for the given columns.

Further Reading

This appendix has introduced signal representations from an algebraic perspective, as for example in [67] or [17]. It is worth noting that the evolution of our understanding of signal representations has been closely related to the cost and availability of computing resources. Over the years, Moore's law has had the effect of lowering the cost of computation and, as a consequence, signal representations that require additional computation have been more widely adopted. Initially, classical signal representations with closed-form mathematical definitions (for example, the discrete-time Fourier transform) and fast implementations were preferred. Wavelets and representations based on filterbanks could be seen as examples of transformations where the elementary basis had to be computed. In more recent years, overcomplete sets have become popular tools for signal representations. With an overcomplete set, an infinite number of representations is possible, and the task of finding the best one under a specific criterion requires computation. Finally, the choice of elementary basis itself can be derived from data, which can be achieved using dictionary learning methods.

Problems

A.1 Let S_1 and S_2 be two subspaces of $\mathbb{R}^N$.

a. Prove that $S_1 \cap S_2$, the intersection of S_1 and S_2, is a subspace of $\mathbb{R}^N$.

b. Assume $S_1 \neq \{\mathbf{0}\}$, $S_2 \neq \{\mathbf{0}\}$ and $S_1 \cap S_2 = \{\mathbf{0}\}$; then prove that $S_1 \cup S_2$ is not a subspace of $\mathbb{R}^N$.

A.2 Define the following time series (for $k \geq 1$):

$$x(1) = 1 \quad \text{and} \quad x(k+1) = x(k) + 1, \quad k > 1.$$

Use this to define a set, S, of vectors in $\mathbb{R}^2$ indexed by a non-negative integer k:

$$S = \{\mathbf{x}_k\}_{k \in \mathbb{Z}, k \geq 1},$$

where

$$\mathbf{x}_k = [x(k) \quad x(k+1)]^\mathsf{T}, \quad k \geq 1, \quad \text{and}$$

$$\mathbf{x}_0 = [-1, 1]^\mathsf{T}.$$

a. Plot in the 2D plane $\mathbf{x}_0$, $\mathbf{x}_1$, $\mathbf{x}_2$ and $\mathbf{x}_3$.

b. Prove that for all $k_1, k_2 \geq 0$, $k_1 \neq k_2$, we have that $\mathbf{x}_{k_1}, \mathbf{x}_{k_2}$ form a basis for $\mathbb{R}^2$.

c. Design an orthogonal basis for $\mathbb{R}^2$, taking as a starting point a pair of vectors $\mathbf{x}_0$ and $\mathbf{x}_k$, with $k \neq 0$.

d. We wish to create a dictionary $\mathcal{S}_D \subset \mathcal{S}$ by selecting vectors in $\mathcal{S}$. Our goal in designing such a dictionary is to minimize the maximum coherence, i.e.,

$$\max_{k_1, k_2 \in \mathcal{S}_D} \frac{1}{\|\mathbf{x}_{k_1}\| \, \|\mathbf{x}_{k_2}\|} \langle \mathbf{x}_{k_1}, \mathbf{x}_{k_2} \rangle$$

where $\|\mathbf{x}_k\|$ is the norm of $\mathbf{x}_k$ and $\langle \cdot, \cdot \rangle$ denotes the inner product. Prove that the minimum coherence dictionary of size d will have the form

$$\mathcal{S}_D = \{\mathbf{x}_0\} \cup \{\mathbf{x}_{k_1}, \mathbf{x}_{k_2}, \ldots, \mathbf{x}_{k_{d-1}}\},$$

that is, $\mathbf{x}_0$ should always be part of such a minimum coherence dictionary.

A.3 Define $\mathbf{T}^\mathsf{T} = [\mathbf{a}_1, \mathbf{a}_2, \mathbf{a}_3]$, where the $\mathbf{a}_i$ are column vectors of dimension 3, such that $\mathbf{T}$ is an orthogonal matrix. Let $\mathbf{X}$ and $\mathbf{Y}$ be 3×3 matrices such that

$$\mathbf{Y} = \mathbf{T}\mathbf{X}\mathbf{T}^\mathsf{T}.$$

This operation can be interpreted as a projection of $\mathbf{X}$ onto an orthogonal basis set with N basis vectors. What is N? Write the basis vectors as a function of the $\mathbf{a}_i$.

A.4 Let $\mathbf{A}$ be a real-valued $N \times N$ orthogonal matrix. Define $\mathbf{D}$, a $2 \times N$ matrix where each of the two rows is taken from $\mathbf{A}$ (any two different rows from $\mathbf{A}$ can be selected). We view $\mathbf{D}$ as a dictionary for $\mathbb{R}^2$ where each column vector is a possible basis vector and we denote each column vector by $\mathbf{u}_i$, for $i = 0, \ldots, N-1$, $\mathbf{u}_i \in \mathbb{R}^2$. Denote by $\|\cdot\|_2$ the l_2 norm of a vector.

a. Let $\mathbf{x} \in \mathbb{R}^2$ be an arbitrary vector. Prove that, for any $\mathbf{x}$, $\mathbf{y} = \mathbf{D}^\mathsf{T}\mathbf{x}$ provides a representation of $\mathbf{x}$ in terms of the atoms of the dictionary, i.e., we can reconstruct $\mathbf{x}$ as $\mathbf{x} = \mathbf{D}\mathbf{y}$.

b. Prove that the representation in part **a** is not unique (i.e., there are other $\mathbf{y}'$ from which $\mathbf{x}$ can be reconstructed, so that $\mathbf{x} = \mathbf{D}\mathbf{y}'$).

c. Express $\|\mathbf{y}\|_2$ for $\mathbf{y}$ defined in part **a** as a function of $\|\mathbf{x}\|_2$. What kind of overcomplete representation is provided by $\mathbf{D}$?

A.5 Let $\mathbf{A}$ and $\mathbf{B}$ be two different real-valued $N \times N$ orthogonal matrices. Let $\mathbf{x} \in \mathbb{R}^N$ be an arbitrary vector and let $\mathbf{y}_a = \mathbf{A}^\mathsf{T}\mathbf{x}$ and $\mathbf{y}_b = \mathbf{B}^\mathsf{T}\mathbf{x}$. Define a new matrix $\mathbf{T}$ with $2N$ columns of size N, containing all the columns of $\mathbf{A}$ and $\mathbf{B}$. Denote by $\|\cdot\|_2$ the l_2 norm of a vector.

a. Compute $\|\mathbf{y}_a\|_2^2 - \|\mathbf{y}_b\|_2^2$.

b. Prove that $\mathbf{T}$ is a tight frame and show how $\mathbf{x}$ can be recovered from $\mathbf{T}^T\mathbf{x}$.

c. Let $\mathbf{x}$ denote a vector having an exact sparse representation with l_0 norm of 2 in the dictionary defined by the columns of $\mathbf{T}$ (assume this is the optimal l_0 solution). Prove that $\|\mathbf{y}_a\|_0 + \|\mathbf{y}_b\|_0 \geq 4$, where $\| \cdot \|_0$ represents the l_0 norm.

A.6 Let E be the vector space of polynomials of a variable x with degree up to 3. One vector (polynomial) in this space is defined as $p(x) = a_0 + a_1x + a_2x^2 + a_3x^3$, where all $a_i \in \mathbb{R}$. The sum of two polynomials and multiplication by a scalar are defined in the usual way, i.e., $p(x) + q(x)$ and $\alpha p(x)$.

a. Define $p_0(x) = 1$, $p_1(x) = x$, $p_2(x) = x^2$ and $p_3(x) = x^3$.
Prove that $\{p_0(x),\ p_1(x),\ p_2(x),\ p_3(x)\}$ forms a basis for E.

b. Are the following sets of vectors S_i subspaces of E? Justify your answers with a proof or a counterexample.
- $S_1 = \left\{q(x) \in E \mid q(x) = a_2x^2 + a_3x^3,\ a_2, a_3 \in \mathbb{R}\right\}$
- $S_2 = \{q(x) \in E \mid q(x) = 1 + a_1x,\ a_1 \in \mathbb{R}\}$
- $S_3 = \{q(x) \in E \mid q(x) = a_0q_0(x) + a_1q_1(x),\ a_0, a_1 \in \mathbb{R}\}$, where $q_0(x)$ and $q_1(x)$, two polynomials in E, are given.

c. Define $q_0(x) = 1$, and then for $i = 1, 2, 3$, define $q_i(x) = \alpha xq_{i-1}(x) + \beta$. For which values of α and β does $\{q_0(x),\ q_1(x),\ q_2(x),\ q_3(x)\}$ form a biorthogonal basis for E? For $\beta = 0$, find the dual basis.

A.7 Consider the space of discrete-time finite-energy sequences $l_2(\mathbb{Z})$. Let $\delta(n)$ denote the impulse function and, for a given $\theta \in [0, \pi]$, define the set of vectors $S_\theta = \{\phi_{2k,\theta}(n), \phi_{2k+1,\theta}(n), k \in \mathbb{Z}\}$ where

$$\phi_{2k,\theta}(n) = \alpha\delta(n - 2k) + \beta\delta(n - (2k + 1)),$$

$$\phi_{2k+1,\theta}(n) = -\beta\delta(n - 2k) + \alpha\delta(n - (2k + 1)),$$

with $\alpha = \cos(\theta)$, $\beta = \sin(\theta)$.

a. Prove that S_θ is an orthonormal basis for $l_2(\mathbb{Z})$.

b. Define an overcomplete set $S' = S_{\theta_1} \cup S_{\theta_2}$, $\theta_1 \neq \theta_2$. Prove that S' is a tight frame and provide the corresponding frame bounds.

c. For S' as defined above choose θ_1, θ_2 so that the resulting dictionary is best in terms of minimizing the maximum coherence. Justify your answer.

A.8 Denote by $\mathbf{U}$ an $N \times M$ matrix of M column vectors in $\mathbb{R}^N$ representing a dictionary, where $M > N$. Assume all the columns are normalized: $\|\mathbf{u}_i\|_2 = 1$, $\forall i = 1, \ldots, M$. Assume we have a vector $\mathbf{x} \in \mathbb{R}^N$ that we wish to approximate using the dictionary $\mathbf{U}$. Let the approximation obtained via the matching pursuits (MP) algorithm be

$$\hat{\mathbf{x}}_1 = \mathbf{U}_3\mathbf{a}_1,$$

where $\mathbf{U}_3 = [\mathbf{u}_1, \mathbf{u}_2, \mathbf{u}_3]$ is the $N \times 3$ matrix containing the three column vectors chosen by MP.
Further, assume that we have the following information about these three vectors:

$$\mathbf{u}_1^T\mathbf{u}_2 = \alpha = \cos\theta,\quad \mathbf{u}_1^T\mathbf{u}_3 = 0,\quad \mathbf{u}_2^T\mathbf{u}_3 = 0.$$

a. Find $\{\mathbf{u}_1', \mathbf{u}_2', \mathbf{u}_3'\}$, an orthonormal basis for the subspace spanned by $\{\mathbf{u}_1, \mathbf{u}_2, \mathbf{u}_3\}$.

b. Denote $\mathbf{x}_p = \mathbf{U}_3' \mathbf{x}$. Let $\hat{\mathbf{x}}_2 = \mathbf{U}_3 \mathbf{a}_2$ be the vector in the span of $\{\mathbf{u}_1, \mathbf{u}_2, \mathbf{u}_3\}$ such that the error $\|\mathbf{x} - \hat{\mathbf{x}}_2\|_2^2$ is minimized. Find $\mathbf{a}_2$ as a function of $\mathbf{x}_p$.

c. Assume that $\|\mathbf{a}_2\|_0 = 3$. Now we wish to approximate $\mathbf{x}$ by a vector $\hat{\mathbf{x}}_0 = \mathbf{U}_3 \mathbf{a}_0$ such that $\|\mathbf{a}_0\|_0 = 2$, by setting one of the entries of $\mathbf{a}_2$ to zero. Describe the optimal approach to perform this selection (in terms of minimizing the additional l_2 approximation error $\|\hat{\mathbf{x}}_2 - \hat{\mathbf{x}}_0\|^2$) as a function of the values of the entries of $\mathbf{a}_2$. Explain your reasoning carefully. Hint: estimate the additional error produced by each choice in order to justify your approach.

d. Assume we wish to add one more vector $\mathbf{u}_0$ to the dictionary, and we are choosing $\mathbf{u}_0$ to be in the subspace spanned by $\{\mathbf{u}_1, \mathbf{u}_2\}$. Select $\mathbf{u}_0$ so that the coherence properties of the resulting dictionary are as good as possible, on the basis of the information provided (i.e., assuming you know only the vectors $\{\mathbf{u}_1, \mathbf{u}_2, \mathbf{u}_3\}$). Hint: without loss of generality assume that $\theta \in [0, \pi/2]$ and explain your choice as a function of the values of θ in that range.

Appendix B GSP with `Matlab`: The GraSP Toolbox

Benjamin Girault

University of Southern California

The tools and methods described in this book are part of the growing toolbox of graph signal processing. In contrast with classical signal processing, which is well supported by many software packages written over the years, GSP is not. As such, there is a need in the community for similar software packages to handle graphs and graph signals.

The primary goal of such a toolbox is to abstract the complexities of GSP behind well-chosen data structures, and to propose tools to work with those structures. In other words, the goal is to simplify the execution of simple but cumbersome tasks in order to focus on important research questions. In this chapter we introduce such a toolbox, GraSP, developed and maintained by the author of this chapter [21]. GraSP is implemented in `Matlab`, a programming language whose main strength is its matrix algebra capabilities. This toolbox started as a personal project during the author's PhD but has grown into a full featured toolbox that is able to interface well with other tools from the community. It is available to the community as free software.

Two alternative toolboxes are worth mentioning before delving into the description of GraSP. The first one is GSPbox [157], another `Matlab` toolbox for GSP. Unfortunately, the authors have warned that this toolbox is not maintained anymore in favor of its Python counterpart, pyGSP [158]. This second toolbox is an interesting starting point for those programming in Python.

In Section B.1, we first describe how to set up the toolbox in `Matlab`. Then, we present a series of use cases with the ultimate goal of a quick start with the toolbox, both in class or for research. These use cases follow the organization of the book, thus linking ideas introduced in the book to their practical implementation. In addition, after each use case, we explain in more detail a number of useful features of GraSP that the reader already familiar with the toolbox may find useful. Code for these use cases can be found online on the book's web page[1].

B.1 Setting Up `Matlab`

Before using the toolbox, it is necessary to download and install it. The version of GraSP[2] used in this book is 1.3.0. Once it is downloaded and unpacked, the reader will find a folder `GraSP-1.3.0` with the toolbox. In `Matlab`, change the current folder to

---

[1] `http://www.graph-signal-processing-book.org/`
[2] Available here: `https://www.grasp-toolbox.org/`.

GraSP-1.3.0. Then, run <u>grasp_install</u>() (or <u>grasp_install</u>('*without_mex*', **true**) if the first command did not succeed[3]). Finally, in order to make all the functions of GraSP available to `Matlab`, run <u>grasp_start</u>. Note that only <u>grasp_start</u> will need to be run each time `Matlab` starts.

Going Further B.1.1

Upgrade

Occasionally, a new version of `GraSP` will require additional steps. The file RELEASENOTES at the root folder of the `GraSP` archive contains a description of these steps in addition to the changes in the new version. For example, new third party dependencies may be added. In that case, simply run again the function <u>grasp_install</u>, which effectively installs missing dependencies while skipping those already installed.

Automatically starting the toolbox

To avoid having to run manually this function each time `Matlab` is started, create (or edit) the file `startup.m` located in the startup folder of `Matlab` (type **help startup** in `Matlab` for more information), and add these lines:

```
1  addpath('<grasp_root_folder>');
2  grasp_start;
```

here `<grasp_root_folder>` is where `GraSP` root folder is to be found.

B.2   Chapter 1 Use Cases

B.2.1   Generate a Random Graph: Watts–Strogatz Model

In this use case, we want to generate a graph and plot it. The code and the resulting figure are shown in Figure B.1 for a Watts–Strogatz graph model (see Section 1.5 for details on this model). In a nutshell, this code creates a specific `Matlab` structure g that holds the data describing the graph (see Going Further B.2.1 for details on the structure). Then, the graph is plotted using a key function of `GraSP`: <u>grasp_show_graph</u>.

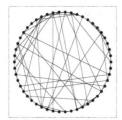

code/appendix_grasp/use_case_random_ws.m

```
1  N = 50;
2  K = 4;
3  beta = 0.4;
4  g = grasp_watts_strogatz(N, K, beta);
5  grasp_show_graph(gca, g);
```

Figure B.1 Generation of a random Watts–Strogatz graph and the resulting plot.

---

[3] Check https://www.mathworks.com/support/requirements/supported-compilers.html for help on fixing MEX files compilation.

Going Further B.2.1

The graph structure in GraSP

GraSP uses a specific Matlab structure to store information about a given graph, e.g., its connectivity, or information about how to plot it. This basic structure is described in `grasp_struct`. This function builds an empty graph by populating a Matlab structure g with the following fields and default values:

```
1  g.A = 0;                         % Adjacency matrix
2  g.A_layout = 0;                  % Adjacency matrix used by grasp_show_graph
3  g.layout = 0;                    % Graph layout (2D or 3D)
4  g.distances = 0;                 % Matrix of distances between any pair of nodes
5  g.node_names = {};               % Cell of strings for node names
6  g.M = 0;                         % Matrix of the graph variation operator
7  g.Q = 0;                         % Matrix of the graph signal inner product
8  g.Z = 0;                         % Matrix of the fundamental matrix of the graph
9  g.fourier_version = 'n.a.';      % Name of the GFT (eigvals, F and Finv)
10 g.eigvals = 0;                   % Graph frequencies of the graph
11 g.F = 0;                         % Fourier matrix
12 g.Finv = 0;                      % Inverse Fourier matrix
13 g.T = 0;                         % Translation operator
14 g.background = '';               % Background image file path
15 g.show_graph_options = struct(); % Options for grasp_show_graph
```

We will explore these fields throughout this appendix to make explicit the contexts in which they are useful. Right now, the following fields are important for us to describe:

- g.A: This is the actual adjacency matrix of the graph (Definition 2.14). It encodes the graph structure necessary to perform graph signal processing tasks.
- g.layout: This $N \times d$ matrix holds the embedding to represent graph nodes, i.e., the position of each of the N nodes in $\mathbb{R}^d$. Only $d = 2$ for a 2D representation and $d = 3$ for a 3D representation are valid. This field is only useful for 2D and 3D representations of graph signals.

Graph size

The number of nodes of a graph can be obtained using

```
1  >> N = grasp_nb_nodes(g);
```

B.2.2 Additional Random Graphs: Classical Models

The random graph models presented in Section 1.5 are all implemented in GraSP. Code Example B.1 shows how to generate them, in addition to random bipartite and random regular graphs. **Random bipartite graphs** are Erdős–Rényi graphs, with edges only between two subsets of N and M nodes, while **random regular graphs** have edges drawn at random with exactly k edges per node. All these graphs are unweighted.

Code Example B.1 Generating random graphs with GraSP.
`code/appendix_grasp/use_case_random_more.m` (N: number of nodes)

```
1 >> % Erdos-Renyi: Uniform random graph model
2 >> p = 0.1; % Edge probability
3 >> er  = grasp_erdos_renyi(N, p, 'directed', false);
4 >> % Barabasi-Albert: Preferential attachment graph model
5 >> m0 = 3; % Size of the initial complete graph
6 >> ba  = grasp_barabasi_albert(N, m0);
7 >> % Random bipartite
8 >> M = 20; % Size of the second set of nodes
9 >> rb  = grasp_random_bipartite(N, M, 'p', p);
10 >> % Random regular
11 >> k = 3; % Degree
12 >> rr = grasp_random_regular(N, k);
```

Going Further B.2.2

Graph symmetrization

Some graph models generate unweighted directed graphs (e.g., `grasp_erdos_renyi`). These can be made undirected (symmetric):

```
1 >> er = grasp_symetrise_unweighted(er);
```

Graph connectedness

Random graphs can lead to disconnected graphs, i.e., graphs with more than one connected component. To extract these, two functions can be used: `grasp_largest_connected_component` to get the largest graph, or `grasp_maximally_connected_subgraphs` to get all of them (in a cell array):

```
1 >> er_lcc = grasp_largest_connected_component(er);
2 >> er_ccs = grasp_maximally_connected_subgraphs(er);
```

For example, computing the positive strong nodal domains of graph g with respect to graph signal **x** (see Section 3.5.5) is done using:

```
1 >> nd = grasp_subgraph(g, find(x > 0));
2 >> nodal_domains = numel(grasp_maximally_connected_subgraphs(nd));
```

B.2.3   Additional Random Graphs: Sensor Network (Geometric Graph)

Geometric graphs are graphs for which each node is naturally localized in a Euclidean space (usually 2D or 3D), i.e., they have coordinates. To generate a random 2D geometric graph with uniform distribution of nodes:

```
1 >> N = 50;
2 >> geom = grasp_plane_rnd(N);
```

This function builds a graph where weights are obtained using a Gaussian kernel of the Euclidean distance.

More generally, if a distance matrix is available in g.distances (see Going Further B.2.3), a complete graph weighted by a Gaussian kernel of the distance[4] can be easily computed (see Section 6.2.1 for details):

```
1  >> g.A = grasp_adjacency_gaussian(g, sigma);
```

Unfortunately, there is no good way to choose sigma. A popular heuristic is to choose one third of the average distance between pairs of nodes [159], which is implemented in grasp_adjacency_gaussian by setting sigma to the string 'auto'. See also the discussion in Section 6.2.2.

Two-moons graph   Another classical geometric graph with edge weights defined using distances is the *two-moons graph* used in machine learning. Each node has a position and is associated with a feature vector. In this context, the distance between two nodes is computed using their feature vectors, rather than their position. This type of graph illustrates the property that not all feature spaces can be linearly separated into two classes, and is useful to show how graph-based methods can help:

```
1  >> N = 50;
2  >> tm = grasp_two_moons_knn(N, N)
```

Going Further B.2.3

Approximately regular geometric graph

For an approximately regular distribution of nodes (such as in Figure B.3), the function grasp_plane_rnd_pseudo_regular(N, M) starts with a uniform distribution of M nodes, and iteratively removes them such that there remain N of them, approximately regularly distributed.

Distance matrix

The distance matrix giving the distance between any pair of nodes can be obtained using the layout of the nodes, i.e., from the embedding in 2D or 3D used to plot the graph. There are two functions that can compute this matrix; the first one assumes that the x and y coordinates of nodes are available in g.layout.

```
1  >> g.distances = grasp_distances_layout(g);
```

However, when coordinates in g.layout are actually geographic coordinates (latitude and longitude), then one can use the following function to compute (approximate) distances between pairs of nodes:

```
1  >> g.distances = grasp_distances_geo_layout(g);
```

Subgraph: edge trimming / sparsification

Working with a full graph may be computationally expensive as the graph may have up to N^2 edges. However, in some cases (e.g., geometric graphs) a number of edges of reduced importance can be removed (see Section 6.1.2). Two approaches are implemented in GraSP:

---

[4] As discussed in Example 1.3 and [20, Section 3.5.2], these weights need to encode similarities between nodes (a higher weight for higher similarity), not distances.

weight thresholding (keeping large weights), and *k* nearest neighbors (keeping for each node the edges to its *k* closest neighbors):

```
1  >> threshold = 0.001;
2  >> g.A = grasp_adjacency_thresh(g, theshold);
3  >>
4  >> k = 5;
5  >> g.A = grasp_adjacency_knn(g, k);
```

The new adjacency matrix is symmetric if the original one is symmetric.

B.2.4 Some Non-random Graphs

Regular graphs As shown in Section 1.3, four graphs can be of interest to draw parallels with classical signal processing: the ring graph (a periodic time series), the path graph (a time series), the grid graph (a pixel grid of an image) and the torus graph (a grid graph where each row or column path graph becomes a cycle).

```
1  >> N = 30;
2  >> path  = grasp_directed_path(N);
3  >> ring  = grasp_directed_cycle(N);
4  >> grid  = grasp_directed_grid(N);
5  >> torus = grasp_directed_torus(N);
```

where N is the number of nodes in the graph (for a path or ring graph) or the size of a border (for a grid or torus). While `grasp_directed_*` builds a directed graph, its counterpart `grasp_non_directed_*` builds an undirected equivalent.

Minnesota road network Working with this classical road network, and with roads (edges) weighted by a Gaussian kernel of the distance, is enabled by:

```
1  >> minn = grasp_minnesota()
```

Going Further B.2.4

Minnesota road network: edges

In addition to roads weighted by a Gaussian kernel of the distance between intersections, one can obtain a variety of edge sets and edge weights using the parameter `'type'`:

```
1  % Roads weighted by a Gaussian kernel of the distance
2  g = grasp_minnesota('type', 'gauss_road')
3  % Complete graph with a Gaussian kernel of the distance
4  g = grasp_minnesota('type', 'gauss_dist')
5  % Complete graph with a Gaussian kernel of the shortest path dist.
6  g = grasp_minnesota('type', 'shortest_path')
7  % Roads weighted by the distance (not a similarity!)
8  g = grasp_minnesota('type', 'road_dist')
```

Graph edges versus edges drawn by `grasp_show_graph`

GraSP allows for a set of edges, drawn using `grasp_show_graph`, that is different from the actual set of graph edges defined by the adjacency matrix g.A. Using this feature is controlled by g.A_layout: if it is not a valid adjacency matrix (e.g., g.A_layout=[]), then g.A is used, otherwise g.A_layout is used. For example, the Minnesota graph always holds the unweighted road network in g.A_layout.

B.2.5 Plotting a Graph Signal on a Path Graph

Here we are interested in visualizing a graph signal, on a simple graph: the path graph. This use case effectively builds Figure 1.8. In Figure B.2, a graph signal is created as a vector x of eight scalar values alternating between 200 and 50 on a path graph. Plotting x is then achieved by setting the parameter *'node_values'* to x in the call to `grasp_show_graph` (cf. Section B.8).

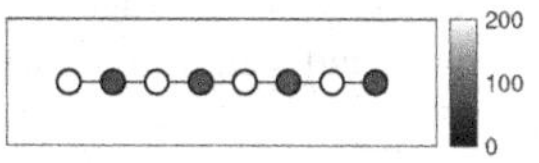

code/appendix_grasp/use_case_path.m

```
1  g = grasp_non_directed_path(8);
2  x(1:2:8) = 200;
3  x(2:2:8) = 50;
4  grasp_show_graph(gca, g,...
5                    'node_values', x,...
6                    'layout_boundaries', .2,...
7                    'show_colorbar', 1,...
8                    'value_scale', [0 200]);
```

Figure B.2 Creating an undirected path and a graph signal on this path, and plotting them.

B.2.6 Importing a Graph

For some applications, the graph we want to use is already given to us as a list of edges. In this section, we assume that we have access to an adjacency matrix describing the graph (from the list of edges). Figure B.3 shows the code to create the GraSP graph structure g and the resulting plot. We further assume here that a layout (here a 2D embedding) for the graph's nodes is given; however, this is not necessary to obtain a valid graph structure (see Going Further B.2.5).

This use case also showcases how the behavior of `grasp_show_graph` can be customized to suit one's need. In Figure B.3 we have represented the edge weights using a gray color scale (*'edge_colormap'*) with a specific edge thickness (*'edge_thickness'*), and enabled the corresponding colorbar below the figure (*'edge_colorbar'*). Finally, the graph boundaries are automatically computed according to Line 15. These parameters, and many others are documented in Section B.8.

Finally, notice that `grasp_importcsv` can be used instead of Lines 1 through 13 in code/appendix_grasp/use_case_import_graph.m

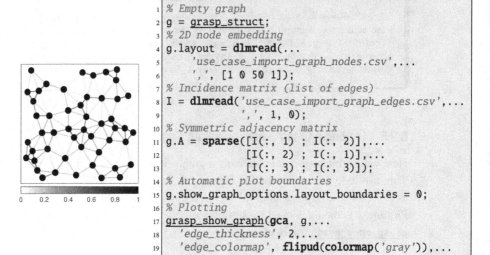

```matlab
1  % Empty graph
2  g = grasp_struct;
3  % 2D node embedding
4  g.layout = dlmread(...
5      'use_case_import_graph_nodes.csv',...
6      ',', [1 0 50 1]);
7  % Incidence matrix (list of edges)
8  I = dlmread('use_case_import_graph_edges.csv',...
9      ',', 1, 0);
10 % Symmetric adjacency matrix
11 g.A = sparse([I(:, 1) ; I(:, 2)],...
12              [I(:, 2) ; I(:, 1)],...
13              [I(:, 3) ; I(:, 3)]);
14 % Automatic plot boundaries
15 g.show_graph_options.layout_boundaries = 0;
16 % Plotting
17 grasp_show_graph(gca, g,...
18     'edge_thickness', 2,...
19     'edge_colormap', flipud(colormap('gray')),...
20     'edge_colorbar', true,...
21     'edge_color_scale', [0 1]);
```

Figure B.3 Importing a weighted graph and plotting it. In this example, the edge color is a function of the edge weight. The correspondence between color and weight is shown in the colorbar below the figure.

Figure B.3 (see Going Further B.2.5). Here, these lines effectively illustrate the process of building a valid graph structure.

Going Further B.2.5

Saving graphs and graph signals

To avoid recomputing the graph structure of a given graph each time `Matlab` is started, or to export the necessary data for reproducibility purposes, it is desirable to save this structure. To that end, `grasp_exportcsv` (to save the adjacency matrix and the node embedding) can be used together with `grasp_exportcsv_signal` (to save a given signal) and `grasp_importcsv` (to import the graph structure in `Matlab`). This is especially useful when dealing with random graphs in order to avoid working on a completely different realization from one session to another, or to share the specific realization used in a scientific communication. For a graph structure g and a graph signal x, use the following syntax to export and import CSVs of the graph structure:

```matlab
1  >> grasp_exportcsv(g, '<node_file.csv>', '<edge_file.csv>');
2  >> grasp_exportcsv_signal(g, x, 'signal_file.csv>');
3  >> g = grasp_importcsv('<node_file.csv>', '<edge_file.csv>');
4  >> x = csvread('<signal_file.csv>', 1, 1);
```

where `<node_file.csv>`, `<edge_file.csv>` and `<signal_file.csv>` are the paths to the files containing the list of nodes, the list of edges and the graph signal, respectively. These functions only save the nodes and their edges and allow for interoperability with alternative tools since CSV files are easy to read and parse. However, if saving all the fields is required (e.g., to keep the computed GFT), then the standard Matlab functions **save** and **load** should be used instead. These save and load .mat files, which are only readable by Matlab. For a graph g and a signal x:

```
1  >> save('<graph_file.mat>', 'g');
2  >> save('<graph_signal_file.mat>', 'x');
3  >> load('<graph_file.mat>');
4  >> load('<graph_signal_file.mat>');
```

Computing a node embedding

If a graph lacks a node embedding, <u>grasp_show_graph</u> will automatically compute one each time it is called, which can become expensive. To avoid this, two functions are available and can be called, using:

```
1  >> g.layout = grasp_layout(g);
2  >> g.layout = grasp_layout_spectral(g);
```

provided g has a valid adjacency matrix g.A. The first function uses Graphviz[a], and requires this tool to be installed. If Graphviz is unavailable, the second function can be called to compute a 2D spectral embedding of the nodes.

[a] https://graphviz.org/

B.2.7 Building a Graph from Scratch

This section shows how to build the graph of Figure 1.4. The general idea is simple: we start with an empty graph, add the desired number of nodes, define their embedding and finally add the directed edges. To do so, we define the adjacency matrix g.A and the embedding matrix g.layout. Figure B.4 shows how this process works in GraSP. Note that having the embedding matrix g.layout is not strictly necessary, in which case building the graph amounts then to just filling the adjacency matrix g.A (see Section B.2.6).

If setting g.layout before g.A, a useful intermediate step is to plot the graph in the top subfigure of Figure B.4, with a node index shown next to it. This is implemented in Line 10, with 'node_text' set to 'ID' (see Section B.8).

Notice also in Figure B.4 how directed and non-directed edges appear: each time there is a directed edge in both directions with the same weight, a non-directed edge is plotted by <u>grasp_show_graph</u> (without the arrowhead).

Going Further B.2.6: Subgraph – Node Set

For various reasons, one may be faced with the task of selecting a subgraph of a graph (see, e.g., Section 5.4.3), which involves removing any node not in a given set and its incident edges. Given such a set, called **node_subset**, this can be done using:

```
1  >> g = grasp_subgraph(g, node_subset);
```

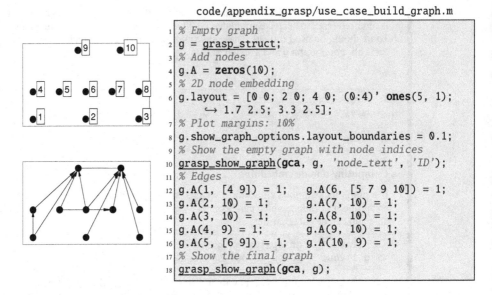

code/appendix_grasp/use_case_build_graph.m

```
1  % Empty graph
2  g = grasp_struct;
3  % Add nodes
4  g.A = zeros(10);
5  % 2D node embedding
6  g.layout = [0 0; 2 0; 4 0; (0:4)' ones(5, 1);
         ↪ 1.7 2.5; 3.3 2.5];
7  % Plot margins: 10%
8  g.show_graph_options.layout_boundaries = 0.1;
9  % Show the empty graph with node indices
10 grasp_show_graph(gca, g, 'node_text', 'ID');
11 % Edges
12 g.A(1, [4 9]) = 1;     g.A(6, [5 7 9 10]) = 1;
13 g.A(2, 10) = 1;        g.A(7, 10) = 1;
14 g.A(3, 10) = 1;        g.A(8, 10) = 1;
15 g.A(4, 9) = 1;         g.A(9, 10) = 1;
16 g.A(5, [6 9]) = 1;     g.A(10, 9) = 1;
17 % Show the final graph
18 grasp_show_graph(gca, g);
```

Figure B.4 Creating a graph by manually setting the node embedding and the adjacency matrix.

B.3   Chapter 2 Use Cases

We now look at one of the most important building blocks of GSP: graph filters. The use cases corresponding to this chapter show how to implement nodal domain graph filters. We use the same graph as Figure B.3 and we would like to implement the following graph filter:

$$\mathbf{H} = [\mathbf{I} + 10\mathcal{T}]^{-1}, \tag{B.1}$$

where $\mathcal{T}$ is the random walk Laplacian of the graph. This filter is essentially a smoothing filter removing large variations along edges of the graph. In each use case, we employ the graph signal defined using x = grasp_delta(g, 10), i.e., a signal with value 1 on node 10 and zero elsewhere.

Our goal is to cover various exact and approximate approaches to implementing this filter with GraSP. We will further explore this topic in Section B.4 using spectral domain graph filters. Note that any filter can be applied using grasp_apply_filter(graph, filter, input_signal), independently of its design.

B.3.0   The Random Walk Laplacian

Throughout this section and the next, we will be using the random walk Laplacian $\mathcal{T}$ as the fundamental graph operator. Therefore, in order to define and use graph filters, we need to set g.Z to this particular matrix. Mathematically, the random walk Laplacian is defined as $\mathcal{T} = \mathbf{D}^{-1}\mathbf{L}$, where $\mathbf{D}$ is the degree matrix and $\mathbf{L}$ is the combinatorial Laplacian (see Section 2.3). In GraSP, this can be achieved with:

```
1 g.Z = grasp_degrees(g) ^ (-1) * grasp_laplacian_standard(g);
```

On the other hand, `grasp_laplacian_normalized` would give us the normalized Laplacian if we were using it as the fundamental graph operator.

B.3.1 Linear Graph Filter

This first approach is straightforward: we just have to compute the matrix inverse in (B.1). In Figure B.5, we plot both the matrix of the graph filter and the output graph signal.

Code Example B.2 Linear filter: `code/appendix_grasp/use_case_filters.m`

```
1 % Filter
2 h_mat = grasp_filter_struct()
3 h_mat.type = 'matrix';
4 h_mat.data = (eye(grasp_nb_nodes(g)) + 10 * g.Z) ^ (-1);
5 % Matrix plot
6 imshow(h_mat.data);
7 % Output plot
8 grasp_show_graph(gca, g,...
9                  'node_values', grasp_apply_filter(g, h_mat, x),...
10                 'value_scale', [0 0.2]);
```

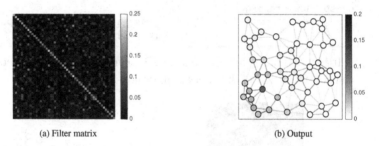

(a) Filter matrix   (b) Output

Figure B.5 Figures obtained using Code Example B.2. (a) Matrix representation of (B.1) and (b) its output graph signal when applied to the input signal **x**.

B.3.2 Polynomial Graph Filter

Another approach is to use a graph filter defined in the nodal domain, as shown in Section 2.4. As examples, we consider the following **graph polynomial filters**:

$$\mathbf{H}_1 = p_1(\mathcal{T}) = \mathbf{I} - \frac{1}{2}\mathcal{T}, \qquad \mathbf{H}_2 = p_2(\mathcal{T}) = \mathbf{I} - \mathcal{T} + \frac{1}{4}\mathcal{T}^2.$$

We choose to implement these, as their behavior on graph signals is similar to that of **H** in (B.1). In Code Example B.3, notice that the polynomials are encoded as vectors with the polynomial coefficients ordered by decreasing power. This follows the Matlab encoding of polynomials, and allows `h.data` to be manipulated using built-in functions such as **polyval** or **polyvalm**.

Notice also that the polynomial graph filter structure encodes only the filter taps (the coefficients of the polynomial). Indeed, to be generic, a polynomial graph filter p must be defined only through these, and applying a graph filter $p(\mathcal{T})$ to a graph signal $\mathbf{x}$ requires a call to `grasp_apply_filter` with both the graph and the graph filter to obtain the output of the filter $p(\mathcal{T})\mathbf{x}$.

Code Example B.3 Polynomial filters: `code/appendix_grasp/use_case_filters.m`

```
1  % Filter 1
2  h_poly_1 = grasp_filter_struct;
3  h_poly_1.type = 'polynomial';
4  h_poly_1.data = [-1/2 1];
5  % Matrix plot
6  imshow(grasp_apply_filter(g, h_poly_1));
7  % Output plot
8  grasp_show_graph(gca, g,...
9                   'node_values', grasp_apply_filter(g, h_poly_1, x),...
10                  'value_scale', [0 0.2]);
11
12 % Filter 2
13 h_poly_2 = grasp_filter_struct;
14 h_poly_2.type = 'polynomial';
15 h_poly_2.data = [1/4 -1 1];
16 % Matrix plot
17 imshow(grasp_apply_filter(g, h_poly_2));
18 % Output plot
19 grasp_show_graph(gca, g,...
20                  'node_values', grasp_apply_filter(g, h_poly_2, x),...
21                  'value_scale', [0 0.2]);
```

In Figure B.6, notice how the degree-2 polynomial filter $\mathbf{H}_2$ extends to one more hop, and how the matrix of the filter has more non-zero coefficients.

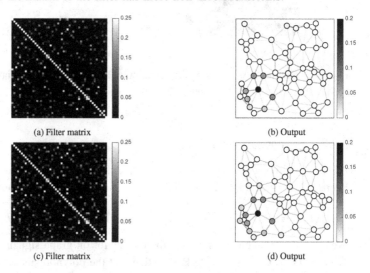

(a) Filter matrix   (b) Output

(c) Filter matrix   (d) Output

Figure B.6 Figures obtained using Code Example B.3. (a) Matrix representation of $\mathbf{H}_1$ and (b) its output graph signal when applied to the input signal $\mathbf{x}$. (c)-(d): Same for $\mathbf{H}_2$.

B.4 Chapter 3 Use Cases

B.4.1 Graph Fourier Transform Computation and Visualization

Here, we would like to compute and visualize the random-walk-Laplacian-based graph Fourier transform. To this end, we use the graph of Figure B.3 and store it in variable g. The function to compute the graph Fourier transform is grasp_eigendecomposition. It follows the formalism of [36] to parameterize the GFT, i.e., it needs two ingredients: (i) the quadratic graph signal variation operator $\mathbf{M}$ such that $\mathbf{x}^\top \mathbf{M} \mathbf{x}$ is the variation of graph signal $\mathbf{x}$ and (ii) the graph signal inner product $\mathbf{Q}$ matrix such that $\mathbf{y}^\top \mathbf{Q} \mathbf{x}$ is the inner product between graph signals $\mathbf{x}$ and $\mathbf{y}$. In the case of the random walk Laplacian we have $\mathbf{M} = \mathbf{L}$, the combinatorial Laplacian, which is the default value for $\mathbf{M}$ in grasp_eigendecomposition, and $\mathbf{Q} = \mathbf{D}$, the degree matrix (see also Section 3.4.3).

Code Example B.4 shows the code to compute the GFT and visualize it. Figure B.8 and Figure B.7 were obtained using grasp_show_fouriermodes to plot a number of graph Fourier modes as graph signals, and grasp_show_transform to stack on a line all the graph Fourier modes that have been plotted (see Section B.9 for details).

Code Example B.4 `code/appendix_grasp/use_case_gft.m`

```
1  % Compute the LRW-based GFT
2  g = grasp_eigendecomposition(g, 'inner_product', 'degree');
3  % Show several GFT modes
4  max_value = max(max(abs(g.Finv(:, 1:9))));
5  grasp_show_fouriermodes(g,...
6                          'modes', 1:9,...
7                          'titles', 0,...
8                          'cmap', flipud(colormap('gray')),...
9                          'node_size', 150,...
10                         'value_scale', max_value * [-1 1]);
11 % Show the complete GFT, stacked
12 grasp_show_transform(gcf, g);
```

Going Further B.4.1

The GFT in GraSP graph structure

The graph structure in GraSP uses three important fields related to this transform:

- g.eigvals: This vector holds the sorted graph frequencies as defined by a graph Fourier transform.
- g.F: This is the forward graph Fourier transform matrix.
- g.Finv: This is the inverse graph Fourier transform matrix.
- g.fourier_version: This gives the name of the graph Fourier transform (less important).

A typical GraSP user does not need to populate these fields with the GFT manually. These are automatically filled by the function grasp_eigendecomposition.
However, when experimenting with alternative GFTs that are not readily available in GraSP, one can simply set the four fields listed above manually, as long as the GFT and its inverse are linear transforms.

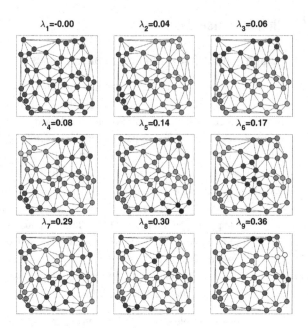

Figure B.7 Visualization of the random-walk-Laplacian-based GFT of the graph of Figure B.3. First nine graph Fourier modes in 2D.

B.4.2   Kernelized Filter

In this approach, the filter is defined as a function (the **kernel**) of the fundamental graph operator $\mathbf{Z}$ (here $\mathcal{T}$). Note that this approach is valid only because $h(x) = (1 + 10x)^{-1}$ is an analytical function.[5] Since such a filter can be written in the spectral domain as $h(\mathbf{\Lambda})$, where $\mathbf{\Lambda}$ is the diagonal matrix of eigenvalues (or graph frequencies), we can also plot its frequency response, as shown in Figure B.9(a). The function call `grasp_apply_filter`(frequency_set, filter)[6] can be used to obtain the frequency response. Noticeably, Figure B.9(b) shows that there is no difference in output compared with Figure B.5(b)

Code Example B.5 Kernelized filter: `code/appendix_grasp/use_case_filters.m`

```
1  % Filter
2  h_kern = grasp_filter_struct;
3  h_kern.type = 'kernel';
4  h_kern.data = @(x) (1 + 10 * x) .^ (-1);
5  % Frequency response
6  plot(0:0.001:2, grasp_apply_filter(0:0.001:2, h_kern));
7  % Output plot
8  grasp_show_graph(gca, g,...
9                   'node_values', grasp_apply_filter(g, h_kern, x),...
10                  'value_scale', [0 0.2]);
```

---

[5] Analytical functions are functions that can be written as an infinite polynomial: $f(x) = \sum_{k=0}^{\infty} f_k x^k$.

[6] This feature can be used whenever `filter` has a frequency response.

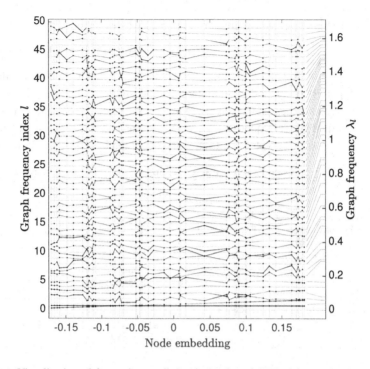

Figure B.8 Visualization of the random-walk-Laplacian-based GFT of the graph of Figure B.3. Full graph Fourier transform.

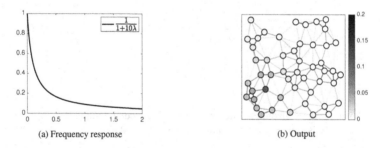

(a) Frequency response   (b) Output

Figure B.9 Figures obtained using Code Example B.5, with (a) the frequency response and (b) the output of the filter when applied to **x**.

B.4.3 Convolutive Filter

As in classical signal processing, a convolutive filter is a filter that convolves its input with a given graph signal (see Section 3.3.4 for more insights). In graph signal processing, this convolution is defined in the spectral domain as the dot product between the spectral components of both signals (with **U** the inverse GFT matrix):

$$\mathbf{x} * \mathbf{y} = \mathbf{U}(\widetilde{\mathbf{x}} \odot \widetilde{\mathbf{y}}). \tag{B.2}$$

Equation (B.2) is known as the **convolution theorem**.

We can define a graph signal **h** such that $\widetilde{\mathbf{h}} = h(\lambda)$ and the graph filter defined by

(B.1) is a convolutive graph filter that convolves its input with **h**:

$$\mathbf{Hx} = h(\mathcal{T})\mathbf{x} = \mathbf{U}h(\mathbf{\Lambda})\mathbf{U}^\top\mathbf{x} = \mathbf{U}\operatorname{diag}(h(\lambda))\widetilde{\mathbf{x}} = \mathbf{U}\left(\widetilde{\mathbf{h}}\odot\widetilde{\mathbf{x}}\right) = \mathbf{h}*\mathbf{x}.$$

Line 4 of Code Example B.6 implements this signal **h**, with the resulting plots shown in Figure B.10. Note that the implementation of graph signal convolutions in `GraSP` does not require the assumption of graph signals having equal values for spectral components of equal graph frequencies (Remark 3.5). Instead, any graph signal in $\mathbf{h} \in \mathbb{C}^N$ can define a convolutive graph filter, even when the graph has equal graph frequencies.

Code Example B.6 `code/appendix_grasp/use_case_filters.m`

```
1  % Filter
2  h_conv = grasp_filter_struct;
3  h_conv.type = 'convolution';
4  h_conv.data = g.Finv * ((1 + 10 * g.eigvals) .^ -1);
5  % Convolution signal plot
6  grasp_show_graph(gca, g,...
7                   'node_values', h_conv.data,...
8                   'value_scale', [-0.3 0.5]);
9  % Output plot
10 grasp_show_graph(gca, g,...
11                  'node_values', grasp_apply_filter(g, h_conv, x),...
12                  'value_scale', [0 0.2]);
```

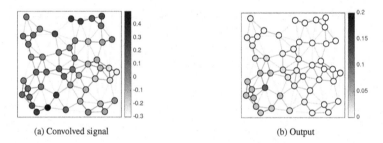

(a) Convolved signal   (b) Output

Figure B.10 Convolutive filter approach using Code Example B.6: (a) graph signal with which the input is convolved, and (b) output when applied to **x**.

B.4.4 Polynomial Filter (Approximation)

Except for the rough polynomial approximations to (B.1) in Section B.3, the filters defined until now suffer from high numerical complexity because they require either a costly matrix inverse, or computation of the graph Fourier transform.

Instead, we can use polynomial approximation to lower the cost of applying a filter. The function `grasp_filter_kernel_to_poly` implements several approaches to polynomial approximation of a given kernelized filter h_kern (see Going Further B.4.2 for details). In this use case, the approximation is performed as a polynomial interpolation of the tuples $(\lambda_l, h(\lambda_l))$, thus rewriting $\mathbf{H} = h(\mathcal{T})$ as the polynomial $\mathbf{H} \simeq p(\mathcal{T})$. Figure B.11(a) shows the resulting frequency response in comparison with the true response of **H**. Notice also that the error on the output in Figure B.11(b) is minimal.

Code Example B.7 `code/appendix_grasp/use_case_filters.m`

```
1  % Filter
2  h_poly = grasp_filter_kernel_to_poly(h_kern,...
3                  'algorithm', 'graph_fit_lsqr',...
4                  'poly_degree', 5,...
5                  'graph', g);
6  % Frequency response
7  plot(0:0.001:2, h_kern.data(0:0.001:2), 'b',...
8      g.eigvals, grasp_apply_filter(g.eigvals, h_poly), ':rx');
9  % Output plot
10 grasp_show_graph(gca, g,...
11                 'node_values', grasp_apply_filter(g, h_poly, x),...
12                 'value_scale', [0 0.2]);
```

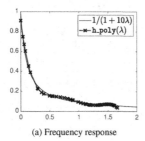

(a) Frequency response

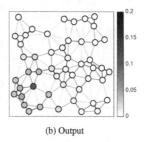

(b) Output

Figure B.11 Polynomial approximation of the filter defined in (B.1): (a) its frequency response compared with that of **H**, and (b) its output when applied to **x**.

Going Further B.4.2

Polynomial approximations of graph filters

Several algorithms are offered that compute a polynomial approximation to a kernelized graph filter `h_kern`. These are selected using the parameter `'algorithm'` of `grasp_filter_kernel_to_poly`:

- `'graph_fit_lsqr'`: least squares approximation of the frequency response on the graph frequencies (arbitrary polynomial degree),
- `'regular_interpolation'`: regular interpolation of the frequency response of `h_kern` using Matlab's **polyfit** (arbitrary polynomial degree),
- `'graph_interpolation'`: Lagrange interpolation polynomial (polynomial degree N),
- `'taylor'`: Taylor interpolation polynomial (arbitrary polynomial degree),
- `'puiseux'`: Puiseux interpolation polynomial (arbitrary polynomial degree).

B.4.5 Chebyshev Polynomial Filter (Approximation)

Another approach to approximating a kernelized graph filter is to perform a Chebyshev approximation of its kernel (see Section 5.5.2 for details). In GraSP, the function `grasp_filter_cheb` implements this approximation using the toolbox chebfun.[7] Sim-

---

[7] http://www.chebfun.org/.

ilarly to polynomial approximation, it is easy to plot the frequency response of these approximations. Figure B.12(a) shows the responses for two approximations of degrees 3 and 5 for the graph filter of (B.1). Notice how the degree-3 approximation introduces some inaccuracies on the output (Figure B.12(b)) compared with the degree-5 approximation (Figure B.12(c)).

Code Example B.8 `code/appendix_grasp/use_case_filters.m`

```
1  % Filter
2  h_cheby3 = grasp_filter_cheb(h_kern.data, 3);
3  h_cheby5 = grasp_filter_cheb(h_kern.data, 5);
4  % Frequency response
5  plot(0:0.001:2, h_kern.data(0:0.001:2), 'k',...
6        0:0.001:2, grasp_apply_filter(0:0.001:2, h_cheby3), '-r',...
7        0:0.001:2, grasp_apply_filter(0:0.001:2, h_cheby5), '-b');
8  % Output plots
9  grasp_show_graph(gca, g,...
10        'node_values', grasp_apply_filter(g, h_cheby3, x),...
11        'value_scale', [0 0.2]);
12 grasp_show_graph(gca, g,...
13        'node_values', grasp_apply_filter(g, h_cheby5, x),...
14        'value_scale', [0 0.2]);
```

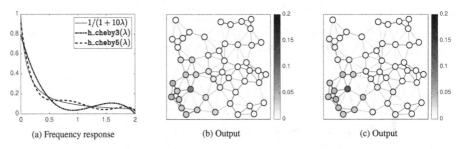

(a) Frequency response   (b) Output   (c) Output

Figure B.12 Chebyshev polynomial approximations of degrees 3 and 5 of the filter defined in (B.1): (a) Frequency responses compared to **H**, and output when applied to **x** ((b) degree-3, (c) degree-5).

B.5   Chapter 4 Use Cases

B.5.1   Node Sampling

We will now consider the node sampling as described in Chapter 4. More precisely, in the context of the Paley–Wiener space of Definition 4.4 (p. 135), we would like to highlight the relationship between the number of iteratively sampled nodes and the increase in the dimension of the Paley–Wiener space from which we can recover signals.

To that end, we first define a function that returns an ordered list of sampled nodes, using the sampling method of [57], described in Section 4.4. More precisely, we select the power of 3 for our spectral proxies, and compute the corresponding optimal

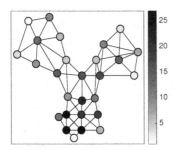

Figure B.13 Optimal node sampling using spectral proxies: sampling order. The colormap on the right shows the mapping from node shading to the order in which the corresponding node was selected.

sampling set (Line 11 in Code Example B.9). Finally, on the GFT visualization in Figure B.14, we highlight for each GFT mode, from the lowest to the highest frequencies, the corresponding sampled node.

Code Example B.9 `code/appendix_grasp/use_case_sampling.m`

```
1  % Load our toy graph
2  grasp_start_opt_3rd_party('usc_graphs');
3  load('toy_graph.mat');
4  % Compute the combinatorial Laplacian-based GFT
5  toy_graph = grasp_eigendecomposition(toy_graph);
6  % Compute the sampling set
7  grasp_start_opt_3rd_party('usc_ssl_sampling');
8  N = grasp_nb_nodes(toy_graph);
9  power = 3;
10 L_k = toy_graph.M ^ power;
11 sampling_set = compute_opt_set_inc((L_k + L_k') / 2, power, N)';
12 % Plot the sampling set order on the graph
13 [~, IX] = sort(sampling_set);
14 grasp_show_graph(gca, toy_graph, 'node_values', IX);
15 % Assign each sampled node to a GFT mode (row)
16 mask = sparse((1:N)', sampling_set, ones(N, 1), N, N);
17 % Plot the GFT matrix with the sampled nodes
18 grasp_show_transform(gcf, toy_graph, 'highlight_entries', mask);
```

Going Further B.5.1

Starting optional third party toolboxes

As shown in Code Example B.9, some tools from the community are conveniently made available through GraSP. While some are readily available after calling `grasp_start`(), most are not. Indeed, naming conventions between toolboxes may lead to conflicting function names. To avoid unnecessary headaches, especially as the number of available toolboxes grows, GraSP does not start them all. Instead, they need to be started individually. To get a list of available toolboxes, use:

```
1  >> grasp_start_opt_3rd_party()
2  #1     GrTheory
3  #2     MatlabBGL
4  #3     anneal
5  #4     gspbox
6  ...
```

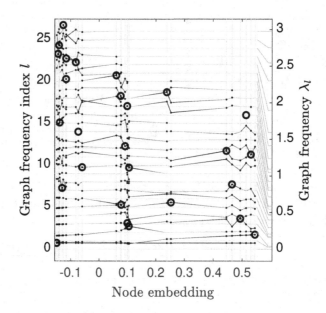

Figure B.14 Optimal node sampling using spectral proxies of power 3 with the combinatorial Laplacian, for the graph in Figure B.13. In this figure we highlight with a circle, for each graph frequency (vertical scale), the node that is sampled to include that frequency in Paley–Wiener space.

and either use the name of a toolbox to start it or the numerical identifier on the left in the listing above:[a]

```
1 >> grasp_start_opt_3rd_party('sgwt')
```

The set of toolboxes from the community is meant to evolve over time on the basis of requests from users of `GraSP`. To request such an addition, either send an email to the author, send a bug request with a link to the toolbox or edit the file `grasp_dependencies_list.m` in `3rdParty/` with the new dependency at the end of the file and submit a pull request.

Bibliography

As a convenience, `GraSP` provides a function to list the scientific communications linked with `GraSP`, and the third party toolboxes that have been started:

```
1 >> grasp_bibliography()
2 GraSP:
3        https://doi.org/10.1109/ICASSP.2017.8005300 -- https://gforge.inria.
         ↪ fr/projects/grasp/
4 sgwt:
5        http://dx.doi.org/10.1016/j.acha.2010.04.005
```

A number of functions also display the scientific communications used for computations when they are executed (e.g., `grasp_eigendecomposition`).

---

[a] The name of the toolbox is preferred, in order to ensure compatibility with future versions of GraSP.

B.6 Chapter 5 Use Cases

B.6.1 Extended Application: SGWT with the Random Walk Laplacian

The spectral graph wavelet transform (SGWT) in [74] has a dedicated toolbox written for Matlab and available as a third party toolbox within GraSP. However, it does not allow for the use of fundamental graph operators other than the combinatorial and normalized symmetric Laplacians. In this extended application, we propose to look at how the SGWT toolbox can be easily reimplemented using GraSP but without restriction on the GFT.

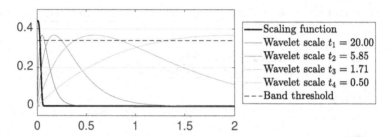

Figure B.15 Mexican hat filter design for the SGWT, with $J = 4$ wavelet scales, and the associated scaling function. The band threshold is used in Figure B.17.

For simplicity, we will build an SGWT based on a Mexican hat design (see Code Example B.10, Line 12 for the scaling function and Line 14 for the mother wavelet). The article [74] describes a more complex cubic spline design, achieving better localization guarantees, but we leave its implementation as an exercise (see Problem B.1).

The first step in implementing a filterbank is to define the filters. This is done in Code Example B.10, where we obtain $J + 1$ kernelized graph filters, and we plot their frequency response in Figure B.15. In this section, we choose the random walk Laplacian as the fundamental graph operator, such that the maximum graph frequency is 2. For alternative fundamental graph operators, this maximum graph frequency will need to be adjusted accordingly.

Code Example B.10 SGWT filter design: `code/appendix_grasp/use_case_sgwt.m`

```
1 % Parameters
2 M = 10; % Chebyshev polynomial approximation degree
3 lambda_max = 2; % Maximum eigenvalue
4 K = 20; % How much of the [0,lambda_max] is covered by wavelets
5 J = 4; % Number of wavelet scales
6 lambda_min = lambda_max / K; % [lambda_min, lambda_max] -> wavelets
7 scales = logspace(log10(2 / lambda_min), log10(1 / lambda_max), J);
8 % Scaling function
9 scaling_function = grasp_filter_struct;
10 scaling_function.type = 'kernel';
11 scaling_function.data = ...
12     @(x) 1.2 * exp(-1) * exp(-(x / (0.4 * lambda_min)) .^ 4);
```

```matlab
13  % Wavelets
14  mother_wavelet = @(x) x .* exp(-x);
15  wavelets = repmat(grasp_filter_struct, J, 1);
16  for j = 1:J
17      wavelets(j).type = 'kernel';
18      wavelets(j).data = @(x) mother_wavelet(scales(j) * x);
19  end
```

For an efficient implementation of those filters, polynomial approximation is desired. In order to do so, the authors of [74] use Chebyshev polynomial approximations. As observed in Section B.4.5, in GraSP this is achieved with Code Example B.11.

Code Example B.11 Chebyshev approximation: `code/appendix_grasp/use_case_sgwt.m`

```matlab
1  % Graph filter Chebyshev approximation
2  M = 15;
3  filterbank = repmat(grasp_filter_struct, J + 1, 1);
4  filterbank(1) = grasp_filter_cheb(scaling_function.data, M,...
5                               'chebfun_splitting', 'on');
6  for j = 1:J
7      filterbank(j + 1) = grasp_filter_cheb(wavelets(j).data, M,...
8                               'chebfun_splitting', 'on');
9  end
```

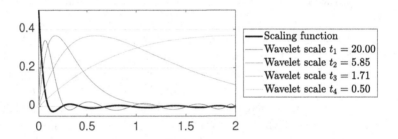

Figure B.16 Chebyshev approximation of the frequency responses shown in Figure B.15.

We have just built our first filterbank and we can now apply it to a graph and to graph signals. In practice, the spectral graph wavelet transform is an orthogonal projection of an input graph signal $\mathbf{x}$ onto a set of graph signals defined by the filterbank above. The impulse responses of those filters is

$$w_{i,s_j}(\mathbf{x}) = \mathbf{g}_{i,s_j}^\top \mathbf{x} \quad \text{with} \quad \mathbf{g}_{i,s_j} = \mathbf{G}_{s_j}\delta_i \tag{B.3}$$

where $\mathbf{G}_{s_j}$ is the graph filter obtained by applying the kernelized graph filter (or its Chebyshev approximation) to the random walk Laplacian, at wavelet scale s_j (or to the scaling function if $j = 1$). In GraSP, computing these impulse responses is achieved by calling `grasp_apply_filter`(graph, filter) for each filter. Iterating through all scales, we can then obtain the matrix of the SGWT, as shown in Code Example B.12. Note that up to now the implementation has not depended on the graph; Code Example B.12 is the first piece of code that explicitly requires the graph structure g.

Code Example B.12 SGWT matrix: `code/appendix_grasp/use_case_sgwt.m`

```
1  g = grasp_eigendecomposition(g, 'inner_product', 'degree');
2  G = zeros((J + 1) * N, N);
3  for j = 1:(J + 1)
4    G((j - 1) * N + (1:N), :) = grasp_apply_filter(g, filterbank(j))';
5  end
```

We have then $\mathbf{w}(\mathbf{x}) = \mathbf{Gx}$, with $\mathbf{G}$ the matrix of the spectral graph wavelet transform with respect to the random walk Laplacian GFT. Going one step further, we can analyze the behavior of each graph signal $\mathbf{g}_{i,s_j}$, and how localization of those signals is affected by the scale s_j. To do so, we use the function `grasp_show_transform` of GraSP to plot the matrix $\mathbf{G}$ (instead of the GFT matrix as in Section B.4.1).

In Code Example B.13, we first map each filter of the SGWT to a graph frequency, the one with maximum frequency response of the associated filter. Note, however, that this may not be a graph frequency of the graph since this maximum may not correspond to an eigenvalue of the random walk Laplacian. On the figure, we then map all basis vectors $\{\mathbf{g}_{i,s_j}\}_i$ at a given scale s_j to the same graph frequency on the right axis (Line 23). We also define the passband of the filters according to the value of cut. The bands are represented in Figure B.15 along with the band threshold (anything above the threshold is within the band), and in Figure B.17 they are represented as shaded areas of varying length on the right axis.

Code Example B.13 SGWT matrix plot: `code/appendix_grasp/use_case_sgwt.m`

```
1    % Mapping between scales and bands
2    sc_freqs = zeros(J + 1, 1);
3    bands    = zeros(J + 1, 2);
4    for j = 2:(J + 1)
5    sc_freqs(j) = fminbnd(@(x) -wavelets(j - 1).data(x),...
6    0, lambda_max);
7    bands(j, 1) = fminbnd(@(x) abs(wavelets(j - 1).data(x) - cut),...
8    0, sc_freqs(j));
9    bands(j, 2) = fminbnd(@(x) abs(wavelets(j - 1).data(x) - cut),...
10   sc_freqs(j), lambda_max);
11   end
12   bands(1, 2) = fminbnd(@(x) abs(scaling_function.data(x) - cut)
↪    ,...
13   0, sc_freqs(2));
14   % Reordering of nodes according to a 1D embedding
15   [~, node_ordering] = sort(g.Finv(:, 2));
16   for j = 1:(J + 1)
17   G((j - 1) * N + (1:N), :) = G((j - 1) * N + node_ordering, :);
18   end
19   % Plot
20   grasp_show_transform(gcf, g,...
21   'embedding', g.Finv(:, 2),...
22   'transform_matrix', G,...
23   'graph_frequencies', kron(sc_freqs, ones(N, 1)),...
24   'bands', bands,...
25   'support_scatter_mode', 'var_width_gray');
```

The next step is to reorder the atoms according to the node ordering shown on the horizontal axis of Figure B.17. To do so, we use the property that this ordering is the

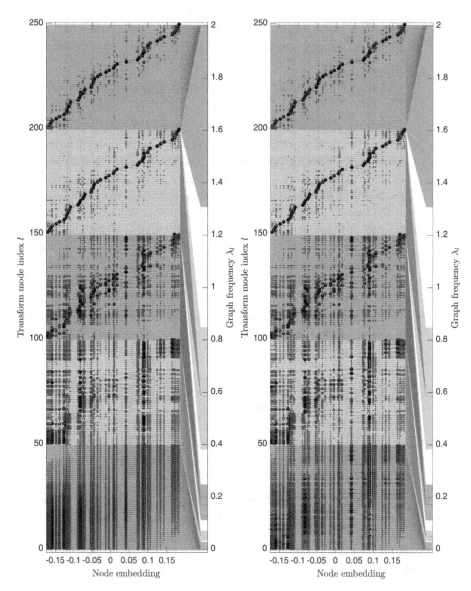

Figure B.17 Exact SGWT (left) and Chebyshev approximation of it (right). Each scale is identified by a shade of gray on the graph frequency axis, representing the frequency band for the filter at that scale.

same as the ordering of the first non-constant eigenvector of the random walk Laplacian, and we thus reorder the rows of **G**, for each scale.

Finally, we can call `grasp_show_transform`. Figure B.17 shows the resulting plot of both the accurate filter design (left) and the approximate one (right). The left scale corresponds to the row number of the matrix **G** (with the first first row at the bottom). We can observe that as the scale decreases to become finer, the support of the SGWT atoms decreases since the number of dots and their sizes decrease. This is consistent with the localized graph signal decomposition that the SGWT is implementing.

More precisely, in these figures, we are showing the support of each atom in a slightly different way than that in Figure B.8: each non-zero entry of the transform matrix **G** is represented by a dot of varying size and shade of gray, thus representing its magnitude. This allows for a visual assessment of the amplitude of the SGWT atoms. It also allows for an assessment of the impact of the Chebyshev approximation: coarser scales, and especially the scaling function, are impacted by the approximation, as indicated by the larger approximation error seen in Figure B.16.

B.7   Chapter 6 Use Cases

B.7.1   Graph Learning

Another approach to defining a graph is to learn its structure from a set of graph signals. For this use case, we use the Gauss–Markov random field (GMRF) approach described in Section 6.3.3. We use the associated toolbox for `Matlab`.

Here, the graph signals we want to model as a GMRF are temperature measurements included in the third party "toolbox" `'bgirault_molene_dataset'`. These are interesting as they highlight several pitfalls to avoid. First of all, temperatures measured by weather stations at various instants define several graph signals. We could compute their empirical covariance matrix but, unfortunately, because of their temporal drift and cyclo-stationarities (night–day cycle), the ergodicity assumption needed to compute the empirical mean graph signal and the covariance matrix does not hold. Instead, we first need to compute the graph signals of the temperature increments between two measurements.

These temperature increments are the random graph signals we wish to model. Their empirical covariance matrix can now be computed, as shown in Code Example B.14. We also compute the mean temperature increment (thus checking that it is sufficiently

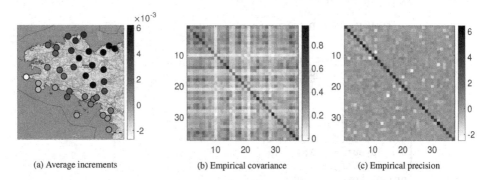

(a) Average increments   (b) Empirical covariance   (c) Empirical precision

Figure B.18 Statistics of the random graph signal of temperature increments. Notice the close-to-zero average increment, the lack of structure in the covariance matrix, and the sparsity of the precision matrix.

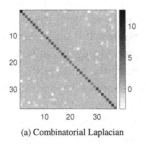

(a) Combinatorial Laplacian

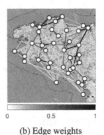

(b) Edge weights

Figure B.19 Learned graph. In (b), only edges with weight greater than 0.1 are plotted.

close to zero) and the precision matrix, which is the target of the learning process. The results are shown in Figure B.18.

Code Example B.14 Covariance estimation:
`code/appendix_grasp/use_case_learning.m`

```
1  % Select the graph
2  grasp_start_opt_3rd_party('bgirault_molene_dataset');
3  load('grasp_molene_data.mat');
4  g = molene_graphs{2};
5  % Compute the temperature increments
6  X = g.data{1};
7  dX = X(:, 2:end) - X(:, 1:(end - 1));
8  % Mean and covariance
9  mu = mean(dX')';
10 S = cov(dX');
11 % Plot the covariance and mean
12 grasp_show_graph(gca, g, 'node_values', mu, 'show_colorbar', 1);
13 imagesc(S);
14 % Plot the precision matrix
15 imagesc(S ^ -1);
```

Now that we have access to a random graph signal covariance matrix, we can perform graph learning under a GMRF assumption. For this, we use the function `estimate_cgl` of the *Graph Laplacian Learning* toolbox (`'usc_graph_learning'`). The results are shown in Figure B.19.

Code Example B.15 Graph learning: `code/appendix_grasp/use_case_learning.m`

```
1  % Learn the combinatorial Laplacian
2  grasp_start_opt_3rd_party('usc_graph_learning');
3  N = grasp_nb_nodes(g);
4  L = estimate_cgl(S, ones(N) - eye(N));
5  % Plot the Laplacian matrix
6  imagesc(L);
7  % Plot the learnt graph
8  g.A = diag(diag(L)) - L;
9  g.A_layout = g.A;
10 g.A_layout(g.A < 0.1) = 0;
11 grasp_show_graph(gca, g,...
12                  'show_edges', true,...
13                  'edge_colorbar', true,...
14                  'edge_color_scale', [0 1]);
```

B.8   Reference for grasp_show_graph Parameters

One of the most important functions of
GraSP is grasp_show_graph. This function plots a graph g on a given Matlab
graphical axis and, as such, provides the
primary mean of visualizing the graph:

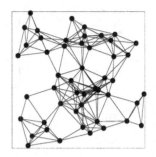

```
1  >> grasp_show_graph(gca, g);
```

An example is shown in Figure B.20.
The rest of this section is a breakdown of
all the parameters of grasp_show_graph
into distinct categories.

Figure B.20 Example of a graph plotted using
grasp_show_graph.

Background Image and Axes Shape

The options described here control the general shape of the figure. First, it may be useful
for some graphs to plot a background image, e.g., geographic graphs may well need a
map in the background. This is controlled by the 'background' option, which holds
the path to an image. Additionally, when the Boolean 'background_grayscale' is set
to true, this image is converted to a black and white image, which can be useful for
distinguishing the colors on and of the graph itself. See Figure B.21 for an example.

Control over the axes, shape and view
is optionally provided by three parameters. 'axis_style' calls Matlab function **axis**, and defaults to 'equal', i.e.,
one unit in either the x or the y direction
appears with the same length on the figure.

The parameter 'layout_boundaries'
can be used to set automatically the
boundaries of the axes with a given margin (e.g., value 0.05 for a 5% margin),

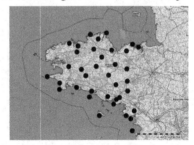

Figure B.21 Graph with a background image.

or to set the boundaries to some arbitrary region in the coordinate system. The default scheme is to restrict to $[0, 10] \times [0, 10]$. When using a background image, the
option 'background_boundaries' is used to map the edges of the image to the coordinate system. This is necessary to get a correct crop of the image when using
'layout_boundaries', and to keep the nodes and the background image consistent.
Refer to Figure B.22 for an example.

Finally, 'viewpoint3D' is useful only for 3D plots and uses the Matlab built-in function **view** to set the camera location, i.e., from where in the 3D space the graph is observed (see the Matlab documentation of **view** to set 'viewpoint3D').

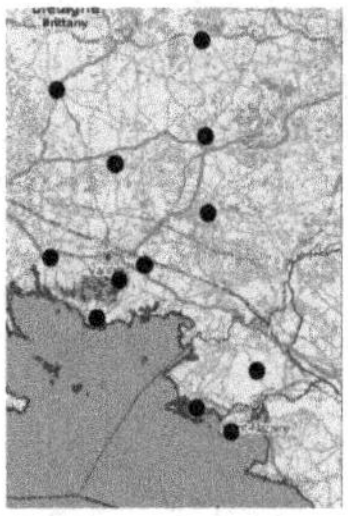

```
1  bg_bound  = [-0.65 -0.07 ; 5.91 6.33] * 10 ^ 6;
2  lay_bound = [-0.35 -0.21 ; 5.95 6.15] * 10 ^ 6;
3  grasp_show_graph(gca, g,...
4                   'axis_style', 'equal',...
5                   'background', 'brittany.png',...
6                   'background_grayscale', true,...
7                   'background_boundaries', bg_bound,...
8                   'layout_boundaries', lay_bound);
```

Figure B.22 Cropped view of Figure B.21.

Node Labels

Plotting a graph leads to losing the mapping between specific nodes and the plotted dots on the figure. To avoid this loss, `grasp_show_graph` provides several options for displaying labels next to each node for identification. The main option is `'node_text'`, which enables these labels. Using the value 0 uses the labels set in the cell array `g.node_names` of the graph structure `g` (or the node ID if this cell array is empty). Custom labels can be used by setting `'node_text'` to a cell array of strings of the correct size.

Other options controlling the labels are `'node_text_shift'`, which shifts the leftmost end of the label (centered on the node with zero shift), `'node_text_fontsize'`, which controls the font size, `'node_text_background_color'`, which controls the background color and `'node_text_background_edge'`, which controls the color of the edge of the label box. An example is shown in Figure B.23.

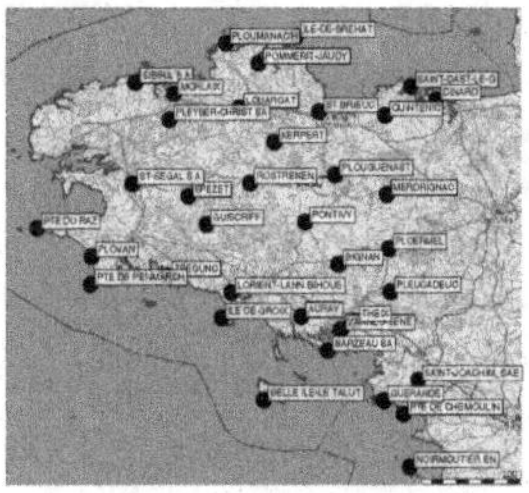

```
1  grasp_show_graph(gca, g,...
2          'node_text', 0,...
3          'node_text_fontsize', 12,...
4          'node_text_shift', [5000 5000],...
5          'node_text_background_color', [1 1 1 0.7],...
6          'node_text_background_edge', [0 0 0]);
```

Figure B.23 Graph with node labels.

Highlighting Nodes

In some contexts, it may be useful to highlight one or more nodes of interest on the graph. In `grasp_show_graph`, this is achieved using a circle around those nodes (see Figure B.24). The option `'highlight_nodes'` is set to a vector containing the IDs of the nodes required to be highlighted. Then, `'highlight_nodes_size'` controls the circle size, `'highlight_nodes_width'` controls its thickness and `'highlight_nodes_color'` controls its color.

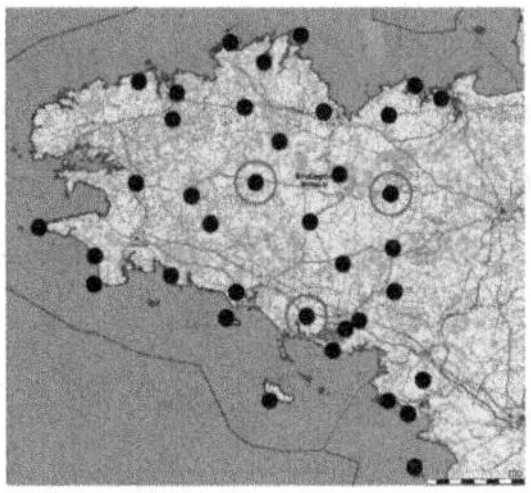

```
1 hlcol = 0.4 * [1 1 1];
2 grasp_show_graph(gca, g,...
3                  'highlight_nodes', [1 2 3],...
4                  'highlight_nodes_size', 3000,...
5                  'highlight_nodes_width', 3,...
6                  'highlight_nodes_color', hlcol);
```

Figure B.24 Highlighting (i.e., circling) nodes.

Edges

The adjacency matrix used to draw the edges of a graph is actually g.A_layout when g.A_layout is non-empty, and g.A otherwise. This allows one to filter out edges that are not important, as has been done in Figure B.25.

Controlling whether to show all edges is achieved using the boolean 'show_edges'. The color of all edges can be tweaked using 'edge_color', and their thickness by 'edge_thickness'.

The graphs used in GSP generally have weighted edges (i.e., they are weighted graphs). As such, plotting the edges if they exist may not reveal the full complexity of the graph as weights are lost in the representation. The function grasp_show_graph addresses this issue by allowing one to draw colored edges, whose colors depend on their weights. This is done using 'edge_colormap' to set the colormap (see the documentation of Matlab function **colormap**, and 'edge_color_scale' to map the lowest and highest colors to a range of weights. Finally, 'edge_colorbar' draws a colorbar below the figure.

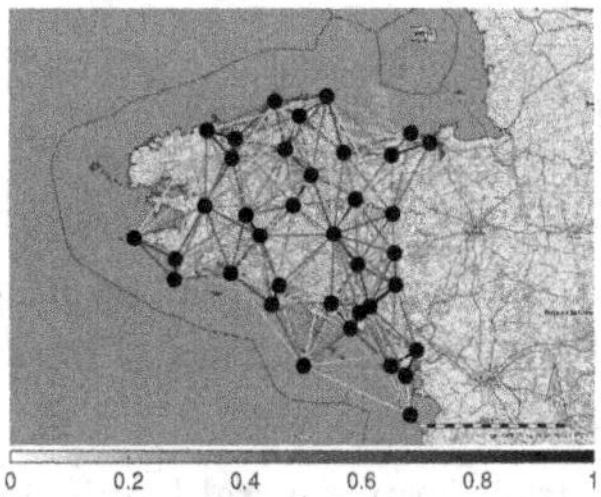

```
1 g.A_layout = grasp_adjacency_thresh(g, 0.2);
2 ecm        = g.show_graph_options.color_map;
3 grasp_show_graph(gca, g,...
4                  'show_edges', true,...
5                  'edge_colormap', ecm,...
6                  'edge_colorbar', true,...
7                  'edge_color_scale', [0 1],...
8                  'edge_thickness', 2);
```

Figure B.25 Edge weight visualization.

Directed Edge Arrow Ends

Some graphs are directed and, as such, the directions of its directed edges need to be shown on the figure. The function grasp_show_graph achieves this by drawing arrows pointing to the target node (see Figure B.26). These arrows' shapes can be modified with several options. The function arrow_max_tip_back_fraction sets how much the very tip of the arrow (the end of the arrowhead) is pulled away from the target node

(0 means that the tip is on the node center, anything greater introduces a gap between the node center and the tip of the arrow). The arrowhead scales with the length of the edge on the figure, and, to avoid very large arrowheads, `arrow_max_head_fraction` sets the maximum length of the arrowhead. Finally, the thickness of the arrowhead can be specified using `arrow_width_screen_fraction`.

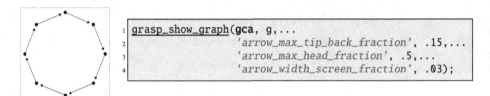

```
1  grasp_show_graph(gca, g,...
2               'arrow_max_tip_back_fraction', .15,...
3               'arrow_max_head_fraction', .5,...
4               'arrow_width_screen_fraction', .03);
```

Figure B.26 Directed edges.

B.9   Reference for `grasp_show_transform` Parameters

The classical graph Fourier modes representation shown in Figure B.7 can be impractical owing to the large space required to display all modes (recall that there are as many modes as there are nodes in the graph). An alternative approach is to embed the nodes in a one-dimensional space, i.e., representing the nodes as if they were irregular samples of a temporal signal. This allows one to stack the graph Fourier modes in 2D, i.e., on a sheet of paper. More specifically, let us assume we have the 1D embedding $x_i \in \mathbb{R}$ for each node $i \in \{1, \ldots, N\}$. Without loss of generality, we can assume that x_i increases with i (see Box 2.1). We can then plot the N graph Fourier modes using the points

$$(x_i, k + [\mathbf{u}_k]_i) \tag{B.4}$$

linked by straight lines on a single figure. If we also assume that all graph Fourier modes satisfy $|[\mathbf{u}_k]_i| \leq 0.5$, $k \in \{0, \ldots, N-1\}$, this plot is without overlap of any of the stacked 1D representations of the graph Fourier modes. This is implemented in `grasp_show_transform`, and an example is shown in Figure B.27.

In addition to representing the graph Fourier transform, several key features are implemented in `grasp_show_transform` to study the properties of the GFT. First, the attentive reader will have noticed that the right part of Figure B.27 is different from that described above. Indeed, regularly stacking the graph Fourier modes as in (B.4) leaves out a crucial piece of information needed to fully understand the graph spectral domain: the graph frequencies. To intuitively show those graph frequencies, the 1D representation of each graph Fourier mode is mapped on the right-axis scale to its graph frequency. In particular, this strongly highlights one key property of the GFT: *in general, the graph frequencies are not equispaced.*

We now categorize the parameters of `grasp_show_transform` and describe how and when to use them.

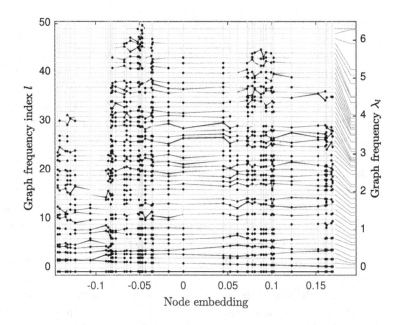

Figure B.27 Example of a graph Fourier transform plotted using `grasp_show_transform`.

Node Embedding

By default, `grasp_show_transform` embeds the nodes into a 1D space based on the first non-constant eigenvector of the random walk Laplacian, since it has been shown that it has good embedding properties. However, this setting can be overwritten by explicitly giving a specific embedding using the parameter `'embedding'`.

For some graphs, the specific clustering of nodes into groups of nodes can be obtained. If a clustering is compatible with the default embedding, or the custom embedding, i.e., the nodes can be also clustered in the same way in the embedding space, then passing this clustering to the function using the parameter `'clusters'` triggers a rendering of vertical bars that separate the clusters.

Graph Transform

The primary usage of `grasp_show_transform` is to plot the full graph Fourier transform. However, it is straightforward to extend the approach to plotting any graph transform. This is implemented by the parameter `'transform_matrix'`, which holds the direct transform matrix; this is similar to what was done in Section B.6.1. A custom graph transform also requires a custom mapping of the transform's atoms to graph frequencies. This is achieved by providing a vector of length M to the parameter `'graph_frequencies'` that holds for each entry m the graph frequencies of the mth basis vector of the transform.

Support and Amplitude Visualization

The support of a transform is a key property for understanding the localization of the transform. The parameter `'epsilon_support'` sets the amplitude condition for an entry of the transform matrix to be in the support, defaulting to 0.05 after amplitude normalization.

Amplitude normalization is the process through which entries of the transform matrix are normalized using `'amplitude_normalization'`, either with a constant normalization factor such that the maximum amplitude is 1 (`'overall_max_abs'`) or such that the maximum amplitude (respectively the ℓ_2 norm) of each basis vector equals 1 (`'l2'`, respectively `'max_abs'`).

To control the density of the plot and the overlap between each mode, the parameter `'amplitude_scale'` can be set to 1 for maximum coverage with no overlap, or to a smaller value for margins between modes or to a larger value for overlap (a value of 2 corresponds to maximum coverage with no more than two modes overlapping anywhere).

Finally, the size of the scatter plot showing the support (the dots in Figure B.27) is controlled by the parameter `'support_scatter_size'`. These scatter plots can alternatively hold information about the value of the entry following a number of schemes set by `'support_scatter_mode'`:

`'constant'`	default as shown in Figure B.27
`'var_size'`	dots have varying size depending on the amplitude
`'var_gray'`	dots have varying shade of gray depending on the amplitude
`'var_color'`	dots have varying color depending on the value (red for positive, blue for negative, white for 0, where the intensity of blue or red is a function of the amplitude)
`'var_width_color'`	combination of `'var_size'` and `'var_color'`
`'var_width_gray'`	combination of `'var_size'` and `'var_gray'` (see Figure B.17).

Spectral Bands

Just as in Figure B.17, spectral bands can be highlighted with shaded colors. These bands are set using the function `'bands'`, assigning it a $K \times 2$ matrix where each of the K rows holds the lower and higher bounds of the band. Colors different from the default `Matlab` colors can be set using `'bands_colors'` with a $K \times 3$ matrix, where each row is a vector with three color values (red, green, blue or RGB).

Miscellaneous

Finally, `'highlight_entries'` can be set to a boolean matrix used to highlight some entries of `'transform_matrix'` (see Figure B.14). To be valid, both matrices need to have the same size.

Problems

B.1 Implement the spline approach in Section B.6.1, instead of the Mexican hat approach. In other words, the scaling function h and the mother wavelet g should satisfy

$$h(x) = \gamma \, \exp\left(-\left(\frac{x}{0.6\lambda_{\min}}\right)^4\right), \qquad g(x) = \begin{cases} x_1^{-\alpha} x^\alpha & \text{for } x < x_1, \\ s(x) & \text{for } x_1 \leqslant x \leqslant x_2, \\ x_2^\beta x^{-\beta} & \text{for } x_2 < x, \end{cases}$$

with $s(x)$ the third-order polynomial such that $g(x)$ and $g'(x)$ are both continuous at x_1 and x_2, and γ is equal to the maximum of g. Use the parameters $\alpha = \beta = 2$, $x_1 = 1$ and $x_2 = 2$. Compare the Chebyshev polynomial approximations of this spline approach with the Mexican hat approach.

B.2 Study the SGWT of Section B.6.1 with respect to the toy graph of Figure B.13.
a. Compute the distance matrix between nodes using grasp_distances_layout.
b. Compute the adjacency matrix as a Gaussian kernel of the distance with the parameter $\sigma = 1.5$ using grasp_adjacency_gaussian.
c. Using grasp_eigendecomposition, compute the GFT based on the random walk Laplacian, i.e., with the combinatorial Laplacian matrix as the parameter 'matrix' and the degree matrix as the parameter 'inner_product'.
d. How do the clusters of the toy graph influence the properties of the SGWT?

References

[1] G. Strang, "The discrete cosine transform," *SIAM Review*, vol. 41, no. 1, pp. 135–147, 1999. https://doi.org/10.1137/S0036144598336745

[2] W. B. Pennebaker and J. L. Mitchell, *JPEG: Still Image Data Compression Standard*. Springer Science & Business Media, 1992.

[3] C. Tomasi and R. Manduchi, "Bilateral filtering for gray and color images." in *Proc. International Conference on Computer Vision (ICCV)*, vol. 98, no. 1, 1998, p. 2. https://doi.org/10.1109/ICCV.1998.710815

[4] L. Page, S. Brin, R. Motwani, and T. Winograd, "The pagerank citation ranking: Bringing order to the web." Stanford InfoLab, Tech. Rep., 1999. http://ilpubs.stanford.edu:8090/422/

[5] L. Liu and U. Mitra, "Policy sampling and interpolation for wireless networks: A graph signal processing approach," in *Proc. 2019 IEEE Global Communications Conference (GLOBECOM)*. IEEE, 2019, pp. 1–6. https://doi.org/10.1109/GLOBECOM38437.2019.9013907

[6] E. D. Kolaczyk and G. Csárdi, *Statistical Analysis of Network Data with R*. Springer, 2014. https://doi.org/10.1007/978-1-4939-0983-4

[7] F. R. Chung, *Spectral Graph Theory*. American Mathematical Society, 1997, no. 92. https://doi.org/http://dx.doi.org/10.1090/cbms/092

[8] P. Van Mieghem, *Graph Spectra for Complex Networks*. Cambridge University Press, 2010. https://doi.org/10.1017/CBO9780511921681

[9] D. I. Shuman, S. K. Narang, P. Frossard, A. Ortega, and P. Vandergheynst, "The emerging field of signal processing on graphs: Extending high-dimensional data analysis to networks and other irregular domains," *IEEE Signal Processing Magazine*, vol. 30, no. 3, pp. 83–98, 2013. https://doi.org/10.1109/MSP.2012.2235192

[10] A. Sandryhaila and J. M. F. Moura, "Big data analysis with signal processing on graphs: Representation and processing of massive data sets with irregular structure," *IEEE Signal Processing Magazine*, vol. 31, no. 5, pp. 80–90, 2014. https://doi.org/10.1109/MSP.2014.2329213

[11] A. Ortega, P. Frossard, J. Kovačević, J. M. F. Moura, and P. Vandergheynst, "Graph signal processing: Overview, challenges, and applications," *Proceedings of the IEEE*, vol. 106, no. 5, pp. 808–828, 2018. https://doi.org/10.1109/JPROC.2018.2820126

[12] L. Stanković, D. Mandic, M. Daković, M. Brajović, B. Scalzo, S. Li, and A. G. Constantinides, "Data analytics on graphs Part I: Graphs and spectra on graphs," *Foundations and Trends® in Machine Learning*, vol. 13, no. 1, pp. 1–157, 2020. https://doi.org/10.1561/2200000078-1

[13] ——, "Data analytics on graphs Part II: Signals on graphs," *Foundations and Trends® in Machine Learning*, vol. 13, no. 2-3, pp. 158–331, 2020. https://doi.org/10.1561/2200000078-2

[14] ——, "Data analytics on graphs Part III: Machine learning on graphs, from graph topology to applications," *Foundations and Trends® in Machine Learning*, vol. 13, no. 4, pp. 332–530, 2020. https://doi.org/10.1561/2200000078-3

[15] A. Sandryhaila and J. M. F. Moura, "Discrete signal processing on graphs," *IEEE Transactions on Signal Processing*, vol. 61, no. 7, pp. 1644–1656, 2013. https://doi.org/10.1109/TSP.2013.2238935

[16] T. H. Cormen, C. E. Leiserson, R. L. Rivest, and C. Stein, *Introduction to Algorithms*. MIT Press, 2009.

[17] M. Vetterli, J. Kovačević, and V. K. Goyal, *Foundations of Signal Processing*. Cambridge University Press, 2014. https://doi.org/10.1017/CBO9781139839099

[18] G. H. Golub and C. F. Van Loan, *Matrix Computations*. Johns Hopkins University Press, 2012, vol. 3.

[19] G. W. Stewart, *Matrix Algorithms: Eigensystems*. SIAM, 2001, vol. 2. https://doi.org/10.1137/1.9780898718058

[20] B. Girault, "Signal processing on graphs – contributions to an emerging field," PhD dissertation, Ecole Normale Supérieure de Lyon, 2015. https://tel.archives-ouvertes.fr/tel-01256044

[21] B. Girault, S. S. Narayanan, A. Ortega, P. Gonçalves, and E. Fleury, "GraSP: A Matlab toolbox for graph signal processing," in *Proc. 2017 IEEE International Conference on Acoustics, Speech and Signal Processing (ICASSP)*. IEEE, 2017, pp. 6574–6575. https://doi.org/10.1109/ICASSP.2017.8005300

[22] J. A. Deri and J. M. F. Moura, "Spectral projector-based graph Fourier transforms," *IEEE Journal of Selected Topics in Signal Processing*, vol. 11, no. 6, pp. 785–795, 2017. https://doi.org/10.1109/JSTSP.2017.2731599

[23] G. Agnarsson and R. Greenlaw, *Graph Theory: Modeling, Applications, and Algorithms*. Pearson/Prentice Hall, 2007.

[24] M. Püschel and J. M. Moura, "Algebraic signal processing theory: Foundation and 1D time," *IEEE Transactions on Signal Processing*, vol. 56, no. 8, pp. 3572–3585, 2008. https://doi.org/10.1109/TSP.2008.925261

[25] M. Püschel and J. M. F. Moura, "Algebraic signal processing theory: Cooley–Tukey type algorithms for DCTs and DSTs," *IEEE Transactions on Signal Processing*, vol. 56, no. 4, pp. 1502–1521, 2008. https://doi.org/10.1109/TSP.2007.907919

[26] F. R. Gantmacher, *The Theory of Matrices*. American Mathematical Soc., 2000, vol. 131.

[27] N. Saito, "How can we naturally order and organize graph Laplacian eigenvectors?" in *Proc. 2018 IEEE Statistical Signal Processing Workshop (SSP)*. IEEE, 2018, pp. 483–487. https://doi.org/10.1109/SSP.2018.8450808

[28] B. Girault and A. Ortega, "What's in a frequency: New tools for graph Fourier transform visualization," *arXiv preprint arXiv:1903.08827*, 2019. https://arxiv.org/abs/1903.08827

[29] E. Isufi, A. Loukas, A. Simonetto, and G. Leus, "Autoregressive moving average graph filtering," *IEEE Transactions on Signal Processing*, vol. 65, no. 2, pp. 274–288, 2016. https://doi.org/10.1109/TSP.2016.2614793

[30] N. Tremblay, P. Gonçalves, and P. Borgnat, "Design of graph filters and filterbanks," in *Cooperative and Graph Signal Processing*. Elsevier, 2018, pp. 299–324. https://doi.org/10.1016/B978-0-12-813677-5.00011-0

[31] D. Cvetković, P. Rowlinson, and S. Simić, *An Introduction to the Theory of Graph Spectra.* Cambridge University Press, 2001. https://doi.org/10.1017/CBO9780511801518

[32] A. Sandryhaila and J. M. F. Moura, "Discrete signal processing on graphs: Frequency analysis," *IEEE Transactions on Signal Processing*, vol. 62, no. 12, pp. 3042–3054, 2014. https://doi.org/10.1109/TSP.2014.2321121

[33] P. Milanfar, "A tour of modern image filtering: New insights and methods, both practical and theoretical," *IEEE Signal Processing Magazine*, vol. 30, no. 1, pp. 106–128, 2013. https://doi.org/10.1109/MSP.2011.2179329

[34] ——, "Symmetrizing smoothing filters," *SIAM Journal on Imaging Sciences*, vol. 6, no. 1, pp. 263–284, 2013. https://doi.org/10.1137/120875843

[35] R. Sinkhorn, "A relationship between arbitrary positive matrices and doubly stochastic matrices," *Annals of Mathematical Statistics*, vol. 35, no. 2, pp. 876–879, 1964. https://www.jstor.org/stable/2238545

[36] B. Girault, A. Ortega, and S. S. Narayanan, "Irregularity-aware graph Fourier transforms," *IEEE Transactions on Signal Processing*, vol. 66, no. 21, pp. 5746–5761, 2018. https://doi.org/10.1109/TSP.2018.2870386

[37] K.-S. Lu, A. Ortega, D. Mukherjee, and Y. Chen, "Perceptually inspired weighted MSE optimization using irregularity-aware graph fourier transform," in *Proc. 2020 IEEE International Conference on Image Processing*, 2020. https://doi.org/10.1109/ICIP40778.2020.9190876

[38] L. W. Beineke, R. J. Wilson, P. J. Cameron *et al.*, *Topics in Algebraic Graph Theory.* Cambridge University Press, 2004.

[39] Y. C. Eldar, *Sampling Theory: Beyond Bandlimited Systems.* Cambridge University Press, 2015. https://doi.org/10.1017/CBO9780511762321

[40] S. Chen, R. Varma, A. Sandryhaila, and J. Kovacevic, "Discrete signal processing on graphs: Sampling theory," *IEEE Transactions on Signal Processing*, vol. 63, pp. 6510–6523, 2015. https://doi.org/10.1109/TSP.2015.2469645

[41] S. Chen, R. Varma, A. Singh, and J. Kovačević, "Signal representations on graphs: Tools and applications," *arXiv preprint arXiv:1512.05406*, 2015. https://arxiv.org/abs/1512.05406

[42] B. Girault, "Stationary graph signals using an isometric graph translation," in *Proc. 2015 23rd European Signal Processing Conference (EUSIPCO).* IEEE, 2015, pp. 1516–1520. https://doi.org/10.1109/EUSIPCO.2015.7362637

[43] N. Perraudin and P. Vandergheynst, "Stationary signal processing on graphs," *IEEE Transactions on Signal Processing*, vol. 65, no. 13, pp. 3462–3477, 2017. https://doi.org/10.1109/TSP.2017.2690388

[44] B. Girault, S. S. Narayanan, and A. Ortega, "Towards a definition of local stationarity for graph signals," in *Proc. 2017 IEEE International Conference on Acoustics, Speech and Signal Processing (ICASSP).* IEEE, 2017, pp. 4139–4143. https://doi.org/10.1109/ICASSP.2017.7952935

[45] N. Cressie, *Statistics for Spatial Data.* John Wiley & Sons, 2015.

[46] A. Serrano, B. Girault, and A. Ortega, "Graph variogram: A novel tool to measure spatial stationarity," in *Proc. 2018 IEEE Global Conference on Signal and Information Processing (GlobalSIP).* IEEE, 2018, pp. 753–757. https://doi.org/10.1109/GlobalSIP.2018.8646692

[47] A. G. Marques, S. Segarra, G. Leus, and A. Ribeiro, "Stationary graph processes and spectral estimation," *IEEE Transactions on Signal Processing*, vol. 65, no. 22, pp. 5911–5926, 2017. https://doi.org/10.1109/TSP.2017.2739099

[48] ——, "Sampling of graph signals with successive local aggregations," *IEEE Transactions on Signal Processing*, vol. 64, no. 7, pp. 1832–1843, 2016. https://doi.org/10.1109/TSP.2015.2507546

[49] Y. Tanaka, "Spectral domain sampling of graph signals," *IEEE Transactions on Signal Processing*, vol. 66, no. 14, pp. 3752–3767, 2018. https://doi.org/10.1109/TSP.2018.2839620

[50] I. Pesenson, "Sampling in Paley–Wiener spaces on combinatorial graphs," *Transactions of the American Mathematical Society*, vol. 360, no. 10, pp. 5603–5627, 2008. https://doi.org/10.1090/S0002-9947-08-04511-X

[51] S. K. Narang, A. Gadde, E. Sanou, and A. Ortega, "Localized iterative methods for interpolation in graph structured data," in *Proc. 2013 IEEE Global Conference on Signal and Information Processing*. IEEE, 2013, pp. 491–494. https://doi.org/10.1109/GlobalSIP.2013.6736922

[52] X. Wang, P. Liu, and Y. Gu, "Local-set-based graph signal reconstruction," *IEEE Transactions on Signal Processing*, vol. 63, no. 9, pp. 2432–2444, 2015. https://doi.org/10.1109/TSP.2015.2411217

[53] Y. Tanaka, Y. C. Eldar, A. Ortega, and G. Cheung, "Sampling signals on graphs: From theory to applications," *IEEE Signal Processing Magazine*, vol. 37, no. 6, pp. 14–30, 2020. https://doi.org/10.1109/MSP.2020.3016908

[54] Y. Bai, F. Wang, G. Cheung, Y. Nakatsukasa, and W. Gao, "Fast graph sampling set selection using Gershgorin disc alignment," *IEEE Transactions on Signal Processing*, vol. 68, pp. 2419–2434, 2020. https://doi.org/10.1109/TSP.2020.2981202

[55] H. Shomorony and A. S. Avestimehr, "Sampling large data on graphs," in *Proc. 2014 IEEE Global Conference on Signal and Information Processing (GlobalSIP)*. IEEE, 2014, pp. 933–936. https://doi.org/10.1109/GlobalSIP.2014.7032257

[56] A. Sakiyama, Y. Tanaka, T. Tanaka, and A. Ortega, "Eigendecomposition-free sampling set selection for graph signals," *IEEE Transactions on Signal Processing*, vol. 67, no. 10, pp. 2679–2692, 2019. https://doi.org/10.1109/TSP.2019.2908129

[57] A. Anis, A. Gadde, and A. Ortega, "Efficient sampling set selection for bandlimited graph signals using graph spectral proxies," *IEEE Transactions on Signal Processing*, vol. 64, no. 14, pp. 3775–3789, 2016. https://doi.org/10.1109/TSP.2016.2546233

[58] M. Tsitsvero, S. Barbarossa, and P. Di Lorenzo, "Signals on graphs: Uncertainty principle and sampling," *IEEE Transactions on Signal Processing*, vol. 64, no. 18, pp. 4845–4860, 2016. https://doi.org/10.1109/TSP.2016.2573748

[59] G. Puy, N. Tremblay, R. Gribonval, and P. Vandergheynst, "Random sampling of bandlimited signals on graphs," *Applied and Computational Harmonic Analysis*, vol. 44, no. 2, pp. 446–475, 2018. https://doi.org/10.1016/j.acha.2016.05.005

[60] S. Chen, R. Varma, A. Singh, and J. Kovačević, "Signal recovery on graphs: Fundamental limits of sampling strategies," *IEEE Transactions on Signal and Information Processing over Networks*, vol. 2, no. 4, pp. 539–554, 2016. https://doi.org/10.1109/TSIPN.2016.2614903

[61] Y. C. Eldar, "Sampling with arbitrary sampling and reconstruction spaces and oblique dual frame vectors," *Journal of Fourier Analysis and Applications*, vol. 9, no. 1, pp. 77–96, 2003. https://doi.org/10.1007/s00041-003-0004-2

[62] A. Agaskar and Y. M. Lu, "A spectral graph uncertainty principle," *IEEE Transactions on Information Theory*, vol. 59, no. 7, pp. 4338–4356, 2013. https://doi.org/10.1109/TIT.2013.2252233

[63] B. Pasdeloup, V. Gripon, R. Alami, and M. G. Rabbat, "Uncertainty principle on graphs," in *Vertex–Frequency Analysis of Graph Signals.* Springer, 2019, pp. 317–340. https://doi.org/10.1007/978-3-030-03574-7_9

[64] B. Pasdeloup, R. Alami, V. Gripon, and M. Rabbat, "Toward an uncertainty principle for weighted graphs," in *Proc. 2015 23rd European Signal Processing Conference (EUSIPCO).* IEEE, 2015, pp. 1496–1500. https://doi.org/10.1109/EUSIPCO.2015.7362633

[65] D. Van De Ville, R. Demesmaeker, and M. G. Preti, "When Slepian meets Fiedler: Putting a focus on the graph spectrum," *IEEE Signal Processing Letters*, vol. 24, no. 7, pp. 1001–1004, 2017. https://doi.org/10.1109/LSP.2017.2704359

[66] M. Crovella and E. Kolaczyk, "Graph wavelets for spatial traffic analysis," in *Proc. INFO-COM 2003, 22nd Annual Joint Conference of the IEEE Computer and Communications Societies.* IEEE, 2003, pp. 1848–1857. https://doi.org/10.1109/INFCOM.2003.1209207

[67] M. Vetterli and J. Kovačević, *Wavelets and Subband Coding.* Prentice Hall, 1995. https://waveletsandsubbandcoding.org/

[68] S. K. Narang and A. Ortega, "Lifting based wavelet transforms on graphs," in *Proc. 2009 Asia-Pacific Signal and Information Processing Association Annual Summit and Conference*, 2009, pp. 441–444. http://hdl.handle.net/2115/39737

[69] I. Daubechies and W. Sweldens, "Factoring wavelet transforms into lifting steps," *Journal of Fourier Analysis and Applications*, vol. 4, no. 3, pp. 247–269, 1998. https://doi.org/10.1007/BF02476026

[70] D. B. Tay and A. Ortega, "Bipartite graph filter banks: Polyphase analysis and generalization," *IEEE Transactions on Signal Processing*, vol. 65, no. 18, pp. 4833–4846, 2017. https://doi.org/10.1109/TSP.2017.2718969

[71] E. J. Stollnitz, T. D. DeRose, A. D. DeRose, and D. H. Salesin, *Wavelets for Computer Graphics: Theory and Applications.* Morgan Kaufmann, 1996.

[72] P. Schröder and W. Sweldens, "Spherical wavelets: Efficiently representing functions on the sphere," in *Proc. 22nd Annual Conference on Computer Graphics and Interactive Techniques*, 1995, pp. 161–172. https://doi.org/10.1145/218380.218439

[73] N. Tremblay and P. Borgnat, "Subgraph-based filterbanks for graph signals," *IEEE Transactions on Signal Processing*, vol. 64, no. 15, pp. 3827–3840, 2016. https://doi.org/10.1109/TSP.2016.2544747

[74] D. K. Hammond, P. Vandergheynst, and R. Gribonval, "Wavelets on graphs via spectral graph theory," *Applied and Computational Harmonic Analysis*, vol. 30, no. 2, pp. 129–150, 2011. https://doi.org/10.1016/j.acha.2010.04.005

[75] I. Pesenson, "Variational splines and paley–wiener spaces on combinatorial graphs," *Constructive Approximation*, vol. 29, no. 1, pp. 1–21, 2009. https://doi.org/10.1007/s00365-007-9004-9

[76] N. Leonardi and D. Van De Ville, "Tight wavelet frames on multislice graphs," *IEEE Transactions on Signal Processing*, vol. 61, no. 13, pp. 3357–3367, 2013. https://doi.org/10.1109/TSP.2013.2259825

[77] D. I. Shuman, C. Wiesmeyr, N. Holighaus, and P. Vandergheynst, "Spectrum-adapted tight graph wavelet and vertex-frequency frames," *IEEE Transactions on Signal Processing*, vol. 63, no. 16, pp. 4223–4235, 2015. https://doi.org/10.1109/TSP.2015.2424203

[78] O. Teke and P. P. Vaidyanathan, "Extending classical multirate signal processing theory to graphs – Part I: Fundamentals," *IEEE Transactions on Signal Processing*, vol. 65, no. 2, pp. 409–422, 2016. https://doi.org/10.1109/TSP.2016.2617833

[79] ——, "Extending classical multirate signal processing theory to graphs – Part II: M-channel filter banks," *IEEE Transactions on Signal Processing*, vol. 65, no. 2, pp. 423–437, 2016. https://doi.org/10.1109/TSP.2016.2620111

[80] A. Anis and A. Ortega, "Critical sampling for wavelet filterbanks on arbitrary graphs," in *Proc. 2017 IEEE International Conference on Acoustics, Speech and Signal Processing (ICASSP)*, 2017, pp. 3889–3893. https://doi.org/10.1109/ICASSP.2017.7952885

[81] S. K. Narang and A. Ortega, "Perfect reconstruction two-channel wavelet filter banks for graph structured data," *IEEE Transactions on Signal Processing*, vol. 60, no. 6, pp. 2786–2799, 2012. https://doi.org/10.1109/TSP.2012.2188718

[82] ——, "Compact support biorthogonal wavelet filterbanks for arbitrary undirected graphs," *IEEE Transactions on Signal Processing*, vol. 61, no. 19, pp. 4673–4685, 2013. https://doi.org/10.1109/TSP.2013.2273197

[83] D. B. Tay, Y. Tanaka, and A. Sakiyama, "Critically sampled graph filter banks with polynomial filters from regular domain filter banks," *Signal Processing*, vol. 131, pp. 66–72, 2017. https://doi.org/10.1016/j.sigpro.2016.07.003

[84] Y. Tanaka and A. Sakiyama, "*M*-channel oversampled graph filter banks," *IEEE Transactions on Signal Processing*, vol. 62, no. 14, pp. 3578–3590, 2014. https://doi.org/10.1109/TSP.2014.2328983

[85] D. I. Shuman, M. J. Faraji, and P. Vandergheynst, "A multiscale pyramid transform for graph signals," *IEEE Transactions on Signal Processing*, vol. 64, no. 8, pp. 2119–2134, 2015. https://doi.org/10.1109/TSP.2015.2512529

[86] P. Burt and E. Adelson, "The laplacian pyramid as a compact image code," *IEEE Transactions on Communications*, vol. 31, no. 4, pp. 532–540, 1983. https://doi.org/10.1109/TCOM.1983.1095851

[87] R. R. Coifman and M. Maggioni, "Diffusion wavelets," *Applied and Computational Harmonic Analysis*, vol. 21, no. 1, pp. 53–94, 2006. https://doi.org/10.1016/j.acha.2006.04.004

[88] D. I. Shuman, "Localized spectral graph filter frames: A unifying framework, survey of design considerations, and numerical comparison," *IEEE Signal Processing Magazine*, vol. 37, no. 6, pp. 43–63, 2020. https://doi.org/10.1109/MSP.2020.3015024

[89] D. A. Spielman and S.-H. Teng, "Spectral sparsification of graphs," *SIAM Journal on Computing*, vol. 40, no. 4, pp. 981–1025, 2011. https://doi.org/10.1137/08074489X

[90] D. A. Spielman and N. Srivastava, "Graph sparsification by effective resistances," *SIAM Journal on Computing*, vol. 40, no. 6, pp. 1913–1926, 2011. https://doi.org/10.1137/080734029

[91] K.-S. Lu and A. Ortega, "Fast graph Fourier transforms based on graph symmetry and bipartition," *IEEE Transactions on Signal Processing*, vol. 67, no. 18, pp. 4855–4869, 2019. https://doi.org/10.1109/TSP.2019.2932882

[92] L. Le Magoarou, R. Gribonval, and N. Tremblay, "Approximate fast graph Fourier transforms via multilayer sparse approximations," *IEEE Transactions on Signal and Information Processing over Networks*, vol. 4, no. 2, pp. 407–420, 2017. https://doi.org/10.1109/TSIPN.2017.2710619

[93] J. Zeng, G. Cheung, and A. Ortega, "Bipartite approximation for graph wavelet signal decomposition," *IEEE Transactions on Signal Processing*, vol. 65, pp. 5466–5480, 2017. https://doi.org/10.1109/TSP.2017.2733489

[94] E. Pavez, H. E. Egilmez, and A. Ortega, "Learning graphs with monotone topology properties and multiple connected components," *IEEE Transactions on Signal Processing*, vol. 66, no. 9, pp. 2399–2413, 2018. https://doi.org/10.1109/TSP.2018.2813337

[95] F. Dörfler and F. Bullo, "Kron reduction of graphs with applications to electrical networks," *IEEE Transactions on Circuits and Systems*, vol. 60, no. 1, pp. 150–163, 2013. https://doi.org/10.1109/TCSI.2012.2215780

[96] G. H. Chen and D. Shah, "Explaining the success of nearest neighbor methods in prediction," *Foundations and Trends in Machine Learning*, vol. 10, no. 5-6, pp. 337–588, 2018. https://doi.org/10.1561/2200000064

[97] S. T. Roweis and L. K. Saul, "Nonlinear dimensionality reduction by locally linear embedding," *Science*, vol. 290, no. 5500, pp. 2323–2326, 2000. https://doi.org/10.1126/science.290.5500.2323

[98] S. Shekkizhar and A. Ortega, "Graph construction from data by non-negative kernel regression," in *Proc. 2020 IEEE International Conference on Acoustics, Speech and Signal Processing (ICASSP)*. IEEE, 2020, pp. 3892–3896. https://doi.org/10.1109/ICASSP40776.2020.9054425

[99] A. P. Dempster, "Covariance selection," *Biometrics*, pp. 157–175, 1972. https://doi.org/10.2307/2528966

[100] C. Uhler, "Geometry of maximum likelihood estimation in gaussian graphical models," *Annals of Statistics*, vol. 40, no. 1, pp. 238–261, 2012. https://www.jstor.org/stable/41713634

[101] J. Friedman, T. Hastie, and R. Tibshirani, "Sparse inverse covariance estimation with the graphical lasso," *Biostatistics*, vol. 9, no. 3, pp. 432–441, 2008. https://doi.org/10.1093/biostatistics/kxm045

[102] N. Meinshausen and P. Bühlmann, "High-dimensional graphs and variable selection with the Lasso," *Annals of Statistics*, vol. 34, no. 3, pp. 1436 – 1462, 2006. https://doi.org/10.1214/009053606000000281

[103] X. Dong, D. Thanou, P. Frossard, and P. Vandergheynst, "Learning laplacian matrix in smooth graph signal representations," *IEEE Transactions on Signal Processing*, vol. 64, no. 23, pp. 6160–6173, 2016. https://doi.org/10.1109/TSP.2016.2602809

[104] V. Kalofolias, "How to learn a graph from smooth signals," in *Proceedings of the 19th International Conference on Artificial Intelligence and Statistics*, ser. Proc. of Machine Learning Research, vol. 51. PMLR, 09–11 May 2016, pp. 920–929. http://proceedings.mlr.press/v51/kalofolias16.html

[105] M. Slawski and M. Hein, "Estimation of positive definite M-matrices and structure learning for attractive Gaussian–Markov random fields," *Linear Algebra and its Applications*, vol. 473, pp. 145–179, 2015. https://doi.org/10.1016/j.laa.2014.04.020

[106] B. Lake and J. Tenenbaum, "Discovering structure by learning sparse graphs," in *Proceedings of the Annual Meeting of the Cognitive Science Society*, 2010, pp. 778–784. http://hdl.handle.net/1721.1/112759

[107] H. E. Egilmez, E. Pavez, and A. Ortega, "Graph learning from data under laplacian and structural constraints," *IEEE Journal of Selected Topics in Signal Processing*, vol. 11, no. 6, pp. 825–841, 2017. https://doi.org/10.1109/JSTSP.2017.2726975

[108] S. Segarra, A. G. Marques, G. Mateos, and A. Ribeiro, "Network topology inference from spectral templates," *IEEE Transactions on Signal and Information Processing over Networks*, vol. 3, no. 3, pp. 467–483, 2017. https://doi.org/10.1109/TSIPN.2017.2731051

[109] X. Dong, D. Thanou, M. Rabbat, and P. Frossard, "Learning graphs from data: A signal representation perspective," *IEEE Signal Processing Magazine*, vol. 36, no. 3, pp. 44–63, 2019. https://doi.org/10.1109/MSP.2018.2887284

[110] G. Mateos, S. Segarra, A. G. Marques, and A. Ribeiro, "Connecting the dots: Identifying network structure via graph signal processing," *IEEE Signal Processing Magazine*, vol. 36, no. 3, pp. 16–43, 2019. https://doi.org/10.1109/MSP.2018.2890143

[111] Y. Li, R. Yu, C. Shahabi, and Y. Liu, "Diffusion convolutional recurrent neural network: Data-driven traffic forecasting," in *International Conference on Learning Representations*, 2018. https://openreview.net/forum?id=SJiHXGWAZ

[112] A. Hasanzadeh, X. Liu, N. Duffield, K. R. Narayanan, and B. Chigoy, "A graph signal processing approach for real-time traffic prediction in transportation networks," *arXiv preprint arXiv:1711.06954*, 2017. https://arxiv.org/abs/1711.06954v1

[113] E. Drayer and T. Routtenberg, "Detection of false data injection attacks in smart grids based on graph signal processing," *IEEE Systems Journal*, 2019. https://doi.org/10.1109/JSYST.2019.2927469

[114] K. He, L. Stankovic, J. Liao, and V. Stankovic, "Non-intrusive load disaggregation using graph signal processing," *IEEE Transactions on Smart Grid*, vol. 9, no. 3, pp. 1739–1747, 2016. https://doi.org/10.1109/TSG.2016.2598872

[115] W. Huang, T. A. Bolton, J. D. Medaglia, D. S. Bassett, A. Ribeiro, and D. Van De Ville, "A graph signal processing perspective on functional brain imaging," *Proceedings of the IEEE*, vol. 106, no. 5, pp. 868–885, 2018. https://doi.org/10.1109/JPROC.2018.2798928

[116] J. D. Medaglia, W. Huang, E. A. Karuza, A. Kelkar, S. L. Thompson-Schill, A. Ribeiro, and D. S. Bassett, "Functional alignment with anatomical networks is associated with cognitive flexibility," *Nature Human Behaviour*, vol. 2, no. 2, p. 156, 2018. https://doi.org/10.1038/s41562-017-0260-9

[117] L. Rui, H. Nejati, S. H. Safavi, and N.-M. Cheung, "Simultaneous low-rank component and graph estimation for high-dimensional graph signals: Application to brain imaging," in *Proc. 2017 IEEE International Conference on Acoustics, Speech and Signal Processing (ICASSP)*. IEEE, 2017, pp. 4134–4138. https://doi.org/10.1109/ICASSP.2017.7952934

[118] M. Ménoret, N. Farrugia, B. Pasdeloup, and V. Gripon, "Evaluating graph signal processing for neuroimaging through classification and dimensionality reduction," in *Proc. 2017 IEEE Global Conference on Signal and Information Processing (GlobalSIP)*. IEEE, 2017, pp. 618–622. https://doi.org/10.1109/GlobalSIP.2017.8309033

[119] F. Monti, M. Bronstein, and X. Bresson, "Geometric matrix completion with recurrent multi-graph neural networks," in *Advances in Neural Information Processing Systems*, 2017, pp. 3697–3707. https://arxiv.org/abs/1704.06803

[120] S. Lee and A. Ortega, "Efficient data-gathering using graph-based transform and compressed sensing for irregularly positioned sensors," in *Proc. 2013 Asia-Pacific Signal and Information Processing Association Annual Summit and Conference*. IEEE, 2013, pp. 1–4. https://doi.org/10.1109/APSIPA.2013.6694166

[121] G. Shen and A. Ortega, "Transform-based distributed data gathering," *IEEE Transactions on Signal Processing*, vol. 58, no. 7, pp. 3802–3815, 2010. https://doi.org/10.1109/TSP.2010.2047640

[122] R. Wagner, H. Choi, R. Baraniuk, and V. Delouille, "Distributed wavelet transform for irregular sensor network grids," in *Proc. IEEE/SP 13th Workshop on Statistical Signal Processing, 2005*. IEEE, 2005, pp. 1196–1201. https://doi.org/10.1109/SSP.2005.1628777

[123] M. F. Duarte, G. Shen, A. Ortega, and R. G. Baraniuk, "Signal compression in wireless sensor networks," *Philosophical Transactions of the Royal Society A*, vol. 370, no. 1958, pp. 118–135, 2012. https://doi.org/10.1098/rsta.2011.0247

[124] H. E. Egilmez and A. Ortega, "Spectral anomaly detection using graph-based filtering for wireless sensor networks," in *Proc. 2014 IEEE International Conference on Acoustics, Speech and Signal Processing (ICASSP)*. IEEE, 2014, pp. 1085–1089. https://doi.org/10.1109/ICASSP.2014.6853764

[125] G. Cheung, E. Magli, Y. Tanaka, and M. K. Ng, "Graph spectral image processing," *Proceedings of the IEEE*, vol. 106, no. 5, pp. 907–930, 2018. https://doi.org/10.1109/JPROC.2018.2799702

[126] J.-Y. Kao, A. Ortega, D. Tian, H. Mansour, and A. Vetro, "Graph based skeleton modeling for human activity analysis," in *Proc. 2019 IEEE International Conference on Image Processing (ICIP)*. IEEE, 2019, pp. 2025–2029. https://doi.org/10.1109/ICIP.2019.8803186

[127] M. Püschel and J. M. F. Moura, "The algebraic approach to the discrete cosine and sine transforms and their fast algorithms," *SIAM Journal on Computing*, vol. 32, no. 5, pp. 1280–1316, 2003. https://doi.org/10.1137/S009753970139272X

[128] H. E. Egilmez, Y.-H. Chao, and A. Ortega, "Graph-based transforms for video coding," *IEEE Transactions on Image Processing*, vol. 29, pp. 9330–9344, 2020. https://doi.org/10.1109/TIP.2020.3026627

[129] A. Buades, B. Coll, and J.-M. Morel, "A non-local algorithm for image denoising," in *Proc. 2005 IEEE Computer Society Conference on Computer Vision and Pattern Recognition (CVPR)*, vol. 2. IEEE, 2005, pp. 60–65. https://doi.org/10.1109/CVPR.2005.38

[130] H. Takeda, S. Farsiu, and P. Milanfar, "Kernel regression for image processing and reconstruction," *IEEE Transactions on Image Processing*, vol. 16, pp. 349–366, 2007. https://doi.org/10.1109/TIP.2006.888330

[131] A. Gadde, S. K. Narang, and A. Ortega, "Bilateral filter: Graph spectral interpretation and extensions," in *Proc. 2013 IEEE International Conference on Image Processing*. IEEE, 2013, pp. 1222–1226. https://doi.org/10.1109/ICIP.2013.6738252

[132] G. Fracastoro, S. M. Fosson, and E. Magli, "Steerable discrete cosine transform," *IEEE Transactions on Image Processing*, vol. 26, no. 1, pp. 303–314, 2017. https://doi.org/10.1109/TIP.2016.2623489

[133] I. Daribo, D. Florencio, and G. Cheung, "Arbitrarily shaped motion prediction for depth video compression using arithmetic edge coding," *IEEE Transactions on Image Processing*, vol. 23, no. 11, pp. 4696–4708, 2014. https://doi.org/10.1109/TIP.2014.2353817

[134] A. Ortega and K. Ramchandran, "Rate-distortion methods for image and video compression," *IEEE Signal Processing Magazine*, vol. 15, no. 6, pp. 23–50, 1998. https://doi.org/10.1109/79.733495

[135] G. Shen, W.-S. Kim, A. Ortega, J. Lee, and H. Wey, "Edge-aware intra prediction for depth-map coding," in *Proc. 2010 IEEE International Conference on Image Processing (ICIP)*, 2010, pp. 3393–3396. https://doi.org/10.1109/ICIP.2010.5652792

[136] W. Hu, G. Cheung, and A. Ortega, "Intra-prediction and generalized graph Fourier transform for image coding," *IEEE Signal Processing Letters*, vol. 22, no. 11, pp. 1913–1917, 2015. https://doi.org/10.1109/LSP.2015.2446683

[137] W. Hu, G. Cheung, A. Ortega, and O. C. Au, "Multiresolution graph Fourier transform for compression of piecewise smooth images," *IEEE Transactions on Image Processing*, vol. 24, no. 1, pp. 419–433, 2015. https://doi.org/10.1109/TIP.2014.2378055

[138] Y.-H. Chao, A. Ortega, W. Hu, and G. Cheung, "Edge-adaptive depth map coding with lifting transform on graphs," in *Proc. 2015 Picture Coding Symposium (PCS)*. IEEE, 2015, pp. 60–64. https://doi.org/10.1109/PCS.2015.7170047

[139] E. Martínez-Enríquez, J. Cid-Sueiro, F. Diaz-De-Maria, and A. Ortega, "Directional transforms for video coding based on lifting on graphs," *IEEE Transactions on Circuits and Systems for Video Technology*, vol. 28, no. 4, pp. 933–946, 2018. https://doi.org/10.1109/TCSVT.2016.2633418

[140] E. Pavez, A. Ortega, and D. Mukherjee, "Learning separable transforms by inverse covariance estimation," in *Proc. 2017 IEEE International Conference on Image Processing (ICIP)*. IEEE, 2017, pp. 285–289. https://doi.org/10.1109/ICIP.2017.8296288

[141] H. E. Egilmez, Y.-H. Chao, A. Ortega, B. Lee, and S. Yea, "GBST: Separable transforms based on line graphs for predictive video coding," in *Proc. 2016 IEEE International Conference on Image Processing (ICIP)*. IEEE, 2016, pp. 2375–2379. https://doi.org/10.1109/ICIP.2016.7532784

[142] K.-S. Lu and A. Ortega, "A graph laplacian matrix learning method for fast implementation of graph Fourier transform," in *Proc. 2017 IEEE International Conference on Image Processing (ICIP)*. IEEE, 2017, pp. 1677–1681. https://doi.org/10.1109/ICIP.2017.8296567

[143] H. E. Egilmez, O. Teke, A. Said, V. Seregin, and M. Karczewicz, "Parametric graph-based separable transforms for video coding," in *Proc. 2020 IEEE International Conference on Image Processing (ICIP)*. IEEE, 2020, pp. 1306–1310. https://doi.org/10.1109/ICIP40778.2020.9191299

[144] J. Pang and G. Cheung, "Graph laplacian regularization for image denoising: Analysis in the continuous domain," *IEEE Transactions on Image Processing*, vol. 26, no. 4, pp. 1770–1785, 2017. https://doi.org/10.1109/TIP.2017.2651400

[145] Z. Wu and R. Leahy, "An optimal graph theoretic approach to data clustering: Theory and its application to image segmentation," *IEEE Transactions on Pattern Analysis & Machine Intelligence*, no. 11, pp. 1101–1113, 1993. https://doi.org/10.1109/34.244673

[146] J. Shi and J. Malik, "Normalized cuts and image segmentation," *IEEE Transactions on pattern analysis and machine intelligence*, vol. 22, no. 8, pp. 888–905, 2000. https://ieeexplore.ieee.org/document/868688

[147] D. G. Lowe, "Distinctive image features from scale-invariant keypoints," *International Journal of Computer Vision*, vol. 60, no. 2, pp. 91–110, 2004. https://doi.org/10.1023/B:VISI.0000029664.99615.94

[148] A. Gadde, A. Anis, and A. Ortega, "Active semi-supervised learning using sampling theory for graph signals," in *Proc. 20th ACM SIGKDD International Conference on Knowledge Discovery and Data Mining*. ACM, 2014, pp. 492–501. https://doi.org/10.1145/2623330.2623760

[149] V. Gripon, A. Ortega, and B. Girault, "An inside look at deep neural networks using graph signal processing," in *Proc. 2018 Information Theory and Applications Workshop (ITA)*. IEEE, 2018, pp. 1–9. https://hal.inria.fr/hal-01959770/

[150] X. Zhu, Z. Ghahramani, and J. D. Lafferty, "Semi-supervised learning using gaussian fields and harmonic functions," in *Proc. 20th International Conference on Machine Learning (ICML03)*, 2003, pp. 912–919. https://www.aaai.org/Papers/ICML/2003/ICML03-118.pdf

[151] J. Bruna, W. Zaremba, A. Szlam, and Y. LeCun, "Spectral networks and locally connected networks on graphs," *arXiv preprint arXiv:1312.6203*, 2013. https://arxiv.org/abs/1312.6203

[152] M. Defferrard, X. Bresson, and P. Vandergheynst, "Convolutional neural networks on graphs with fast localized spectral filtering," in *Advances in Neural Information Processing Systems*, 2016, pp. 3844–3852. https://arxiv.org/abs/1606.09375

[153] T. N. Kipf and M. Welling, "Semi-supervised classification with graph convolutional networks," *arXiv preprint arXiv:1609.02907*, 2016. https://arxiv.org/abs/1609.02907

[154] F. Gama, J. Bruna, and A. Ribeiro, "Stability properties of graph neural networks," *IEEE Transactions on Signal Processing*, vol. 68, pp. 5680–5695, 2020. https://doi.org/10.1109/TSP.2020.3026980

[155] M. M. Bronstein, J. Bruna, Y. LeCun, A. Szlam, and P. Vandergheynst, "Geometric deep learning: Going beyond Euclidean data," *IEEE Signal Processing Magazine*, vol. 34, no. 4, pp. 18–42, 2017. https://doi.org/10.1109/MSP.2017.2693418

[156] F. Gama, E. Isufi, G. Leus, and A. Ribeiro, "Graphs, convolutions, and neural networks: From graph filters to graph neural networks," *IEEE Signal Processing Magazine*, vol. 37, no. 6, pp. 128–138, 2020. https://doi.org/10.1109/MSP.2020.3016143

[157] N. Perraudin, J. Paratte, D. Shuman, L. Martin, V. Kalofolias, P. Vandergheynst, and D. K. Hammond, "GSPBOX: A toolbox for signal processing on graphs," *arXiv preprint arXiv:1408.5781*, 2014. http://arxiv.org/abs/1408.5781

[158] M. Defferrard, L. Martin, R. Pena, and N. Perraudin, "PyGSP: Graph signal processing in python." https://doi.org/10.5281/zenodo.1003157

[159] O. Chapelle, B. Schölkopf, and A. Zien, *Semi-Supervised Learning*.   MIT Press, 2006. https://doi.org/10.7551/mitpress/9780262033589.001.0001

Index